Reviews of Physiology, Biochemistry and Pharmacology 130

Springer-Verlag Berlin Heidelberg GmbH

Reviews of
130 Physiology Biochemistry and Pharmacology

Editors
M.P. Blaustein, Baltimore H. Grunicke, Innsbruck
D. Pette, Konstanz G. Schultz, Berlin
M. Schweiger, Berlin

Honorary Editor:
E. Habermann, Gießen

With 95 Figures and 5 Tables

Springer

ISSN 0303-4240

ISBN 978-3-662-31187-5 ISBN 978-3-540-70691-5 (eBook)
DOI 10.1007/978-3-540-70691-5

Library of Congress-Catalog-Card Number 74-3674

Originally published by Springer-Verlag Berlin Heidelberg New York in 1997.
Softcover reprint of the hardcover 1st edition 1997

Production: PRO EDIT GmbH, D-69126 Heidelberg
SPIN: 10542541 27/3136-5 4 3 2 1 0 – Printed on acid-free paper

The Impact of Somatosensory Input on Autonomic Functions

A. Sato[1], Y. Sato[2], and R.F. Schmidt[3]

[1] Department of the Autonomic Nervous System, Tokyo Metropolitan Institute of Gerontology, 35-2 Sakaecho, Itabashi-ku, Tokyo 173, Japan
[2] Laboratory of Physiologiy, Tsukuba College of Technology, 4-12-7 Ksuga, Tsukuba 305, Japan
[3] Physiologisches Institut der Universität Würzburg, Röntgenring 9, 97070 Würzburg, Germany

Contents

1 Introduction

The autonomic nervous system controls and regulates visceral organs by one of the two following modes:

1. The first type of organ control is by commands which originate elsewhere in the central nervous system (CNS), for instance as a consequence of emotions, circadian rhythms or the state of consciousness, to be forwarded to the visceral organs via the autonomic efferent nerve fibers. This control is independent of the level of activity of the visceral organs. It may result in phenomena such as psychic sweating, increases in heart rate and blood pressure during emotional excitement, and the anticipatory increase in skeletal muscle blood flow in preparation for exercise.
2. The second mode of control of visceral organs is the regulation of their function by reflexes originating from the peripheral sensory receptors, including visceral, somatic and specific cranial receptors. The information picked up by these receptors is transmitted to the CNS via visceral, somatic, and special sensory fibers and forwarded to the autonomic nervous system which in turn emits control signals to the visceral organs to modify their function as a consequence of the afferent input. Well-known and frequently studied examples of such reflexes originating from visceral sensory receptors include the baroreceptor reflex modifying cardiovascular function, the regulation of continence and micturition of the urinary bladder and the reflex regulation of digestion (for literature, see, e.g., Heymans and Neil 1958; Roman and Gonella 1981; Torrens and Morrison 1987).

In contrast to the impressive body of knowledge concerning the effects of visceral afferent activity on autonomic functions, there is, generally speaking, much less information available on the reflex regulation of visceral organs by somatic afferent activity from the skin, the skeletal muscle and their tendons, and from joints and other deep tissues. The elucidation of the neural mechanisms of somatically induced autonomic reflex responses, usually called somato-autonomic reflexes, is, however, essential to developing a truly scientific understanding of the mechanisms underlying most forms of physical therapy, including spinal manipulation and traditional as well as more modern forms of acupuncture and moxibustion.

In the early 1960's, one of the present authors (A.S.) initiated a line of research aimed at clarifying the spinal and supraspinal pathways of somato-sympathetic reflexes. He was later joined in this endevour by the other two authors (see also p. 263). For these experiments, the use of anesthesia was considered essential to eliminate any factors related to emotion during and following somatic afferent stimulation, since the involvement of emotion presented a most formidable problem in the quest to clarify the neural mechanism of somato-autonomic reflexes. This work was later expanded to elucidate the contributions to these reflexes of the various types of cutaneous and muscle afferent nerve fibers. A review of this work, published by Sato and Schmidt (1973), emphasized three aspects of the somato-sympathetic reflexes: (1) the various types of somatic afferents involved, (2) sympathetic nerve activity, and (3) central pathways.

Actually, this review article became something of a landmark for the neurophysiological study of the somato-autonomic reflexes. It established clearly the roles of the spinal and supraspinal reflex centers and distinguished between the influences of sensory input from myelinated and unmyelinated afferent fibers. It was apparent that in addition to the neurophysiological (electrophysiological) approach, there was a need for a parallel, integrative line of research; one which applied and extended the neurophysiological findings on somato-sympathetic reflexes to the responses of various visceral organs to somatosensory stimuli.

Since then, there has been remarkable progress in our understanding of somato-autonomic reflex regulation of visceral functions, particularly in the following five areas:

1. There has been considerable clarification of the role of natural stimulation of somatic sensory receptors in autonomic reflexes. This involves an entirely new interpretation of the effects of natural stimulation based on a modern understanding of neurophysiological principles derived from electrophysiological techniques. Also there is a better appreciation of the importance of afferent fiber characteristics in determining autonomic reflex reponses in various visceral organs. For instance, since

1973, there has been a great accumulation of knowledge concerning the impact of fine somatosensory afferents on autonomic functions, particularly under pathophysiological conditions, where, due to peripheral and central sensitization (see Sect. 2.3.4), the central effects of the somatosensory volleys are greatly enhanced.

2. With respect to different autonomic pathways innervating various visceral organs, the contributions of the spinal cord and supraspinal structures as reflex centers in the somato-autonomic reflexes have been systematically clarified. Indeed, one of the most important findings was that somato-autonomic reflex centers were located not only in the brain but also in the spinal cord. The importance of the segmental level of the spinal somatic afferent nerves involved in specific autonomic reflexes was first emphasized by Sato and Schmidt (1971). They showed in anesthetized cats with the CNS intact, that larger sympathetic reflex discharges in white rami of spinal origin were induced by electrical stimulation of somatic afferent nerves at the same or adjacent rather than distant segmental levels. They foresaw the possibility that specific autonomic organ functions might be reflexly regulated by stimulation of somatic afferent nerves of certain segments. Additionally, several types of sympathetic and parasympathetic reflex discharges of spinal origin have been demonstrated in many autonomic efferent fibers, some under strong inhibitory control from the brain, and others under weak inhibitory control.
3. Cannon's concept of a generalized sympathetic nerve reaction during an emergency (Cannon 1929) seems somewhat divorced from somatically induced autonomic reflex regulation. Specifically, it has been found that the participation of the supraspinal and spinal reflex components differs from one organ system to another. For example, somatically induced cardiovascular reflex regulation is dominated by supraspinal reflex influences, with the involvement of only a minor spinal reflex component, due to a descending inhibitory effect from the brain. In contrast, somatically induced reflex regulation of urinary bladder and gastric motility are dominated by either spinal reflex influences or supraspinal reflex influences, depending upon the different segmental areas stimulated. Characteristically, noxious stimulation of the limbs has been seen to produce a more generalized sympathetic response, as advocated by Cannon. However, this is due to the fact that limb afferents enter the spinal cord at levels essentially devoid of autonomic preganglionic neurons, so that autonomic responses are necessarily

mediated by supraspinal reflex centers. On the other hand, stimulation of segmental spinal afferents which enter the spinal cord at other levels produces quite a different pattern of response. Where segmental spinal afferents have the opportunity to synapse with spinal preganglionic autonomic neurons, localized somato-autonomic reflexes can be mediated within the spinal cord. Spinally mediated somato-autonomic reflexes may show a very strong segmental organization and, under the appropriate conditions, the effects on target organs may be quite specific. Often, however, in the CNS-intact animal, these spinal reflexes are masked by descending influences from the brain.

4. There is a new trend in research on somato-autonomic reflexes, focusing on the role of the endocrine and immune systems. Two examples of hormonal involvement are (i) catecholamine secretion from the adrenal medulla, whose secretion is controlled by adrenal sympathetic efferent nerve activity, and (ii) pancreatic hormonal secretion, which is partially under the control of pancreatic sympathetic and parasympathetic efferent nerve activity. In addition, some hormones belonging to the hypothalamic-pituitary system act as hormonal efferent pathways in the reflex responses of visceral function to somatic afferent stimulation. It is also well known that immune-related organs receive autonomic innervation. Few studies have as yet been done, but there is evidence that immune function is reflexly influenced by autonomic efferent nerve activity following somatic afferent stimulation.
5. The central reflex pathways involved in somato-autonomic, somato-endocrine, and somato-immune reflexes have been resolved into essentially four classes:
   a) *The axon reflex.* This peripheral "reflex" has no direct autonomic efferent involvement, but produces effects somewhat comparable to those of autonomic activation.
   b) *The spinal (segmental) reflex.* This reflex is elicited when spinal nerves originating at specific segmental levels are stimulated. The segmental afferent nerves modulate visceral organs via autonomic efferent nerves or modulate them indirectly by affecting visceral afferent input.
   c) *The medullary reflex.* This generalized reflex is elicited by stimulation of various spinal nerves, and stimulation of limb afferents evokes a particularly strong reflex effect.

d) *The supramedullary reflex.* This reflex requires the involvement of supramedullary central neurons in specific functions, such as sweating, hormonal secretion and regulation of cerebral blood flow.

It appears to us that there is now an appropriate point in time to summarize and review this vast amount of new literature and to set it into perspective both in regard to the scientific progress which has been achieved, and in regard to the clinical and most particularly the therapeutic consequences which these new findings imply. Most of the work on the mode of operation and the central mechanisms of somato-autonomic interactions reported in this review has been done with anesthetized animals, especially with rats, to eliminate emotional factors arising from somatosensory inflow. However, this review also includes recent relevant work in conscious animals and in humans.

When starting to write this review we spontaneously tended to focus on the questions we have studied, and, as a consequence about 20% of all references in this text are by ourselves and our collaborators. However, the more the review progressed over the last 3 years or so, the more we made a deliberate effort to include the relevant observations of our colleagues from all over the world. We hope that this review will provide a lucid picture of the vital field of somato-autonomic reflex physiology, and we hope that the research summarized here will have some impact on the therapeutic application of somatosensory stimulation to improve visceral functions in humans.

2 Somatic Sources of Afferent Input to the Autonomic Nervous System

2.1 Introduction

Everyday experience has taught animals and humans alike that activation of certain types of cutaneous and deep tissue sensory receptors not only evokes sensations but induces a great variety of motor and autonomic responses, some of them being very welcome, others not at all. As regards the autonomic responses, which are the topic of this review, examples of the welcome type include the vasodilation of an ice-cold hand as a consequence of stimulating its warmth receptors when exposing the hand to an open fire place. On the other hand, unwelcome autonomic effects of somatosensory activity include, amongst many more harmless examples, the severe symptoms of circulatory failure, sweating, nausea, vomiting etc. as a consequence of excruciating visceral pain.

In the general introduction to this review some of the basic principles governing the relationship between the somatosensory and autonomic nervous systems have been outlined. In this chapter we will focus our attention on the input side of this relationship. Actually, when the authors of this review initiated their systematic experimental exploration in mammals of the impact of the somatosensory on the autonomic nervous system in the second half of the 1960s, they started on the assumption that all types of somatosensory receptors have a more or less powerful access to the autonomic nervous system and its effector organs. As will be shown in the subsequent chapters of this review, this basic notion could not be fully confirmed. Truly, many, if not the majority, of somatosensory receptors have potent access to autonomic functions (the very topic of this review) but there are also others whose activation seems to have little if any impact on autonomic functions. In this respect the most striking examples are the primary muscle spindle afferent units and those from

Golgi tendon organs whose central actions seem to concentrate entirely on the motor system.

Against this background, this chapter focuses on the description of the location, structure and function of those somatosensory receptors whose activation has been shown experimentally and clinically to have predictable effects on one or another autonomic effector organ. These receptors may therefore act as important modulators of autonomic nervous activity and effector function either in everyday life or when stimulated to make use of their potential autonomic effects for therapeutic reasons. In addition to the sensory receptors themselves, we will outline the properties of their afferent nerve fibers and of the cutaneous, muscle, articular and mixed nerves in which the somatosensory afferents project to the spinal cord and brain stem. Furthermore, the central, mostly spinal terminations of these afferents will be discussed together with the ascending pathways carrying the sensory messages to the supraspinal autonomic structures.

2.2 Low-Threshold Somatosensory Receptor

This review of the somatosensory receptors in mammals will first put its attention to the low-threshold cutaneous and subcutaneous receptors, i.e., those afferent units which are most readily available for activation by innocuous external stimuli. The low threshold of these sensory receptors is, by the way, certainly one of the main reasons why they have attracted, in recent decades, much more attention than any of the other somatic afferent systems. The receptors in nonprimate mammals will be discussed separately from those of primates, including humans, not only because they have been investigated at different times and, at least in part, with somewhat different methods, but also because the hairy and glabrous skin is so different in nonprimates versus primates that a separate outline appears unavoidable. In addition, those readers who want to make full therapeutic use of the information contained in this review need only to refer to those sections in this chapter particularly concerned with human skin.

Secondly, we will turn our attention to the low-threshold receptors in skeletal muscles and their tendons. These afferent systems are of such uniform structure and function in the various types of mammals that no separate discussion of those in nonprimates versus primates appears justified. The same holds true for the low-threshold articular receptors which will be considered thereafter.

2.2.1
Cutaneous and Subcutaneous Low-Threshold Sensory Receptors

2.2.1.1
Mechanoreceptors

Not surprisingly, in view of the fact that the structure and function of skin differ not only in different animal species but also in different parts of the body surface, such as in hairy versus glabrous skin, a wide variety of sensory structures has been described in the skin of mammals (for a review of the older literature see Darian-Smith (1984b). Nevertheless, a rather uniform picture has emerged over the years both for the innervation of the hairy as well as the glabrous skin in mammals, and on the correlation of the various sensory terminal structures with their physiological function.

2.2.1.1.1
Mechanoreceptors Innervating Glabrous Skin in Nonprimate Mammals

Mechanoreceptors with Myelinated Afferents. As regards nonprimate mammals, the most rigorous studies of such sensory receptors were carried out for the foot pad of the cat. Figure 1 illustrates the most frequently used methods to identify and investigate these sensory receptors as well as their central terminals and projections (Schmidt et al. 1967). In daily life, the cat's foot pads are exposed to two main types of mechanical stimuli: prolonged pressure, as during standing and walking, and short, much weaker stimuli, often of vibratory character, for instance when the foot pads are used for exploring. Well suited for these tasks, three classes of mechanoreceptors have been found, namely PC, RA, and SA units (e.g., Gray 1966; Jänig et al. 1968).

The first group are called PC receptors because their properties are the same as those described for Pacinian corpuscles elsewhere. They are by far the most sensitive mechanoreceptors of the cat's foot pads, 90% of them having a threshold <4 μm for a single stepwise skin indentation, and 50% being excited by <0.8 μm skin indentation (Jänig et al. 1968). Since they have wide receptive field-zones, a single pulse of some 10–20 μm indentation applied to the middle of the large pad will excite all or almost

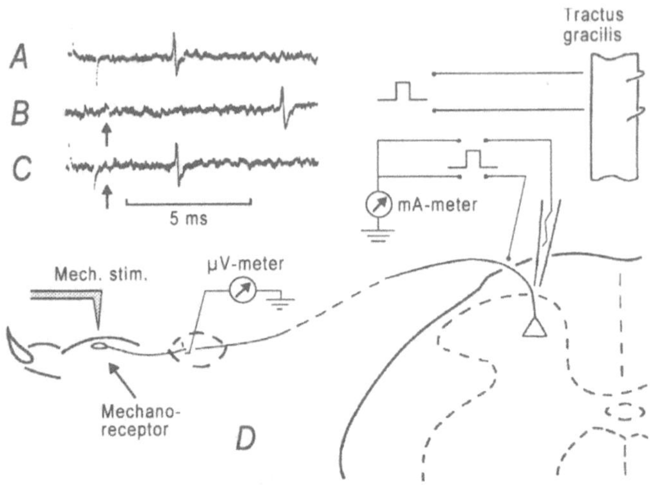

Fig. 1A–D. Stimulating and recording arrangements for the identification and characterization of the peripheral properties and the central (spinal) excitability of cutaneous mechanoreceptors in the central pad of a cat's hind foot. Setups corresponding to the one shown in **D** are being used for other types of sensory receptors in the skin and when working with muscular, articular, or visceral primary afferent units (see Figs. 2, 6–9, 14). Specimen records of action potentials recorded from a fine filament of the plantar nerve **A** after peripheral activation of a mechanoreceptor by a piezo-driven mechanical stimulator, **B** following stimulation of the same afferent fiber in the spinal cord through a platinum wire microelectrode, and **C** following simultaneous peripheral and central stimulation. Collision of the orthodromic with the antidromic action potential prevents the latter from progressing to the recording electrode. If such antidromic identification is not required, the afferent nerve can be cut at the recording site to facilitate the dissection of single fibers and to give better signal to noise ratios of the action potentials. *Arrows* indicate the central stimulation. **D** Antidromic impulses are evoked by stimulating the fiber under observation close to its dorsal horn terminal via a microelectrode. The excitability of the terminal and its changes, e.g., by primary afferent depolarization (PAD) in the course of presynaptic inhibition, can be determined by measuring the threshold stimulus current. *Top right*, stimulation of the dorsal columns (dissected free from the spinal cord and put on wire electrodes) allows us to determine whether or not a collateral of the primary afferent fiber projects via this pathway to the dorsal column nuclei. (Modified from Schmidt et al. 1967)

all PC afferents coming from this region. Prolonged touch stimuli elicit a single impulse from a PC unit both at the beginning and at the end of the skin displacement, i.e., the units adapt instantaneously. But they are extremely sensitive for vibratory stimuli in the range between 30–200 Hz, with an optimum sensitivity at 180–200 Hz (0.7–1.5 μm amplitude of sinusoidal vibration for a 1: 1 following of the receptor, Fig. 2). The PC units are arranged in the subcutaneous fat along the nerves and vessels (Jänig 1970). This system seems particularly suited to transmit vibratory stimuli of small amplitude, and it has been reported that the sensitivity to vibratory stimuli remains high even if the foot is exposed to high constant pressure (Jänig et al. 1968).

The second group are called RA receptors because they adapt rapidly, i.e., within 200–500 ms, to constant pressure stimuli. The receptive fields

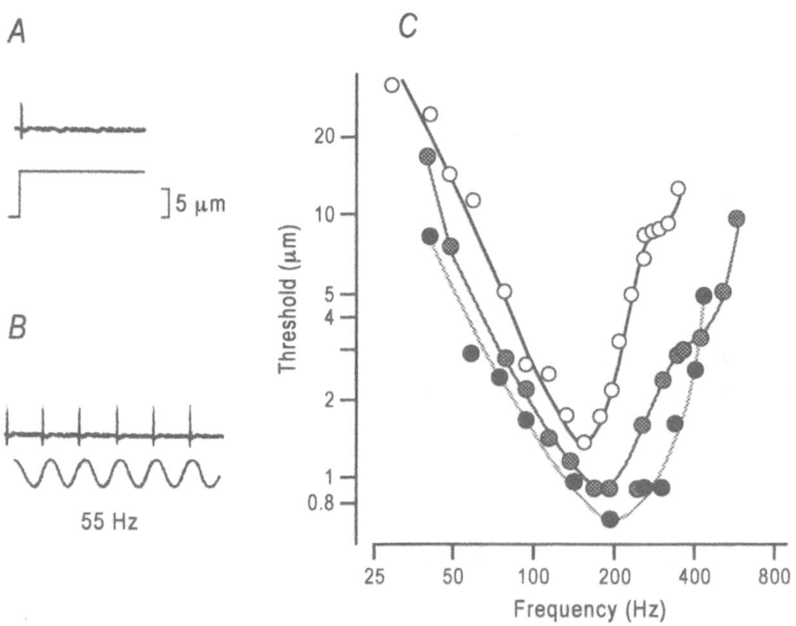

Fig. 2A–C. Responses of subcutaneous PC receptors in the cat's hind footpad to mechanical stimuli. **A** A rectangular stimulus elicits a single action potential, i.e., PC receptors adapt very rapidly to constant mechanical stimuli (pressure). **B** Specimen recording of the repetitive activation of the PC unit by sinusoidal stimulation. **C** The thresholds of three different PC units to sinusoidal stimulation. The graph relates the frequency of a sinusoidal stimulus (abscissa) to the peak-to-peak amplitude required to evoke one impulse for each cycle period (as shown in **B**). Note the logarithmic scaling of the ordinate. The stimuli were applied at the sites of maximum sensitivity for single-pulse stimuli, which were 0.4–0.45 μm. (Modified from Jänig et al. 1968)

of these units are small. They do not have any resting discharge and show no response to sudden cooling of the skin. Their thresholds are much higher than those of the PC units but lower than those of the SA units. Morphological studies of functionally identified RA units revealed that they are located in the dermal papillae (upper dermis), and that, among the nerve fiber terminations found there, the extended, encapsulated, ramified fibers are the most probable candidates for these mechanoreceptors (Jänig 1970).

The slowly adapting SA units form the third and final group of these mechanoreceptors (Jänig 1970; Jänig et al. 1968). Most likely they are of two subtypes with somewhat different properties. The first subtype discharges irregularly in response to stimuli of constant force and has a high dynamic sensitivity. It responds to cooling of the skin, and often has a low irregular spontaneous activity, i.e., its functional properties resemble but are not fully comparable to those of the SA I units in primate glabrous skin. Its histological basis is the Merkel cell nerve fiber complex fixed in the epidermis by desmosomes, its fine structure being similar to that of the touch corpuscle. The second subtype discharges more regularly in response to stimuli of constant force and has a lower dynamic and static sensitivity than the first type; it is not spontaneously active and does not respond to cooling the skin surface. Again, there is some resemblance to the SA II units in primate glabrous skin. Their histological structure, however, remains to be identified.

As pointed out by Iggo and Andres (1982), it has now generally been accepted that the Merkel cells in mammals function as slowly adapting type I units (SA I), although this is much better documented for the touch spots in hairy than in glabrous skin. The Ruffini endings have been identified as the receptors of the slowly adapting type II (SA II) cutaneous mechanoreceptor in the hairy skin of the cat (e.g., Chambers et al. 1972; see also above). Since similar functional properties are found for mechanoreceptors in the monkey and human skin, both hairy and glabrous (see below), it may be assumed that, everywhere this receptor is found, this spindle shaped dermal structure is functionally of the SA II type.

Mechanoreceptors with Unmyelinated Afferents. The glabrous skin of non-primates seems to contain few if any low-threshold mechanoreceptors with unmyelinated afferents. According to Bessou et al. (1971), they are entirely absent from the glabrous skin of the cat's foot pad. Because of this negative result no systematic search seems to have been undertaken in the glabrous skin of other nonprimate mammals.

2.2.1.1.2
Mechanoreceptors Innervating Hairy Skin in Nonprimate Mammals

Mechanoreceptors with Myelinated Afferents. Hairy skin is concerned with relaying information about the changing patterns of contact between the body surface and adjacent objects. This pattern of activity provides information on the disposition of the body and limbs relative to the object world around us. Conveniently, the afferents innervating hairy skin may be divided into rapidly and slowly adapting ones on the basis of their response to stepwise indentation and flexing of the hairs.

Within the rapidly adapting fiber population two subgroups are recognized: (1) Pacinian receptors (PC units, rare in hairy skin, for their histology and function see above and the section on human glabrous skin) and (2) various types of hair follicle receptors which are subdivided on the basis of the type of follicle they innervate. Type G fibers innervate the follicles of guard hairs. Their primary afferents belong to group II. Each fiber innervates guard hairs over a rather extensive receptive-field zone (up to several square centimeters) but each fiber innervates only a small fraction of guard hairs within its field zone. Type G fibers respond not only to movement of guard hairs but also to indentation of the skin of the receptive field zone. With this type of stimulation they are best activated by sinusoidal vibratory stimuli in the frequency range of 20–50 Hz. They are, however, considerably less sensitive to vibration than PC units. Type T units, which innervate two to seven tylotriche hairs in the hairy skin of the cat, may be distributed over several square centimeters and have group II primary afferents (Brown and Iggo 1967). Type D units innervate the down hairs. Their afferents have conduction velocities between 10–30 m/s, i.e., they include both slow group II and fast group III units (Brown and Iggo 1967).They respond to minute movements of the fine down hairs and the guard hairs within the receptive field zone which is similar in area to those of the G and T units.

Within the slowly adapting mechanoreceptor population of hairy skin, SA I units can be discriminated from SA II units (Burgess et al. 1968). Both types of units have group II primary afferents with conduction velocities in the range of 40–70 m/s. The SA I units innervate the small dome-shaped Merkel touch corpuscles, each fiber supplying several neighboring domes (one to five in the cat), and innervating the Merkel cells in the deepest layer of the epidermis (Iggo and Muir 1969). Indentation of the touch corpuscle evokes an initial high frequency burst of discharges which drops within 10–15 ms to a more static level, typically with a highly variabile interspike interval. In contrast, SA II units respond to sustained indentation with a very regular discharge (often starting from a regular resting activity of 3–20 imp/s). Their receptive fields are confined to a single small area of 2–3 mm^2, and stretching the skin surrounding this central sensitive zone usually elicits discharges. The SA II fibers innervating hairy skin terminate in the dermis as Ruffini endings (Chambers et al. 1972). Both SA I and SA II units may be excited by rapid cooling, and the resting activity of SA II fibers may stop when the skin is warmed. The temperature sensitivity of these units is, however, well below that of specific thermoreceptors and so may have little physiological significance.

The quite distinctive sinus hairs, including vibrissae, in the perioral and carpal skin of the nonprimate have a complex innervation which, in the context of this article, deserves only brief mention. For instance, the sinus hairs of the cat are innervated by afferents terminating in Golgi-Mazzoni corpuscles (giving a rapidly adapting response to hair bending), and by slowly adapting units terminating either in Merkel cells (and being direction sensitive) or in lanciform endings (Gottschaldt et al. 1973).

Mechanoreceptors with Unmyelinated Afferents. In contrast to the glabrous skin (see above), in the cat's hairy skin about 50% of the unmyelinated fibers respond selectively to sustained indentation of the skin (for references s. Darian-Smith 1984b). These fibers have small receptive fields and respond sluggishly to sustained indentation but not to movement of hairs or to vibratory stimuli. They frequently discharge for several seconds after the stimulus has ceased. It must be concluded that such units, although not small in numbers, relay little information about most commonly encountered tactile stimuli.

2.2.1.1.3
Mechanoreceptors Innervating Glabrous Skin in Primates and Humans

Mechanoreceptors with Myelinated Afferents in the Human Glabrous Hand. The investigation of the functional properties of human cutaneous sense organs and their relation to the psychophysics of tactile sensibility was greatly facilitated by the introduction of the microneurographic technique by Vallbo and Hagbarth (1968). This technique, which is illustrated in Fig. 3A, allows not only record from individual afferent units with per-

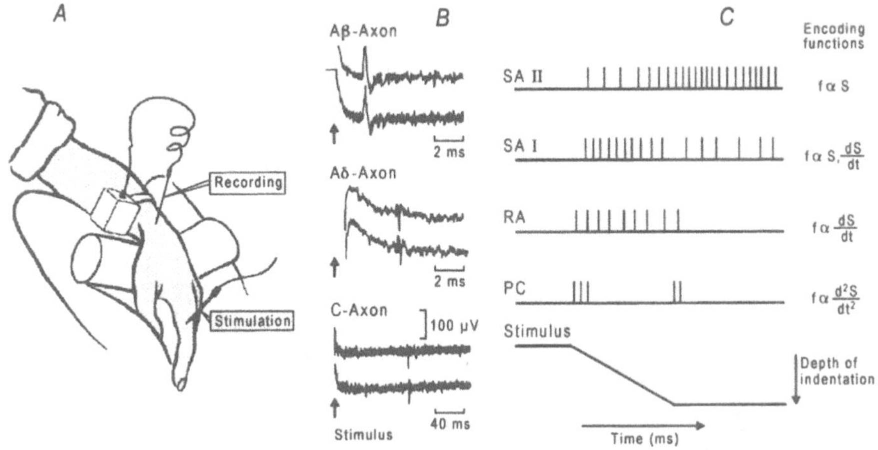

Fig. 3A–C. Microneurography to study human sensory receptors. **A** Setup to record transcutaneously from single primary afferent fibers in the radial nerve following electrical stimulation of their receptive fields. **B** Specimen records of action potentials recorded in this way from the indicated types of primary afferent fibers. Note the change in time scale when recording from C fibers (group IV fibers). **C** Discharge behavior of the various types of low-threshold cutaneous mechanoreceptors in human skin in response to a ramp-shaped mechanical skin indentation (lowermost record). *Right*, the coding function of each type of receptor is listed (*S*, stimulus strength; *t*, time; *dS/dt*, rate of rise, i.e., velocity, of stimulus ramp; dS^2/dt^2, acceleration or deceleration of the stimulus movement; *a*, equation constant). *SA II*, type II slowly adapting mechanoreceptor (Ruffini organ); *SA I*, type I slowly adapting mechanoreceptor (Merkel disc); *RA*, rapidly adapting mechanoreceptor (velocity detector, Meissner corpuscle); *PC*, very rapidly adapting mechanoreceptor (acceleration detector, Pacini corpuscle). (Modified from H. Handwerker in Schmidt 1995)

cutaneously inserted tungsten microelectrodes but also permits the stimulation of the same fiber to elucidate the properties of the percept induced by a defined series of impulses in an identified sensory unit (for references see Vallbo and Johansson 1984; Lamotte et al. 1992).

The glabrous skin of primates, including humans, contains four types of tactile units with functional properties corresponding closely to the properties of the four categories of mechanoreceptors described physiologically in subhuman species (see above). These four types of units, RA, PC, SA I, and SA II (see above for an explanation of the abbreviations) have skin end organs of very different morphology (Fig. 4) and display

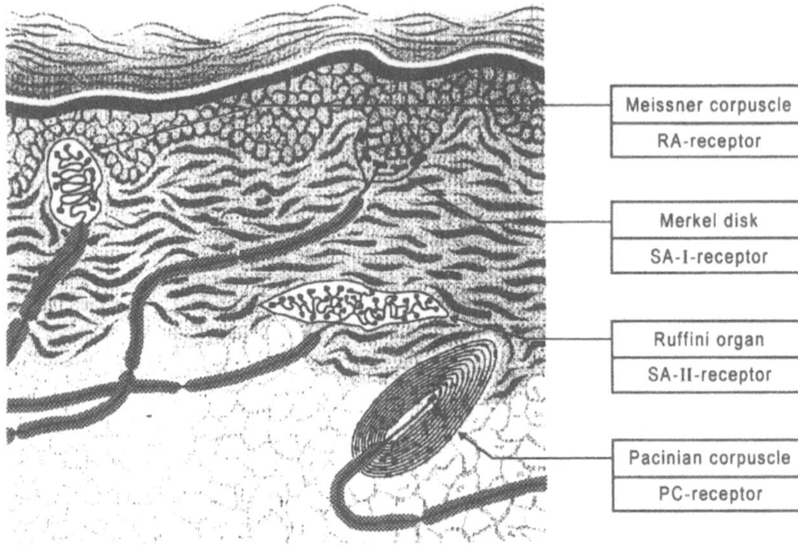

Fig. 4. Primate glabrous skin; different types of specialized nerve fiber terminals and their locations within epidermis and dermis are shown. *Right*, the correlations between structure and function are shown (see also Fig. 3). (Modified from H. Handwerker in Schmidt 1995)

different organizations of their afferent terminals in the spinal cord (see Sect. 2.5.1). The low-threshold mechanoreceptors of the human glabrous skin are most easily differentiated on the basis of two features: adaptation to sustained indentation, and structure of the receptive field. The fast adapting RA units are characterized by small receptive fields with quite distinct borders whereas the fast adapting PC units have large receptive fields with diffuse borders. The slowly adapting SA I and SA II types are similarly differentiated from each other by field characteristics, as with the RA and PC units.

The glabrous skin of a human hand contains some 17 000 tactile units (Vallbo and Johansson 1984; Johansson and Vallbo 1979). Their locations and receptive field sizes are illustrated in Fig. 5. The fast adapting RA and PC units constitute 43% and 13% respectively of the total number of tactile units. Of the slowly adapting units the SA I units constitute about 25% and the SA II units about 19% of all tactile units in the glabrous skin of a human hand.

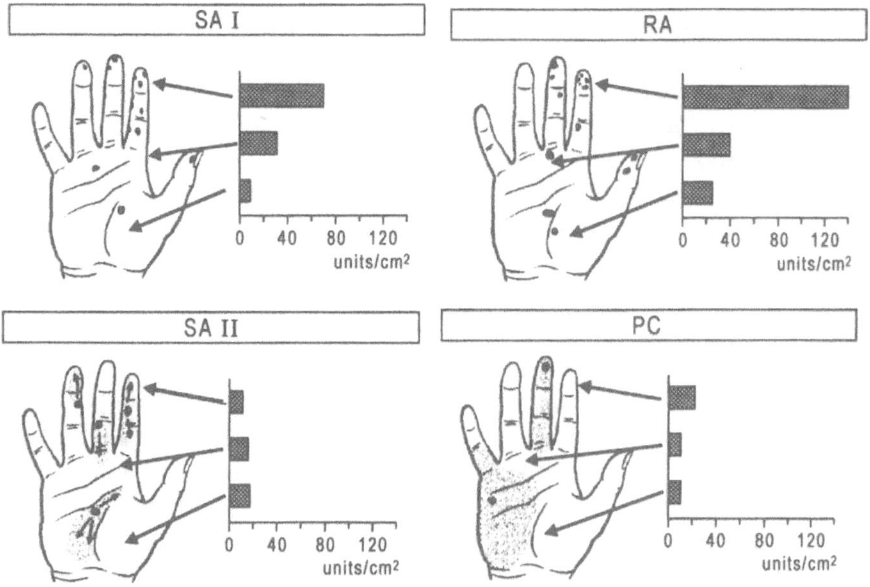

Fig. 5. Receptive fields and innervation density of the four major types of low-threshold mechanoreceptors in the human hand. The receptive fields and the relative innervation densities of the different receptor types were determined with transcutaneous microneurography (see Fig. 3), and the absolute innervation densities were determined by histological nerve fiber counts. (Data from Vallbo and Johansson 1984, arranged by H. Handwerker in Schmidt 1995)

Mechanoreceptors with Unmyelinated Afferents. Y. Zotterman (personal communication) always maintained that low-threshold C afferents from glabrous skin should exist, and that they may be responsible for transmitting the tickling sensation. However, as far as we are aware, such units have not yet been unequivocally identified in primate glabrous skin.

2.2.1.1.4
Mechanoreceptors Innervating Hairy Skin in Primates and Humans

Mechanoreceptors with Myelinated Afferents. This discussion will be brief in view of the fact that a detailed account of the fiber types innervating the nonprimate skin has been given above, and since the fiber types innervating primate and nonprimate mammalian hairy skin are essentially similar, although the frequency of occurrence of each type probably differs (Merzenich and Harrington 1969; Hunt and McIntyre 1960; Burgess et al. 1968; Brown and Iggo 1967; Aoki and Yamamura 1977). As with mechanoreceptive units innervating primate glabrous skin, those innervating hairy skin may be divided into rapidly adapting and slowly adapting ones. Within the rapidly adapting fiber population two subgroups are recognized, namely the PC units, i.e., Pacinian afferents, similar in every respect to those found in primate glabrous skin but even less common, and fibers innervating the different types of hair follicles. Similarly, the slowly adapting fiber population may be subdivided in SA I and SA II types that parallel, in most respects, the analogous types supplying glabrous skin in man (see above).

Mechanoreceptors with Unmyelinated Afferents. In contrast to the findings in hairy skin of cats where many mechanoreceptors with unmyelinated afferents are present (see above), low-threshold mechanoreceptive afferents are rare in the hairy skin of the monkey (Merzenich and Harrington 1969; Kumazawa and Perl 1977). Their functional properties resemble those described above for the corresponding units in the cat's hairy skin.

2.2.1.2 Thermoreceptors

2.2.1.2.1 Cutaneous Thermoreceptors

Definition of Receptor Specificity. Since any biological process is dependent on temperature, it is not always easy to decide whether a certain neural tissue can be defined as "thermosensitive." Besides the quantitative amount of the response to local cooling or warming, one must consider the degree of specificity of such a response. Therefore, certain criteria have been proposed for the definition of cutaneous thermoreceptors (Hensel 1973, 1974, 1981): (a) that they have a static discharge at a constant temperature (T); (b) that they show a dynamic response to temperature changes (dT/dt), with either a positive temperature coefficient (warm receptors) or a negative coefficient (cold receptors); (c) that they do not respond to mechanical stimuli within reasonable limits of intensity.

The variety of cutaneous thermoreceptors can be divided, on the basis of their dynamic responses, into well-defined classes of warm and cold receptors (Fig. 6). Irrespective of the initial temperature, a warm receptor will always respond with an overshoot of its discharge on sudden warming and a transient inhibition on cooling, whereas a cold receptor will respond in the opposite way, namely, with an inhibition on warming and an over-

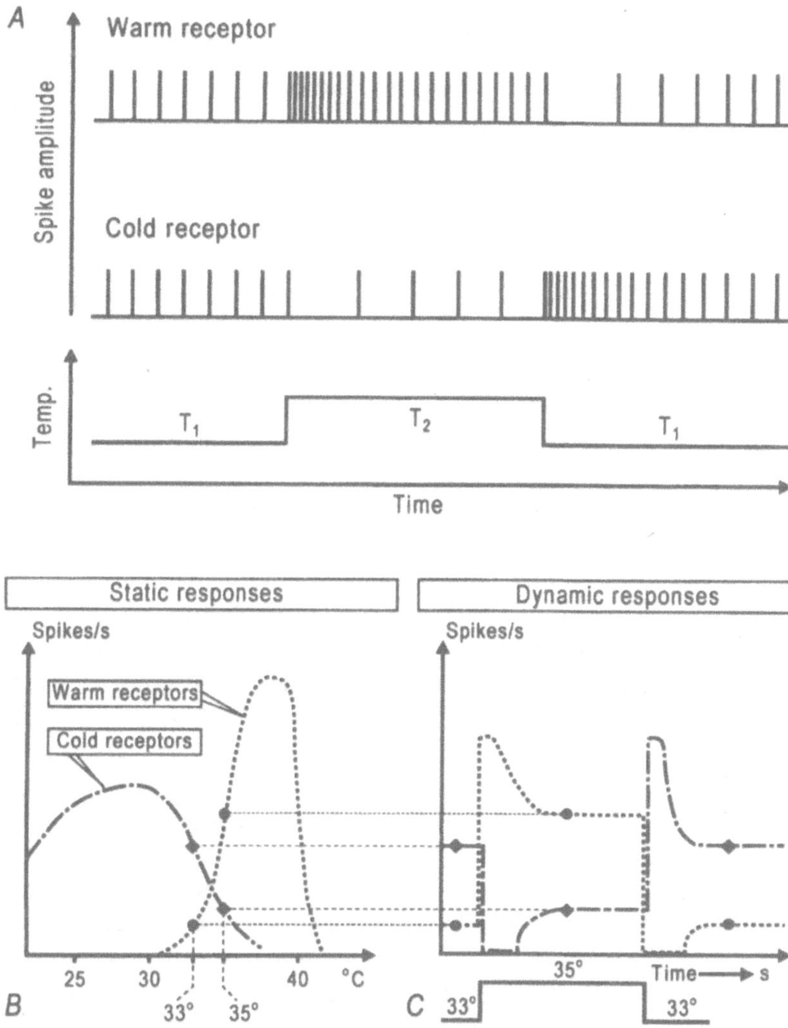

Fig. 6A,B. General properties of primate thermoreceptors. **A** Impulse discharges from a single warm and a single cold receptor during a constant midrange skin temperature *(T_1)* and in response to a rectangular warm stimulus *(T_2)*. **B,C** Static and dynamic behavior of populations of warm and cold receptors. The dynamic responses were obtained during the temperature step shown at the bottom of C. (Modified from Hensel 1981 and H. Handwerker in Schmidt 1995)

shoot on cooling. Besides this dynamic behavior, there are also typical differences in the static frequency curves of both types of cutaneous receptors, in that the temperature of the maximum discharge is much lower for cold receptors than it is for warm receptors.

The pioneering observations on thermoreceptors by Zotterman, Hensel, Iggo and their coworkers were initially carried out mainly with cats (for reviews, see Zotterman 1953; Hensel 1973, 1974, 1981). Unfortunately, in the cat, thermoreceptors are scarce except in the nerves supplying the tongue, facial and scrotal skin. In addition, the thermal stimulation procedures were inconvenient, and, as a result, systematic studies of thermoreceptor responses to transient changes in skin temperature were lacking. As soon as the exploration of thermoreceptors was extended to primates in the 1960s and 1970s, and with the introduction of refined thermal stimulation techniques, substantial advances were made concerning the physiology of warm and cold receptors.

General Properties of Cutaneous Thermoreceptors (Fig. 6). In primates specific cold and warm receptors have been found in abundance both in the glabrous and hairy skin (for literature see Darian-Smith 1984a,b). For instance, in the median nerve of the macaque, at the level of the wrist, cold fibers constitute about one-third of all Aδ myelinated fibers, and warm fibers form a substantial fraction of the unmyelinated fibers in the same nerve.

In the glabrous skin of the hand, the specific low-threshold thermoreceptive fibers are uniquely excited either by warming (warm fibers) or by cooling (cold fibers) the appropriate region of skin (Fig. 6A,C). Starting from normal skin temperatures, warm fibers begin to discharge more rapidly soon after the onset of skin warming, and continue to do so until the warm stimulus stops. By contrast, cold fibers discharging slowly at the resting skin temperature stop firing as the temperature begins to rise and often remain silent until the warm stimulus ends. During the subsequent fall of the skin temperature back to its initial level the cold fiber may once again discharge, sometimes with an increased afterdischarge.

Cooling the skin evokes a reversal of responses in the warm and cold fibers: the warm fibers become silent and the cold fibers respond vigorously. Again, an afterdischarge may be observed in the warm fibers as the skin temperature climbs back to its initial level. As required by Hensel (see above), both warm and cold fibers are unresponsive to mechanical deformation of the skin unless the stimulus is of an intensity sufficient to evoke pain.

The receptive fields of most warm and cold fibers innervating the macaque's palmar and digital skin respond only from a single, minute area, 1 mm or less in diameter, when tested with small hot or cold probes.

In the hairy skin of the monkey, warm and cold units with similar receptive properties are found too, except that their single spot-like receptive fields are commonly somewhat larger in diameter (3–5 mm vs. 1 mm or less) than the fields of thermoreceptors innervating glabrous skin. In addition, a significant number of the units innervating hairy skin, and also the transitional skin between hairy and glabrous skin, have more complex receptive fields that consist of two to four spots separated by zones of unresponsive skin.

2.2.1.2.2
Thermoreceptors in the Central Nervous System and in Other Body Tissues

Changes in body core temperature produce profound changes in the processes of temperature regulation, indicating that body core temperature is being monitored by internal thermoreceptors. Such internal thermoreceptors have indeed been identified in the CNS, namely in the regio praeoptica of the anterior hypothalamus (Brown and Brengelmann 1970; Bligh 1973; Hensel et al. 1973), in the lower brain stem (Jessen 1985) and in the spinal cord (Brück and Wünnenberg 1970; Thauer and Simon 1972; Simon 1974; Jessen 1985). Other sites outside the skin and the CNS where thermoreceptors are likely to occur are the dorsal wall of the abdominal cavity and the skeletal musculature (Simon 1974; Jessen 1985). In the anterior hypothalamus warm sensitive neurons, i.e., neurons which increase their discharge upon warming, seem to be more frequent than cold sensitive neurons. The same seems to be true for the thermoreceptive neurons in the lower brain stem and the spinal cord (Jessen 1985).

2.2.2
Low-Threshold Sensory Receptors in Muscle and Tendon

2.2.2.1
Muscle Spindles and Tendon Organs

This consideration of muscle spindles and tendon organs will be brief for two reasons. First, several recent reviews cover nearly every aspect of the structure and properties of these organs (Matthews 1972, 1981; Hunt 1974; Barker 1974). Second, their activation seems to have very little impact on any autonomic outflow (e.g., Sato and Schmidt 1973; Sato et al. 1981).

2.2.2.1.1
Structure and Function of Muscle Spindles

Muscle Spindles Are Stretch Receptors. Practically every skeletal muscle contains muscle spindles. There are none in the extraocular muscles of some animals, such as the rabbit, cat, dog and horse, but humans and many other mammals have numerous muscle spindles in these muscles as well. Each muscle spindle encloses in a capsule of connective tissue a number of intrafusal muscle fibers. The main sensory innervation of muscle spindles consists of the annulospiral endings that wind around the centers of the intrafusal fibers. Their afferents are thick, myelinated afferents called Ia fibers (see also Table 1). Each spindle receives only a single Ia fiber, which branches to form several annulospiral endings. Many but not all spindles have a second sensory innervation by group II afferents (Table 1). All intrafusal fibers have a motor innervation by thin motoaxons, commonly called γ-motofibers.

Muscles spindles lie in parallel with the extrafusal fibers, and when a muscle is stretched to its resting length most of the primary muscle spindle endings (those of the Ia fibers) discharge. During stretching the discharge rate increases. Isotonic contraction of the extrafusal musculature decreases the tension of the muscle spindle, so that the discharge ceases. These findings imply that the muscle spindle measures primarily the length of the muscle.

In addition to stretching there is a second way to excite the primary muscle-spindle endings, namely be contraction of the intrafusal muscle fibers via the γ-motoneurons. The contractile force exerted by the intrafusal muscle fiber is too small to change the length or tension of a muscle as a whole, even if all the intrafusal fibers in the muscle contract simultaneously. However, the intrafusal contraction stretches the central part of the intrafusal fibers, thus inducing excitation of the primary muscle spindle ending and its afferents, just as when the whole muscle is stretched. These two means of spindle activation, stretching the muscle and intrafusal contraction, can have additive effects. On the other hand, intrafusal contraction can more or less compensate for the effect of extrafusal contraction, so that even during shortening of the muscle the spindles can continue to function as length sensors.

Table 1. Sensory receptors of skeletal muscles and their tendon organs (based on a compilation by M. Wiesendanger in Schmidt and Thews 1995)

Type of receptor	Type of nerve fiber	Localization of terminal	Adequate stimulus	Response pattern	Central effects	Function
Spindle primary sensory ending	Ia	Muscle (in parallel to extrafusal muscle fibers)	Muscle stretch	Phasic and tonic	Monosynaptic excitation of homonymous motoneurons, disynaptic inhibition of antagonistic motoneurons	Myotatic reflex "tendon reflex", (H reflex), length servo, compensation of external disturbances (load), regulation of muscle tone (together with Golgi tendon organs)
Spindle secondary ending	II	Muscle (in parallel to extrafusal muscle fibers)	Muscle stretch	Tonic	Polysynaptic excitation and/or inhibition of homonymous motoneurons	Flexor reflex as a protective reflex; also participation in tonic stretch reflex; contribution to position sense
Golgi tendon organ	Ib	Border area between muscle and tendon (lying in series)	Increase in muscle (tension particularly by contraction)	Phasic and tonic	Disynaptic inhibition of homonymous muscles, disynaptic excitation of antagonists	Tension servo, regulation of muscle tone (together with Ia afferents)
Mainly free nerve endings	II, III, IV "Flexor reflex afferents"	Muscle, tendons, ligaments; also skin, periost, joint capsule	Mechanical and chemical noxious (algesic) stimuli	Phasic and tonic	Excitation of flexor motoneurons, inhibition of extensor motoneurons (both polysynaptic)	Flexor reflex and other protective reflexes

The primary endings of muscle spindles in the cat gastrocnemius-soleus muscle are strongly activated by warming and depressed by cooling (Mense 1978). Group II units with a high background discharge (muscle stretched) behaved like Ia units, whereas those with a low discharge rate (muscle relaxed) exhibited an increase in discharge rate on cooling and a marked depression on warming. With no evidence to the contrary, we may presume that in other mammalian species, including humans, changes in muscle temperature have similar effects on the sensory outflow from muscle spindles.

As reviewed and summarized by Mitchell and Schmidt in 1983, other changes in the extracellular fluid caused by muscular activity, such as changes in the composition, osmolarity, and pH, as well as endogenous inflammatory mediators or algesic substances seem to have little effect on muscle spindles (and for that matter on tendon organs). Judging from these observations, group I and II afferent units in muscle nerves do not seem to contribute to muscular chemo-nociception.

2.2.2.1.2
Structure and Function of Tendon Organs

All the muscles of terrestrial vertebrates are equipped with Golgi tendon organs. These sense organs consist of the tendon fascicles of about ten extrafusal muscle fibers enclosed in a connective-tissue capsule and supplied by one or two thick myelinated nerve fibers called Ib fibers (Table 1). The Golgi tendon organs are stretch receptors which lie in series with the extrafusal muscle fibers. When a muscle is stretched to about its resting length the tendon organs are usually silent. During stretching they begin to discharge, and during isotonic contraction they keep on discharging. During isometric contraction their discharge rate increases greatly, i.e., the Golgi tendon organ measures primarily the actual tension of its muscle (for the basic reflex action of muscle spindles and tendon organs, see Table 1).

2.2.2.2
Noncorpuscular Low-Threshold Sensory Receptors in Muscle and Tendon

Mechanoreceptors Reacting with Short Latencies to Contractions. Contraction-sensitive low-threshold sensory receptors with either group III or group IV afferent fibers have been repeatedly described (for literature see Mense 1986). Many of these contraction sensitive units were also found to be sensitive to stretch, although with a threshold far exceeding that of muscle spindles and Golgi tendon organs (Kaufman et al. 1983; Mense and Stahnke 1983). These units require noxious pressure to produce responses of a magnitude similar to those elicited by contractions. Thus it appears that these units are sensitive predominantly to forces acting in the direction of the long axis of the muscle. Most of the contraction-sensitive group III units showed a sudden onset of their response, slow or no adaptation during repetitive contractions, no afterdischarges, and a graded response to increasing forces of contraction (Fig. 7). Therefore, it is likely that their mechanism of activation is mechanical (Mense and Stahnke 1983), although an additional sensitivity to metabolites cannot be ruled out (Kaufman et al. 1983).

Mechanoreceptors and Chemoreceptors Reacting with Long Latencies to Contractions. Most of the group IV units that were activated by muscular contractions had a delayed response followed by afterdischarges. This time course of response is suggestive of a chemical or combined mechano-chemical mechanism of activation. The possible factors activating these types of units are numerous. Of course, chemical alterations in the contracting muscle were considered the principal stimulus, but up to now all attempts to identify such a chemical substance have failed (for a review see Mense 1986).

About 20% of the group III and IV units from cat gastrocnemius-soleus muscle which were tested with rhythmic contractions belonged to the contraction-sensitive units described in the two preceding paragraphs. During static (sustained) contractions, the proportion of group III and IV receptors being activated was found to be much higher (63% and 68%, respectively; Kaufman et al. 1983). The reason for this difference may be that static contractions are likely to impair the blood supply to the muscle, thus leading to an accumulation of metabolites which may contribute to the excitation of the receptors during exercise.

Both types of contraction-sensitive units are good candidates to signal the amount of work performed by a contracting muscle. It was proposed by Kao (1963) and others that such units should be termed "ergoreceptors."

Low-Threshold Touch-sensitive Units. Such units, which are insensitive to muscle stretch and contraction, have been described in the muscular and tendon tissue of the cat's gastrocnemius-soleus muscle (Mense and Meyer

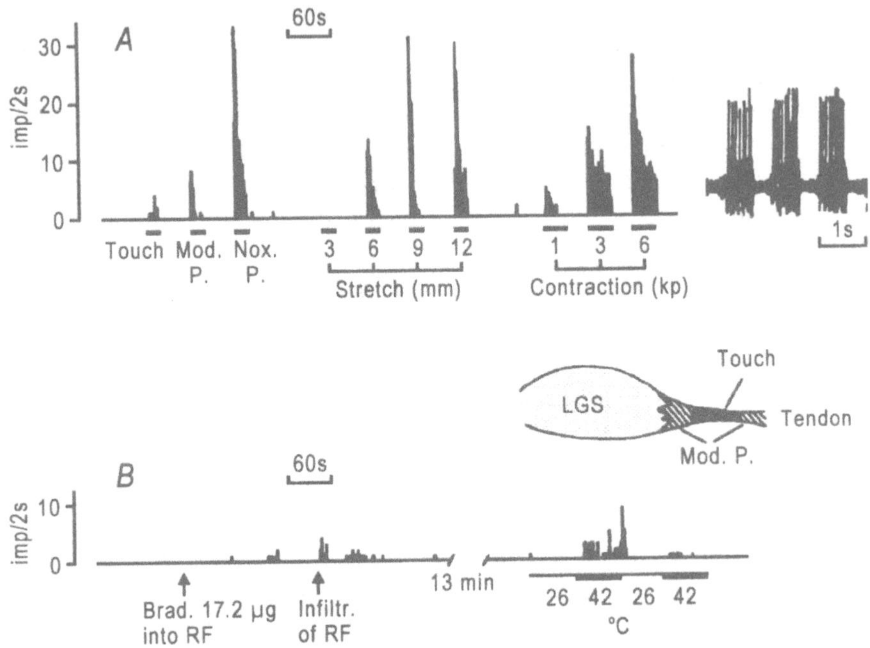

Fig. 7A,B. Response characteristics of muscular low-threshold mechanoreceptors with fine afferent nerve fibers. The figure depicts a contraction-sensitive unit with a group III afferent fiber, conduction velocity 26 m/s. Location of the receptive field (*RF*) on the calcaneal tendon. *LGS*, lateral-gastrocnemius-soleus muscle of the cat. A central RF area with very low threshold *(Touch)* is bordered rostrally and caudally by RFs with higher but still innocuous thresholds *(Mod. P.)*. **A** (from *left* to *right*), responses to local stimulation (touch; *Mod.P., Nox. P.,* moderate and noxious pressure, respectively), to stretch to the indicated length, and to contraction to the indicated force in kilopond (kp). The duration of the stimuli used is indicated by the length of the *bars* underneath the histograms. The specimen record on the *far right* shows the fiber's activity during rhythmic contractions with a force of 6 kp. **B** *Arrows*, single injection of bradykinin *(Brad.)* into or infiltration *(Infiltr.)* of the receptive field *(RF)* with the algesic agent. The response to warming of the tendon (*right*) was not reproducible and therefore not considered to indicate thermoreceptive properties of the unit. *Imp*, impulses. (Modified with permission from Mense and Meyer 1985)

1985). They could not be found in the skeletal muscle of the dog (Kumazawa and Mizumura 1977).

2.2.3
Low-Threshold Sensory Receptors in Articular Tissue

2.2.3.1
Response Properties of Group II Articular Afferents

Group II articular afferents (conduction velocity, 21–65 m/s) usually have no resting discharges and low mechanical thresholds (Fig. 8; see also Fig. 9, upper left-hand diagram). Most of them are excited by gentle local stimuli and by movements in the working range of the knee joint. Although they encode pressure and particularly movement stimuli up to the noxious range, their responses are more closely related to the particular type of stimulus (e.g., movement in a specific direction) than to intensity (innocuous vs. noxious). These results suggest that they presumably serve proprioceptive functions such as deep pressure sensation and kinesthesia (Dorn et al. 1991; Proske et al. 1988; McCloskey 1978; for a more complete review of the literature, see Schaible and Grubb 1993), and that they are probably not involved in joint nociception and pain. Almost none of them

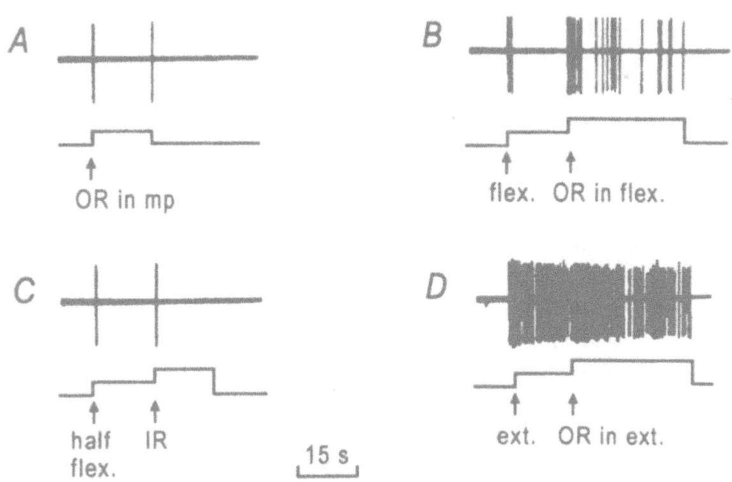

Fig. 8A–D. Responses of a group II unit to passive movements of a normal knee joint. **A** Outward rotation (*OR*) starting from the midrange position. *mp*, resting position. **B** Full flexion (*flex.*) and outward rotation starting from the flexed position. **C** Half flexion and inward rotation (*IR*) starting from half-flexed position. **D** Full extension (*ext.*) and outward rotation starting from extension. (From Dorn et al. 1991, with permission)

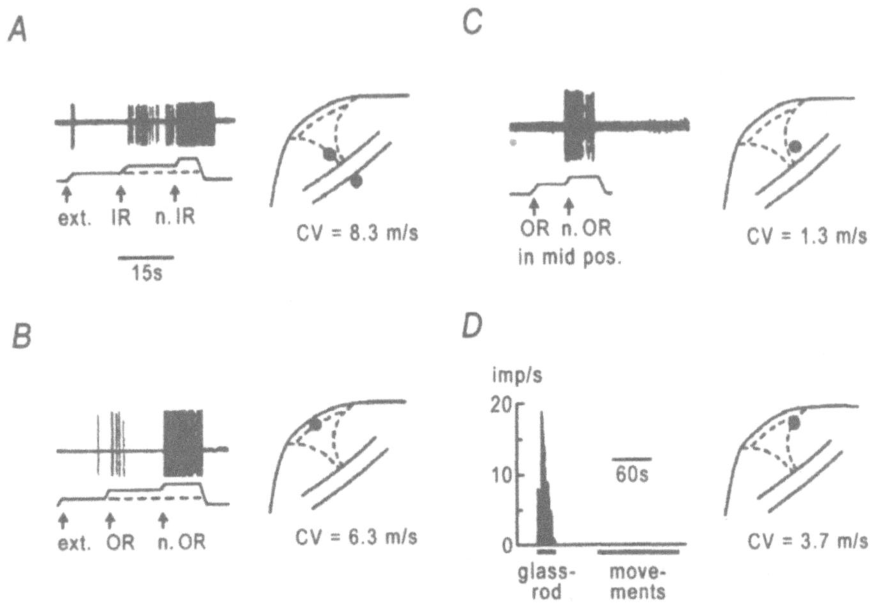

Fig. 9A–D. Response behavior of articular sensory units with fine afferent fibers (group III and IV fibers). Classification according to the response during passive movements of the knee joint. **A** Activated by non-noxious movements. **B** Weakly activated by non-noxious movements. **C** Activated only by noxious movements. **D** Not activated by movements. The *insets* on the right of each record display the receptive fields on the medial aspect of the joint capsule and the conduction velocity (*CV*) of each unit. The time scale in **A** also applies to **B** and **C**. *Ext.*, extension; *IR* and *OR*, innocuous inward and outward rotations, respectively (pronation and supination, respectively); *n.IR* and *n.OR*, noxious inward and outward rotations, respectively; *mid pos.*, mid-position (normal resting position of the joint between flexion and extension); *imp*, impulses. The activity in **D** is displayed as a peristimulus-time histogram; the movement program included noxious flexions, extensions, and rotations. The measurements were performed by H.-G. Schaible and R. F. Schmidt on units of the medial articular nerve of the cat's knee joint

respond to close i.a. injection of algesic substances. Group II afferents are equipped with corpuscular endings of the Ruffini, Golgi, and Pacini type (Boyd and Davey 1968) and these are located in the fibrous capsule, articular ligaments, menisci and adjacent periosteum, but not in the synovial tissue and the cartilage (see Sjölander et al. 1989; Johansson et al. 1991; Schaible and Grubb 1993).

2.2.3.2
Response Properties of Low-Threshold Group III and IV Articular Afferents

In articular group III and group IV afferent units, resting discharges occur in no more than one third of all fibers (see Fig. 12). The frequency of these irregular discharges is low, usually below 0.5 Hz (Schaible and Schmidt 1983a). With regard to their mechanosensitivity, the units with fine afferent nerve fibers can be grouped together according to their movement sensitivity (Fig. 9). The spectrum includes units that are readily (A) or only marginally (B) activated by non-noxious events; units that are only activated by noxious movements (C); units which are not activated by any movements, even extremely noxious ones (D). These movement-insensitive units are excited by local mechanical stimulation, i.e., they have circumscribed receptive fields (for further discussion, see below, Sects. 2.3.3, 2.3.4). In the group III range, the majority of afferents have low thresholds, whereas in the group IV range there a about equal percentages of the low-threshold and the high-threshold (nociceptive) type (see Table 3). As with the group II units dealt with in the preceding paragraph, the low-threshold fine afferent units may subserve proprioceptive functions such as deep pressure sensation and kinesthesia, although little experimental or clinical evidence is available on this matter.

In contrast to the group II afferent units, most of the group III and IV ones are activated and/or sensitized to mechanical stimulation by close i.a. injection of inflammatory mediators and/or algesic substances. With regard to their activation by such substances, afferent units would have to be classified as polymodal provided that tissue concentrations of such substances became sufficiently high under in vivo pathophysiological conditions. Regarding their sensitization by inflammatory mediators, it has to be appreciated that in inflamed tissue there is a sensitization not only of the high- but also of the low-threshold fine afferents, and this sensitization is most likely due to chemical factors (see Sects. 2.3.3 and 2.3.4).

2.2.3.3
Total Sensory Outflow at Rest and During Movement of Normal Joints

Using histological data and the results of recording the discharges of individual fine afferents, an estimate has been made of the number of afferent impulses reaching the spinal cord when the knee joint is at rest

Table 2. Outflow from a normal and an acutely inflamed knee joint of the cat via the medial articular nerve, MAN; for the composition of this nerve see Table 3, p. 40 (based on histological and electrophysiological data from the Institute of Physiology, University of Würzburg; compiled by B. Heppelmann 1990). A comparable outflow, unknown in detail, will take place via the posterior articular nerve, PAN (cf. Grigg et al. 1986)

	Normal joint	Inflamed joint	Change
Resting activity (impulses/30 s)	1,800	11,100	× 6.2
Flexion movement[a] (impulses/30 s)	4,400	30,900	× 7.0
Low-threshold units	240 fibers	550 fibers	× 2.3

[a]In the normal working range of the knee joint.

and in response to a single flexion movement from the midposition of the joint (Heppelmann 1990). Since sufficient data were only available for the MAN, the estimate has been restricted to that particular input (Table 2). In the normal joint 35% of the group III and 36% of the group IV afferents have a mean resting activity of 10 and 19 impulses/min, respectively. Group II fibers are barely active at rest. Altogether this makes for some 1800 impulses per 30 s entering the spinal cord from a knee joint resting in midposition (Heppelmann 1990; Schaible and Schmidt 1985; Dorn et al. 1991). A flexion, extension or rotatory movement in the working range of a normal joint excites 89% of the group II, 45.5% of the group III and 29.5% of the group IV afferents, i.e., they are low-threshold afferents (Heppelmann et al. 1987). In the MAN these percentages amount to 240 afferent fibers having low thresholds and 390 fibers having high ones (see also Table 3; p. 40).. Again, it is not absolutely settled which type of information is signaled to the nervous system by these discharges. Most units respond to more than one type of joint movement, i.e., only in an imprecise way may these discharges contribute to measuring joint movement and joint direction. During a simple flexion movement the afferent volley in the MAN consists of approximately 4400 impulses per 30 s (including the resting discharges).

2.3 Somatic Nociceptors

Nociceptors are afferent units that ideally are exclusively activated by tissue-threatening or tissue-damaging mechanical, thermal or chemical stimuli, and which are capable, by their response behavior, of distinguish between innocuous and noxious events. As an additional feature, nociceptors may have a high stimulation threshold; this feature is useful for the recognition of nociceptors in animal experiments and in microneurographic sessions when mechanical or thermal stimuli are applied, but less so upon chemical stimulation (Perl 1984; Besson and Chaouch 1987). It against the background of these considerations that the term nociceptor being used in this section on somatic nociceptors.

Studies on nociceptors have been performed over many decades, in many species and in a great variety of tissues. It is not surprising, therefore, that a certain degree of confusion exists as to the classification of nociceptors, the most frequently used classification being that of dividing them into mechanical nociceptors, thermal nociceptors, and polymodal nociceptors. Even if such receptor types do indeed exist, it seems risky to propose a strict classification since the response characteristics of nociceptors are variable (phenomena of sensitization and desensitization, see below). In addition, in many studies the sensitivity of nociceptors to all types of stimulation was not always systematically tested, adding a further amount of uncertainty to any rigid classification.

2.3.1 Cutaneous and Subcutaneous Nociceptors

2.3.1.1 Mechanonociceptors (HTM-Aδ-Units and CM Units)

HTM-Aδ-Units. This ubiquitous type of cutaneous nociceptor was first identified and characterized by Perl and his associates in the cat, and has also been studied in rats, monkeys, and humans (for reviews, see Burgess and Perl 1973; Perl 1984; Besson and Chaouch 1987). These nociceptors are nearly all associated with group III afferent fibers, particularly in the slower group III range. Therefore, and because of their high threshold to mechanical stimulation, they are often referred to as HTM-Aδ-units. However, in primates myelinated nociceptors have been found with conduction velocities in the fast group III/slow group II range.

HTM-Aδ-units are not activated by thermal or chemical stimuli and respond only to moderately intense or noxious mechanical stimuli. Their responses increase with the intensity of stimulation. This type of nociceptor makes up about 20% of the cutaneous group III fibers. Their receptive fields are usually large (1–2 cm in diameter), comprising several sensitive zones separated by areas where mechanical stimulation is without effect. The units do not exhibit spontaneous activity, and adapt slowly to a prolonged stimulus. In the monkey hand, the mean threshold of these nociceptors is more than three times higher than that for low-threshold, slowly adapting mechanoreceptors (Campbell et al. 1979).

CM Units. Purely mechanosensitive nociceptors with unmyelinated afferents have only recently been detected in human peroneal nerve (Schmidt et al. 1995). These units were not excited by heating the skin up to the subject's tolerance level. Their mechanical thresholds are similar to those of the CMH units described below. They were encountered about half as frequently as the CMH units.

2.3.1.2
Heat Nociceptors (CH Units)

This rare type of unit has recently been discovered in human skin (Schmidt et al. 1995). The units respond to heating with thresholds ranging from 45–48°C but are not activated by mechanical stimulation even with rather stiff von Frey filaments. This type of CH unit has been reported before in animals, albeit rarely (Georgopoulos 1976; Welk et al. 1984; Baumann et al. 1991).

2.3.1.3
Polymodal Nociceptors (CMH and AMH Units)

C Polymodal Nociceptors (CMH Units). These units with unmyelinated afferent fibers (group IV or C fibers) have been described in the cat, rat, rabbit, monkey, and in humans (for literature see Besson and Chaouch 1987; Handwerker and Kobal 1993). Frequently the term CMH units is used for these nociceptors. They are characterized by their responsiveness to intense thermal and mechanical stimuli as well as to irritant chemicals.

In the absence of stimulation the receptors do not produce any spontaneous activity. The receptive field is generally small; it consists of a cutaneous area with a diameter that can, on occasion, be less than 0.5 mm and that has a uniform sensitivity to mechanical stimulation, however, more complex receptive fields have also been observed.

The most effective stimuli exciting these polymodal C nociceptors are those of the greatest intensity applied by pointed objects. Activities of 50–150 action potentials/s have been observed with intense stimulation. After a large initial discharge, the response adapts and settles to a lesser level that can outlast the stimulation. Another characteristic of polymodal C nociceptors consists of the appearance of "fatigue" following the repetition of a stimulus to the same point in the receptive field.

Heat thresholds of single CMH units vary over a wide range between 40° and 50°C in the hairy and glabrous skin of the monkey and other mammals. No significant differences were found between the thermal thresholds of CMHs in hairy and in glabrous skin. CMH units are frequent in human skin nerves, and their heat and mechanical thresholds are in the same range as those of CMHs in other mammals (for references, see Handwerker and Kobal 1993).

Aδ-Polymodal Nociceptors (AMH Units). These units, often called heat and mechanoreceptive units (AMH units) have been encountered much less frequently than C polymodal units. In the monkey, their response to mechanical stimulation is similar to that of the polymodal C nociceptors (Georgopoulos 1976). The AMH units described hitherto in humans seem to have an excitability spectrum rather similar to that of human CMH units (Adriaensen et al. 1983).

2.3.1.4
Unresponsive Cutaneous C Nociceptors (CM_iH_i Units)

The human skin seems to be innervated by a considerable number of unmyelinated afferent units which in normal skin cannot be excited by strong mechanical stimuli (far beyond the threshold of the CMH units described above), and which also are not stimulated by heating to levels of 48°C or higher, temperatures that also surpass the thresholds of previously described human CMH units (Schmidt et al. 1995). These units were labeled CM_iH_i, with the subscript "i" standing for "insensitive." A certain proportion of these units could be excited either by mustard oil or capsaicin application or were sensitized by these chemical irritants to respond to mechanical and/or heat stimuli.

In the study of Schmidt et al. (1995), about one quarter of all group IV units encountered were of the CM_iH_i type. This proportion is similar to the range of 30% insensitive C units which have been found in the hairy skin in the monkey (Meyer et al. 1991) and also the 26% and 15% recorded, respectively, in vivo and in vitro in the hairy skin in rat (Kress et al. 1992). About one third of the CM_iH_i units were excited by mustard oil, and this proportion of CM_iH_i units can therefore be regarded as chemo-nociceptors. However, chemo-nociception is probably also mediated by other unit types, since about one third of CM and CH units and almost 60% of CHM units were sensitive to mustard oil.

2.3.1.5
Experimental Activation of Cutaneous Nociceptors

Since nociceptors have high thresholds to mechanical, thermal or chemical stimulation, it is somewhat difficult to excite them without activating low-threshold sensory receptors at the same time. All in all, in humans and primates, radiant heat stimuli have been found most suitable to activate nociceptors in isolation since many of them respond to heat (see above). Various devices have been used for this purpose, and the introduction of the CO_2 laser as a means to deliver concentrated thermal energy in short time frames has been proven to be particularly useful (for literature and a discussion of the method see Bromm and Treede 1991). Such heat stimuli simultaneously activate both Aδ (group III) and C (group IV) nociceptive afferents which terminate in the most superficial skin layers, without activating any other sensory structures.

2.3.2
Nociceptors in Muscle and Tendon

Nociceptors in muscle and tendon have been studied almost exclusively in animals (rat, cat, dog, monkey), the most thorough studies having been conducted by Mense and his associates (e.g., Mense 1986, 1991a,b, 1993; Mense and Meyer 1988; Hoheisel et al. 1993). Only preliminary microneurographic recordings from human muscular nociceptors are available, but they are sufficient to suggest that the basic neurobiological mechanisms of nociception in muscle and tendon are similar in all these species.

2.3.2.1
Localization and Structure of Muscle and Tendon Nociceptors

Muscle and tendon nociceptors all seem to have fine afferent fibers in the group III and group IV fiber range, with conduction velocities from 30 m/s to less than 1 m/s. A receptive ending typically consists of several branches or terminals which altogether form the receptor in the physiological sense. The main type of receptive ending of fine muscle and tendon afferents is the noncorpuscular "free" nerve ending. Their typical location is the wall of arterioles and the surrounding connective tissue; the capillaries proper are not supplied with these endings (Stacey 1969). "Free" nerve endings are actually not free or naked in the strict sense, as they are almost completely ensheathed by Schwann cells. Only some areas of the axonal membrane remain uncovered by Schwann cell processes and are directly exposed to the interstitial fluid (Andres et al. 1985). The exposed membrane areas are supplied with mitochondria and vesicles and show other structural specializations characteristic of receptive areas; they are presumed to be the transduction sites of these terminals. At the electron microscope level several different types of endings connected to group III and IV fibers have been described for the calcaneal tendon of the cat (Andres et al. 1985), but no correlation between structure and function of these various terminals is yet possible.

2.3.2.2
Response Properties of Muscle and Tendon Nociceptors

2.3.2.2.1
Muscle Nociceptors in Nonprimate Mammals

Muscle nociceptors which do not respond to everyday stimuli, such as weak local pressure, contractions and stretches within the physiological range, but which are easily excited by noxious intensities of mechanical stimulation were first described unequivocally for the cat's gastrocnemius soleus muscle (for a review see Mense 1993). In addition to being excited by such mechanical stimuli, the majority of these nociceptive units were readily excited by endogenous pain-producing substances such as bradykinin (Bk), serotonin (5-HT), and high concentrations of potassium ions (Fock and Mense 1976; Kumazawa and Mizumura 1977; Kaufman et al. 1982). Bk and 5-HT also have strong actions on blood vessels and appear to influence the vasculature at lower concentrations than nerve endings.

Therefore, these substances have been called vasoneuroactive substances (Sicuteri 1967).

The typical muscle nociceptor responds both to noxious local pressure and to injections of Bk (i.a. or i.m., Fig. 10), but there are also nociceptors which can be activated only by one type of noxious stimulation (mechanical or chemical). This finding may indicate that different types of nociceptors are present in skeletal muscle, as with the skin where mechano-, mechano-heat, and polymodal nociceptors have been identified (see above). The doses of pain-producing substances required in animal experiments to excite muscle nociceptors are similar to those eliciting muscle pain in humans upon intra-arterial or intra- and subcutaneous injection (Coffman 1966; Lindahl 1961). The same is true for the time

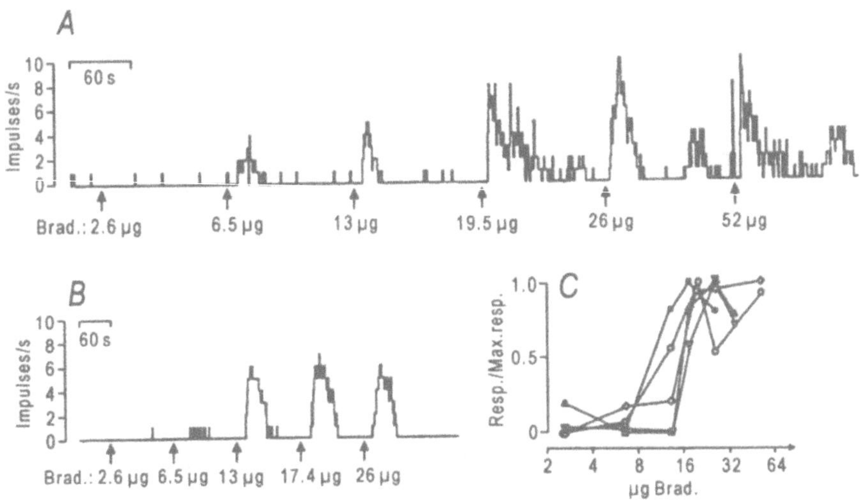

Fig. 10A–C. Discharge characteristics of muscular fine afferent units following stimulation with algesic substances. **A** Responses of a single group IV afferent (conduction velocity, 0.76 m/s) from the cat's gastrocnemius-soleus muscle (GS) to increasing doses of bradykinin *(Brad.)*. Appearance of separate afterdischarges at a higher dosage (19.5–52 µg). *Arrows* indicate start of injection, which was completed in 10–15 s. **B** Group IV unit from lateral GS; conduction velocity, 0.64 m/s. Saturation in the magnitude of response at a bradykinin dosage exceeding 13 µg. **C** Dose–response plots of five group IV afferent units from GS muscle. *Ordinate,* relative magnitude of response in terms of impulses evoked per injection. *Abscissa,* doses of bradykinin applied as single injections at intervals of 2 min (logarithmic scale). (Data from Mense 1977 in Mense 1986, with kind permission)

course of the receptor activation (latency and duration) which – in the case of Bk injections – resembles closely the time course of painful sensations elicited by Bk in humans. These findings increase the probability that chemically induced muscle pain is really due to activation of that set of noncorpuscular ("free") nerve endings that were classified as nociceptors in animal experiments.

Many nociceptive units in muscle have two receptive fields, i.e., they can be activated from two separate areas in the muscle. Some units seem to have one of their receptive fields in the muscle, the other in the skin nearby. The anatomical basis of this feature probably branching of the afferent fiber close to the area of termination (alternatively the branching could occur in the spinal nerves, see below). Silent or sleeping nociceptors have not yet been identified muscle or tendon tissue.

2.3.2.2.2
Muscle Nociceptors in Humans

The response behavior of human muscle nociceptors explored by microneurography seems to be identical to that described above for those in the cat and dog. They respond to noxious local pressure and algesic substances with a discharge frequency and time course identical to those observed in animal experiments. The chemically induced discharges of the human muscle nociceptors parallel very closely the intensity and duration of the subjective pain elicited by the injection of the stimulants. Furthermore, electrical stimulation of muscle nociceptors leads to cramp-like subjective sensations that are characteristic of muscle pain. Although the possibility has to be considered that the recorded nociceptive unit is not stimulated in isolation, but rathere together with other fibers, these data support the assumption that the muscle nociceptors observed in animal experiments really mediate muscle pain.

2.3.3 Articular Nociceptors

2.3.3.1 Fine Structure of Group III and IV Articular Terminals

2.3.3.1.1 Terminal Branches of Fine Articular Afferents

The most detailed studies on the three-dimensional ultrastructure of articular nociceptors have been performed on the knee joint of the cat, combining electron microscopy with stereology and three-dimensional reconstruction (Heppelmann et al. 1989, 1990a,b, 1995). Within the peripheral nerves, each group III fiber has its own (the myelin sheath-producing) Schwann cell, whereas several group IV axons are usually tied together by one Schwann cell forming Remak bundles. This principle is maintained within the transition zone between peripheral nerve and sensory endings, where group III fibers are no longer myelinated but still completely or partly enclosed by the perineurial sheath. When the fibers leave the perineurium to form sensory endings, group III fibers may join group IV fiber bundles or continue as single fibers, whereas nearly all group IV axons keep sharing the accompanying Schwann cell with other group IV axons. During the course of the sensory endings, the bundles become smaller and smaller with a decreasing number of sensory axons. As a result, single group III endings are quite common, whereas single group IV endings are rarely found.

2.3.3.1.2 Ultrastructure of the Terminal Beads and End Bulbs

The sensory axons of group III and group IV fiber endings are characterized by varicose segments, the "sensory beads" (linked together by thin segments of the sensory axon), measuring 5–12 μm in length in group III and 3–8 μm in group IV fibers (Fig. 11). The sensory axons, in particular their beads, are not completely surrounded by the Schwann cell, but exhibit free areas at which the axolemma is separated from the surrounding tissue by nothing than the basal lamina. At such sites, where the axoplasm seems to bulge through the gaps between the Schwann cell processes towards the adjacent tissue, the axolemma is slightly thickened. Membrane receptors may be located there. The axoplasm that underlies the bare areas of axolemma shows a faint filamentous substructure and appears more elec-

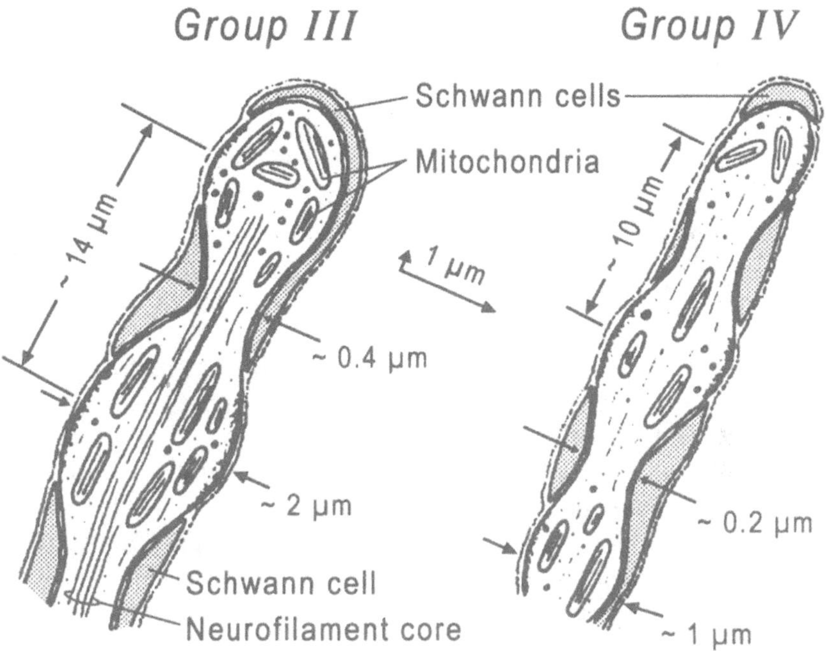

Fig. 11. Terminal portion of a group III (*left*) and group IV (*right*) nerve fiber with approximate dimensions obtained by three-dimensional reconstructions from serial electron microscope sections of the medial capsule of the cat's knee joint. The sensory axons consist of periodically arranged thick and thin segments forming spindle-shaped beads. The axolemma is not completely ensheathed by its accompanying Schwann cells; the bare areas presumably are receptive sites. A central axis of neurofilaments (*Neurofilament core*) exists only in sensory axons of group III fibers. (Data collected by B. Heppelmann, K. Messlinger, and others in the laboratory of R. F. Schmidt; Schmidt et al. 1994)

tron-dense than elsewhere in the sensory axon. Such submembranous axoplasm modifications have been described as an essential component of the "receptor matrix" in several types of nerve endings (see Andres and von Düring 1973). They may represent a special membrane-bound cytoskeleton. The tips of the sensory endings form final beads ("end bulbs"), the ultrastructure of which is in no way different from all other beads, with the exception of the neurofilament core in group III endings (cf. Fig. 11) that is lacking within the last segment. It is most likely that the sensory beads represent the receptive sites of the sensory endings.

2.3.3.2 Response Properties of Group III and IV Articular Nociceptors

2.3.3.2.1 Levels of Responsiveness in Articular Nociceptors and Other Fine Afferents

As outlined above (see Sect. 2.2.3), the mechanosensitivity of fine afferent units from normal joints ranges from those with very low threshold to those insensitive to mechanical stimulation ("sleeping nociceptors"). As can be seen from Fig. 12 (see also Fig. 9), in the medial articular nerve (MAN) of the cat's knee joint, approximately 55% of the group III and 70% of the group IV units either respond only to (potentially) noxious movements or fail completely to respond to movements (Schaible and Schmidt 1983b). Units in the posterior articular nerve of the same joint (PAN) can also be grouped into different categories. But the vast majority of the group IV fibers in this nerve do not respond to mechanical stimulation at all (Grigg et al. 1986). Therefore, it has to be appreciated that at least 60% of all fine afferent fibers innervating the knee joint of the cat have nociceptive properties when classified by mechanical stimuli (Table 3).

Table 3. Number of nociceptive units and distribution of neuropeptides in primary afferents of the medial articular nerve (MAN) (data from the Institute of Physiology, University of Würzburg; compiled in Schmidt et al. 1994)

	n	%
Electrophysiology and morphometry		
MAN afferent fibers	630	100
Nociceptive afferents		
Group II	6	1
Group III	71	11
Group IV	310	49
Total	*387*	*~60*
Non-nociceptive afferents	243	~40
Immunohistochemistry		
Substance P (SP)		18
Neurokinin A (NKA)		8
Calcitonin gene-related peptide (CGRP)		35
Somatostatin (ST)		18

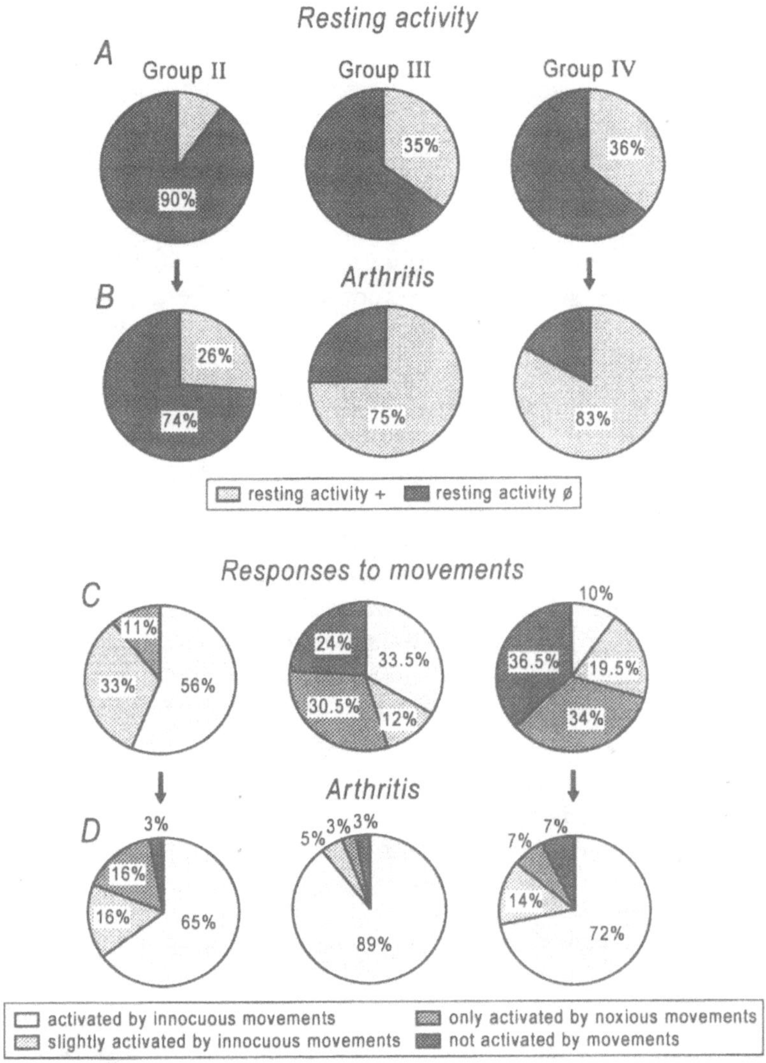

Fig. 12A–D. Sensitization of articular sensory receptors as a consequence of tissue damage. Relative distributions (percentages) of the various afferent units in the knee joint of the cat as classified with regard to their conduction velocities (*groups II, III, IV*), their resting activity (**A,B**), and their responsiveness to passive joint movements (**C,D**). **A,C** Results obtained from normal joints. **B,C** Results obtained during experimental arthritis. Note the rather small changes induced by the inflammation in the group II afferents. The figure does not show that there is a considerable increase in the frequency of the spontaneous and evoked discharges induced by inflammation. (Data from R. F. Schmidt and associates; Schmidt et al. 1994)

2.3.3.2.2
Levels of Unresponsiveness in Silent Units

For silent units, three categories with different "levels of unresponsiveness" have to been recognized resolved: (1) units which do not respond to movements but have a receptive field and a response to close i.a. application of KCl (as an indication of chemical excitability); (2) units that respond neither to movements nor to local mechanical stimulation (no detectable receptive field) but show a typical response to KCl; and (3) units having no mechanosensitivity and no clear response to KCl. Resting activity is practically always absent in all three types of units. Units of all three categories become sensitized in the course of inflammation, some of them to the point that they develop activity at rest and vigorous responses to movements in the working range of the joint.

Actually, the existence of mechanoinsensitive, i.e., silent, primary afferents in the articular nerves of the cat's knee joint slowly but steadily emerged over the years 1983–1988 (Schaible and Schmidt 1983a,b, 1984, 1988a,b; Grigg et al. 1986). In addition, it became obvious that there is most likely, both under physiological and under pathophysiological conditions, a continuity of mechanical thresholds for articular fine afferents ranging all the way from very low to deep into the range of silent units. Under physiological conditions the spectrum of thresholds is rather evenly distributed over this range, whereas under pathophysiological conditions, such as an experimental arthritis, all units become sensitized and the spectrum of thresholds is skewed, i.e., most units can now be activated by stimuli which under normal conditions are completely innocuous.

2.3.3.2.3
Silent Nociceptors in Others Tissue and in Other Species Including Man

Since their original description in articular nerves of the cat, silent primary afferent units have been described in various tissues. They seem to be widespread if not universal in the mammalian somatoafferent systems. A first brief review on this topic published in 1990 by McMahon and Koltzenburg (1990) summarized reports on the existence of silent primary afferents in the urinary bladder of the cat (Häbler et al. 1988) (see also Häbler et al. 1990), and preliminary evidence on silent units in the skin of the rat and monkey which was substantiated by more extensive publications in the following year (rat, Handwerker et al. 1991; monkey, Meyer et al. 1991). In the same year silent afferent units were described in the colon of the cat (Jänig and Koltzenburg 1991).

Those afferent units, which in the normal joint and in other tissues are not excited by any movements, may lack any significant function except under pathological conditions (Schaible and Schmidt 1983b; Schaible and Grubb 1993). Kruger supposed that in the normal tissue those afferents may subserve an efferent rather than an afferent function (Kruger 1987). Fibers that are only activated by noxious mechanical stimuli can be referred to as nociceptors as defined by the specificity theory. The units that are readily excited while the joint is being moved through its normal working range cannot be considered nociceptive (for further discussion on movement-insensitive and completely silent units, see below).

2.3.4 Sensitization and Desensitization of Somatic Nociceptors

2.3.4.1 General Aspects of Sensitization of Nociceptors

2.3.4.1.1 Decrease in Threshold by Inflammatory Mediators

In any consideration of the mode of operation of the peripheral nociceptive system it has to be appreciated that the threshold of somatic nociceptors (and for that matter also of visceral ones) is neither uniform for their total population nor for any given nociceptor always at the same value. In healthy tissue nociceptors either have a high threshold (i.e., they can only be activated by tissue-threatening or tissue-damaging stimuli, see above) or they are, as also discussed above, insensitive even to noxious stimulation, i.e., silent or sleeping nociceptors). However, pathophysiological alterations of the tissue – such as inflammation – sensitize the nociceptors, i.e., their threshold for thermal and/or mechanical stimulation will be reduced so that even innocuous stimuli lead to their excitation. In addition there is an "awakening" of the sleeping nociceptors. At the same time, as exemplified by articular nociceptors below, many nociceptors start to discharge in the absence of stimulation and their original low frequency resting discharges increase in frequency. This sensitization of the normal and the normally silent nociceptors is one of the major factors responsible for the clinical finding that inflamed organs may be painful at rest, and that innocuous stimulation of such organs gives rise to painful sensations.

Most likely, the sensitization is caused by inflammatory mediators (e.g., bradykinin, serotonin) which are released from the lesioned tissue or which are being synthesized at a higher rate as a consequence of the lesion (e.g., prostaglandins, leukotrienes). Through their action on the vessels (vasodilatation and/or increase the permeability of the vessel walls) they contribute to the development of the typical symptoms of inflammation such as swelling, warming and reddening, and at the same time they lead to the sensitization of the fine primary afferents causing the tenderness, hyperalgesia and spontaneous pain discussed in the preceding paragraph.

2.3.4.1.2
Efferent Function of Nociceptive Terminals

In relation to the peripheral sensitization just discussed, it should be pointed out here that many if not all nociceptive units have efferent functions in addition to serving as afferent communication channels. Upon activation of a nociceptive ending, neuropeptides (e.g., substance P, calcitonin gene-related peptide, see Table 3) are released from the activated terminal as well as from those antidromically invaded by the action potential on its way to the spinal cord (hence the old expression "axon reflex" for this phenomenon, see Sect. 3.4). The neuropeptide release, in turn, contributes to the inflammatory tissue processes by amplifying the action of the other locally released or synthesized inflammatory mediators. This amplification is called the neurogenic component of inflammation.

In relation to the topic of this review, it may be permissible to speculate that tissue manipulation, e.g., massage, may excite sensory terminals in the manipulated tissue inducing thereby the release of neuropeptides and possibly other substances. These substances may contribute to the therapeutic effects of the manipulations by exerting not only sensitizing actions on the afferent units but by having additional effects on the surrounding tissue, e.g., vasodilatation and plasma extravasation.

2.3.4.1.3
Central Sensitization as a Consequence of Nociceptor Sensitization

Peripheral sensitization will, upon stimulation, enhance the afferent inflow from the organ which the nociceptors (and, presumably, the low-threshold fine afferents, too) have been sensitized. In addition, the increased afferent inflow induces in the secondary and later afferent neurons receiving this input an increase in excitability. This central sensitization, induced by and acting "on top" of the peripheral one enhances the central actions (such as somatoautonomic reflexes) of any incoming afferent volley far above what may be expected from the size of the afferent volley alone. The mechanism of central sensitization is still poorly understood, but current evidence suggest that not only the activation of various types of glutamate receptors, such as the *N*-methyl-D-aspartate (NMDA) receptor but also increased synthesis and spinal release of neuropeptides are involved (Neugebauer et al. 1993, 1994a,b, 1996; Schaible et al. 1994).

2.3.4.2
General Aspects of Desensitization of Nociceptors

2.3.4.2.1
Desensitization by Nonsteroidal Anti-inflammatory Drugs

In relation to the main theme of this review, three examples of pharmacological desensitization have to be mentioned. The first is the mode of action of nonsteroidal anti-inflammatory drugs (NSAIDs). Such drugs, acetylsalicylic acid (ASA) and indomethacin, are used for the analgesic and antiphlogistic treatment of inflammation. For a long time, it was assumed that not only the antiphlogistic but also the analgesic effect of these drugs was mediated by an action at the inflammation site. This assumption was based on the results of different experimental approaches but not on the recording of single unit activity originating in the inflamed tissue (for a review of the literature see Heppelmann et al. 1986). Meanwhile, such direct observations have shown that resting discharges and movement-evoked activity originating in the inflamed tissue were reduced following the application of these drugs. This depression presumably reflects the analgesic action of these anti-inflammatory substances at the peripheral nerve fiber (Heppelmann et al. 1986; Guilbaud and Iggo 1985).

The time course of these effects is in agreement with clinical experience. In patients, the analgesic effect of aspirin starts within the first hour after the application, and it may be maximal after 1 h and lasts two to several hours. Depressive effects on the resting activity in single units from inflamed knee were found within the first hour after intravenous injection, starting in most units within the first 10 or 20 min. Thus the onset of the depression and the continuous decline of ongoing discharges fit well with the clinical manifestation of pain relief after treatment with anti-inflammatory drugs. In yet another aspect our results fit well with clinical experience: even after long observation times, the resting activity and the movement evoked-discharges were never completely blocked. The extent of sensitization was reduced but sensitization was never abolished within the recording time. This may correspond to the weak analgesic potency of NSAIDs.

2.3.4.2.2
Desensitization by Capsaicin

The second drug to be mentioned is capsaicin, which is well known as a toxic agent selectively impairing and/or destroying slowly conducting afferent fibers. Low doses of capsaicin can be used to activate such afferents and to elicit acute painful sensations. In cutaneous nerves this drug seems to preferentially excite thin myelinated and unmyelinated afferents which are nociceptive, whereas non-nociceptive afferents are rarely activated by the compound (for a review of the literature see He et al. 1990). For joint afferents, a similar selectivity for high-threshold afferents has been found (Fig. 13) (He et al. 1988). However, it has to be appreciated that this selective action is certainly not an absolute one. It is best, although only in a negative

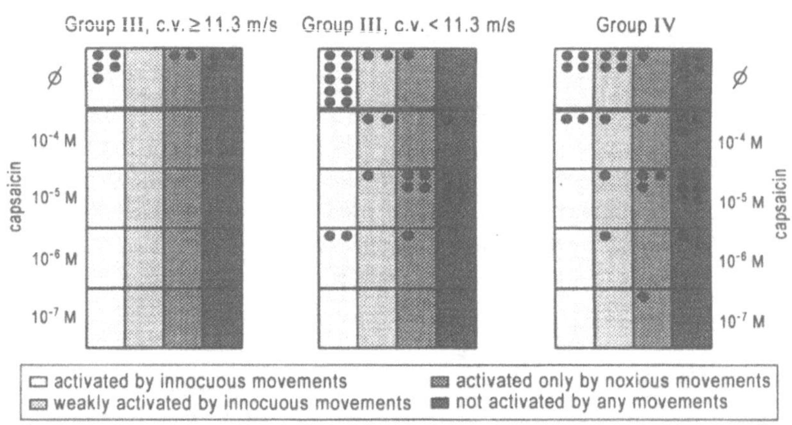

Fig. 13. Sensitivity to capsaicin of a population of 84 fine afferents recorded from the knee joints of 24 cats. The units were classified in three panels in regard to their conduction velocity and to their responses to passive movements of the knee joint (*left* to *right* in each panel, see key at the bottom of the figure; see also Fig. 9). The study included only units that were readily excited by a close i.a. bolus injection of 0.3 ml twice isotonic KCl solution. Units that did not respond to a 0.3-ml bolus injection of 10^{-4} *M* capsaicin were classified as unresponsive (ϕ; *uppermost row* in each panel). (Data obtained by H. He, R. F. Schmidt, and H. Schmittner, assembled in Schmidt et al. 1994)

sense, for group III fibers with conduction velocities greater than 11 m/s which were never excited even by high doses of capsaicin regardless of their mechanical threshold. For the slow group III fibers as well as the group IV ones, many exceptions in both directions were noted, i.e., units with low mechanical threshold being excited by capsaicin, and high-threshold ones remaining unaffected. Nevertheless, it seems worth mentioning that capsaicin is "better" in discriminating between high-threshold and low-threshold fine afferent units than any other substance (e.g., bradykinin, various prostaglandins, serotonin) tested so far in this respect.

The excitation induced by capsaicin in fine articular afferent units is closely coupled to a suppression of the responses to subsequent mechanical and chemical stimuli (He et al. 1990). Such a "desensitizing" effect of capsaicin on afferents is in line with other studies which used different preparations and different types of recording (literature in He et al. 1990). Our own and the other studies suggest that these effects are largely independent of the target tissue of the afferents and also of the species. Our experiments on single identified afferents also showed that the desensitizing effect is evoked in myelinated as well as in unmyelinated afferents, and that this desensitization is predominantly seen in nociceptive units. It also became evident that the suppression of the responses was closely related to an excitatory effect of capsaicin, as responses were not inhibited in units which were not excited by capsaicin or were only excited by high doses. The mechanism of this inhibitory action of capsaicin is not known, but there are good reasons to believe that capsaicin triggers an intraneuronal process which subsequently inhibits the transduction process (He et al. 1990).

2.3.4.2.3
Desensitization by Opioids

The third example is the finding that opiates can exert a desensitizing action at neuronal sites in afferent fibers. Opiate receptors were identified on small-diameter afferent terminals in the spinal cord and on cell soma in dorsal root ganglia (Stein et al. 1990). An inhibitory effect on the calcium-dependent component of the action potential in dorsal root ganglion neurons in culture was described for a range of opiate agonists. The vasodilatation and extravasation in the skin induced by antidromic nerve stimulation were inhibited by various opiates, and this may result from inhibition of neuropeptide release from sensory terminals during electrical stimulation. Under various experimental and clinical conditions a series of opiates and their antagonists were shown to have, respectively, peripheral analgesic and algesic effects (for literature see Russell et al. 1987; Stein et al. 1991; Czlonkowski et al. 1993; Stein 1993).

We had contributed to the unfolding story of peripheral opiate receptors and their potential analgesic action by studying, in 1987, the peripheral effects of a range of μ- and κ-opiate agonists on the spontaneous impulse traffic in small-diameter afferents from the acutely inflamed joint (Russell et al. 1987). At the time, the model was chosen because, as discussed above, the symptoms of inflammation could clearly be reduced by cyclo-oxygenase inhibitors (Heppelmann et al. 1986) or partly mimicked by prostaglandins E1 and E2 (Heppelmann et al. 1985; Schaible and Schmidt 1988a). In this model we expected the antinociceptive effect of opiates to be reflected in a depression of the spontaneous discharges of individual fine primary afferent fibers. As can be seen from Fig. 14, these expectations were fully justified. The spontaneous discharges of nearly all of the tested fine afferent units from inflamed joints were significantly inhibited by one or several opiates in a naloxone-reversible manner. These data provided the first electrophysiological demonstration that opiates may act on opiate receptors located at peripheral sites of primary afferent fibers and hence exert a peripheral analgesic effect. Most of the subsequent evidence on the peripheral analgesic action of opiates is still of a much more indirect nature (Russell et al. 1987; Stein et al. 1991; Czlonkowski et al. 1993; Stein 1993).

2.3.4.3 Special Aspects of the Sensitization of Cutaneous, Muscular, and Articular Nociceptors

2.3.4.3.1 Cutaneous Nociceptors

Human cutaneous nociceptors – like those in animals – sensitize following a mild burn injury, and their lowered threshold and enhanced responsiveness give rise to an increased pain sensation, called primary hyperalgesia, to noxious mechanical and thermal stimuli (for literature, see Willis 1992). If the sensitization is such that lightly brushing or stroking the skin already evokes pain or discomfort, the term allodynia is used to described this pathophysiological pain sensation. Allodynia can be considered a part of primary hyperalgesia. Pain perceived from gentle moving tactile stimuli is synonymously described as brush-, touch- or stroke-

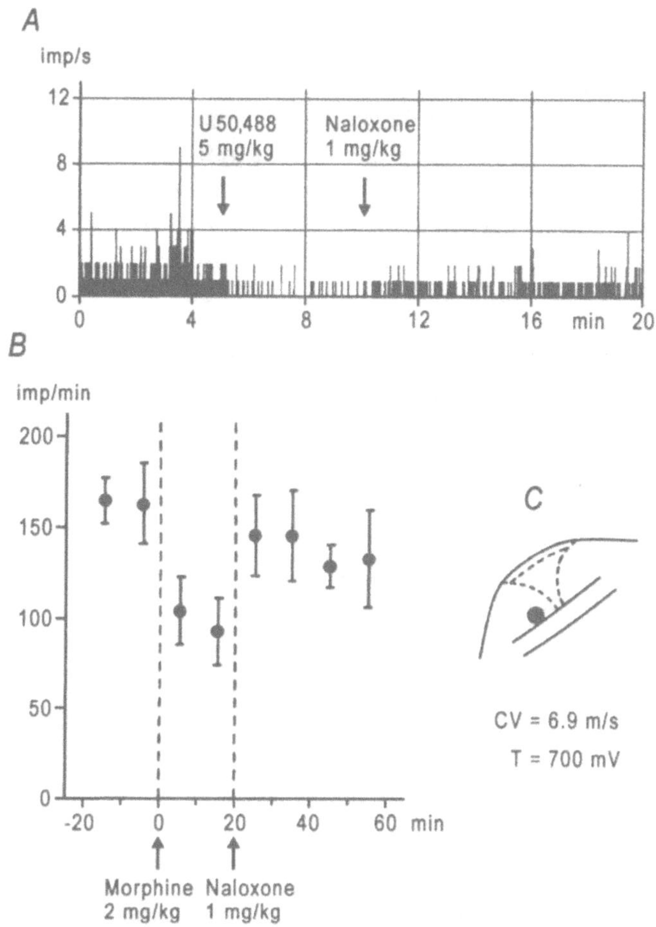

Fig. 14A–C. Effects of close intra-arterial injection of **A** the κ-opiate agonist U50488 and **B** the μ-opiate agonist morphine on the spontaneous discharges of a single group III afferent unit in the medial articular nerve of an acutely inflamed knee joint of the cat. Naloxone reversed the actions of both drugs. **C** Location of the receptive field of this unit on the medial aspect of the knee joint. *T*, electrical threshold of the fiber under observation; *CV*, conduction velocity. (Modified from Russell et al. 1987)

evoked pain. In addition to mild burn, chemical stimuli, such as applying mustard oil to the skin, can be used to evoke allodynia and hyperalgesia. These experimentally induced pain states are similar to those seen in patients with neuropathic pain. In addition to peripheral sensitization, it is likely that central sensitization (see above) also contributes to the hyperalgesia (for literature, see Willis 1992). For instance, during differential compression nerve blocks the allodynia part of brush evoked pain disappeared as soon as the low-threshold myelinated afferent fibers (Aβ-fibers, which normally encode nonpainful tactile sensations) were blocked, and this was taken as evidence that central sensitization is responsible for Aβ-mediated hyperalgesia (Koltzenburg et al. 1994).

Hyperalgesia to warm (heat) or cold stimuli (thermal hyperalgesia) may occur concurrently with, or independent of, spontaneous pain and mechanical hyperalgesia. In turn, heat and cold hyperalgesias may be expressed concurrently or independent of each other. The contribution of peripheral mechanisms is currently supported by reasonable evidence for heat hyperalgesia and weak evidence for cold hyperalgesia in humans. Central hyperexcitability mechanisms are strongly suspected (for literature, see Willis 1992).

Hyperalgesia within the region of injury is called primary hyperalgesia. The cutaneous hyperalgesia that can develop in a wide area of normal skin surrounding a local injury is called secondary hyperalgesia. The evidence presently available, although still controversial, suggests that the secondary hyperalgesia surrounding a cutaneous injury is mainly, if not exclusively a consequence of central sensitization, and not the result of the sensitization of peripheral nerve fibers (for literature, see Willis 1992).

2.3.4.3.2
Nociceptors in Muscle and Tendon

Sensitization by Mechanical Forces. A variety of pathophysiological conditions may give rise to activation and sensitization of nociceptors in muscle and tendon. The first of these which comes to mind is trauma by acute and strong mechanical forces. Such forces not only excite muscle nociceptors mechanically but they disrupt blood vessels and muscle fibers which – as outlined above – will result in an increase in tissue concentrations of endogenous vasoneuroactive substances, thereby sensitizing nociceptors. Similar although somewhat more subtle changes can be observed when a muscle is forced to perform physical work of unaccustomed intensity or duration. A typical way to overload a muscle is to have it perform eccentric contractions or negative work. The characteristics of eccentric contractions are that the external forces acting on the muscle are greater than those produced by the muscle itself, i.e., the muscle performs lengthening contractions (Asmussen 1956; Cavanagh 1988). Walking downhill or downstairs are typical situations in which extensor muscles of the leg contract in this way. After heavy eccentric exercise of this type, biopsy material shows signs of a necrotic inflammation with disruption and swelling of muscle fibers and cellular infiltration of the extracellular space (Asmussen 1956; Round et al. 1987; Friden et al. 1988; Stauber et al. 1990). The damage to the muscle fibers has been explained by assuming that during negative work a smaller number of motor units is active than during positive work of the same intensity. Therefore, the mechanical stress to the Z bands in the sarcomeres and to the connective tissue is higher.

Mechanism and Time Course of Sensitization by Mechanical Forces. Muscle soreness usually occurs with a delay of more than 10 h after the end of heavy unaccustomed exercise. Several hypothesis in regard to its origin have been put forward (for literature see Mense 1993), the most likely being the "torn-fiber" theory of Hough (1902) which states that during exercise muscle fibers and fibers of the connective tissue are being ruptured. This leads to repair processes which are accompanied by local swelling and sensitization of nociceptors, and thus some aspects of inflammation (Staton 1951).

Ischemic Sensitization. Ischemia of a resting muscle is not painful but when a muscle is forced to contract under ischemic conditions pain develops within about one minute. The mechanism underlying this type of ischemic pain is still a matter of controversy. A substance probably more likely involved than others is bradykinin. The kinin is released from plasma proteins during ischemia (Sicuteri et al. 1964) and, because of its strong sensitizing action on nociceptors, is likely to contribute to the pain of intermittent claudication. The nociceptive output during ischemic contractions is probably carried by a small subpopulation of group IV muscle nociceptors which are not or are only weakly activated during contractions without arterial occlusion, but which show strong excitation when the same amount of muscle work is performed under ischemic conditions (Mense and Stahnke 1983).

Inflammatory Sensitization. Myositis, experimentally induced by infiltrating a muscle with carrageenan, produces, both in the cat and in the rat, the typical signs of nociceptor sensitization, namely an increase in the background activity and an increase in the proportion of fine afferents being activated by weak mechanical stimuli (Berberich et al. 1988; Diehl et al. 1988). The background discharge is irregular, often intermittent in nature, with phases of bursting activity alternating with long periods of silence. Such discharges are likely to cause spontaneous pain. The sensitization of the fine afferent units also offers an explanation for the tenderness of inflamed muscle and the pain during movements. According to these results the tenderness is mainly due to a sensitization of group IV receptors, whereas spontaneous pain and the dysesthesias (common in patients with myositis) are probably caused by activity in group III units.

Mode of Generation of Trigger Points. A trigger point is a painful and tender muscle hardening from which local twitch responses and referred pain can be elicited by local pressure (Travell and Simons 1983; Simons 1987). The exact mechanisms underlying the formation of trigger points are unknown. A trigger point is silent in the EMG, therefore a reflex contraction via α- or γ-motoneurons can be excluded. According to Mense (1993), a possible explanation is that a muscle lesion leads to the rupture of the sarcoplasmic reticulum and releases calcium from the intracellular stores. The increased calcium concentration causes sliding of the myosin and actin filaments; the result is a local contracture (myofilament activation without electrical activity) that has a high oxygen consumption and causes hypoxia. An additional factor may be the traumatic release of vasoneuroactive substances which produce a local edema that in turn compresses venules and enhances the ischemia and hypoxia. Because of the hypoxia-induced drop in ATP concentration, the function of the calcium pump in the muscle cell is impaired, and the sarcoplasmic calcium concentration remains elevated. This perpetuates the contracture (Travell and Simons 1983).

2.3.4.3.3
Articular Nociceptors

Changes in Resting Activity. During inflammation resting activity is observed in 75% of group III and 83% of group IV units of the MAN (Fig. 12, lower left panel). The discharges are irregular and sometimes of high frequency. Both the percentages of units with resting activity and the frequencies of their discharges are more than twice as high as in the control sample. Their resting activity might represent the neural correlate of spontaneous pain, and the large number of fibers exhibiting resting activity under inflammatory conditions also implies (and has been confirmed by direct observation, see below) that there are numerous nociceptors (usually silent in the normal joint) that exhibit resting activity under these conditions.

Changes in Evoked Activity. Nearly all fine afferent units from inflamed joints have low thresholds to movement, and most of them respond well to flexion and extension (Fig. 12, lower right panel). The increase in the number of easily excitable afferents corresponds to the clear decrease in the number of units belonging to the other classes. Particularly the number of units which in the normal joint respond only to noxious movements or respond not at all becomes very small as soon as the joint is inflamed.

Time Course of the Sensitization of Articular Afferents During Experimental Arthritis. Remarkable differences exist in the time course of sensitization of the various types of articular afferents (Schaible and Schmidt 1988b,c). In high-threshold afferents, including originally silent ones ("sleeping" nociceptors), the sensitization becomes evident within the second to third hour after induction of inflammation, with a further increase later on, and persists for many hours. In contrast, low-threshold units in the group II and III fiber ranges, develop increased activity mostly in the first hour after the injection of the inflammatory compounds, sometimes starting immediately after the injection (usually low-threshold group II units). These increased responses vanish more or less completely in the later stages of the inflammation.

2.3.4.3.4 Total Sensory Outflow at Rest and During Movement from Inflamed Joints

The percentage of fibers having resting activity and their mean discharge frequency increase considerably in the course of inflammation (Fig. 12). As already mentioned above, in the group III range 75% and in the group IV range 83% of the afferents become spontaneously active with a mean resting activity of 54 and 46 impulses per minute, respectively. Also, some of the group II afferents develop a small resting activity. Both factors contribute to the increase of afferent activity from 1800 to 11 100 impulses reaching the spinal cord within 30 s with the joint in resting position (Table 2) (Dorn et al. 1991; Schaible and Schmidt 1988b,c). The peripheral neurobiological correlate for pain at rest from an inflamed joint seems to consist of these impulses conducted through the MAN together with those in the PAN which most likely undergo a similar sensitization. It appears not too far fetched to assume that similar mechanisms of sensitization are operative not only in acute inflammatory but possibly also in chronic pain processes. Inflammation of the joint reduces the mechanical thresholds of most of the afferent fibers regardless of their threshold under normal conditions (see above). A movement in the normal working range of the joint, such as the test flexion used here, will now excite approximately 550 units in the MAN, and, in addition, the discharge rates of those low-threshold units having already been excited in the normal state are now much higher than previously. All in all, in the MAN the afferent volley in response to the test movement now consists of some 30900 impulses per 30 s, i.e., about seven times more than under normal conditions (Table 2; in individual afferents which have been studied consecutively both under normal and inflamed conditions, the afferent discharges sometimes increased more than 100-fold). This huge volley, together with the corresponding one in the PAN, obviously transmits those messages responsible for the noxious reflexes and the pain perception when moving an inflamed joint.

2.4 Composition of Somatic Nerves

2.4.1 Classification of Primary Somatic Afferent Nerve Fibers

Somatic nerve fibers are frequently classified either by the diameter of the fibers or by their conduction velocity. Two classifications have been proposed. The first classification was originally based on measuring conduction velocities in sensory nerves (Erlanger and Gasser 1937; Gasser 1960; Burgess and Perl 1973). The fastest mass potential on the oscilloscope, i.e., the fastest group of fibers was called the A elevation and the slowest was called the C elevation. Within the A-fiber group, several subgroups were recognized (α–δ, even post-δ). The second classification was based on measuring fiber diameter in deefferented muscle nerves or in cutaneous nerves (Lloyd 1943; Hunt 1954).The four major peaks in the resulting fiber-diameter distribution histograms were labeled groups I–IV in order of decreasing fiber diameter. Appropriate conversion factors can be used to relate fiber diameter and conduction velocity of myelinated fibers to each other (Hursh 1939; Boyd and Davey 1968; Boyd and Kalu 1979).

Both classifications are now used more or less interchangeably irrespective of whether fiber diameter or conduction velocity has been measured (Fig. 15). There is, however, a preference with muscle innervation to label efferent fibers (motor axons, vasomotor fibers) with the letter code and afferent fibers with the Roman numeral code (Table 4).

2.4.2
Composition of Cutaneous Nerves

In transverse sections of cutaneous nerves two distinct sizes of myelinated fibers are obvious: Afferent fibers of medium diameter (group II or Aα,β-fibers) and of small diameter (group III or Aδ-fibers) contribute to two distinct peaks in fiber-size histograms (Fig. 15A). The group II fibers innervate the various types of low-threshold mechanoreceptors, the group III fibers serve thermoreceptors and nociceptors (see above). The range of diameter and mean diameter within groups II and III varies little from cutaneous to cutaneous nerve (Boyd and Davey 1968).

The unmyelinated afferent fibers (group IV or C fibers) form a narrow but large peak in the fiber-size histogram (Fig. 15A). They constitute nearly

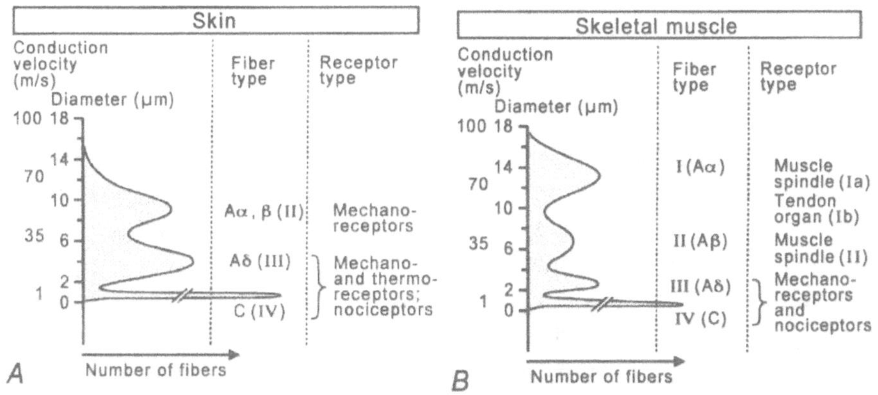

Fig. 15A–B. Fiber diameter and conduction velocity distribution of the various types of primary afferents in **A** cutaneous and **B** skeletal muscle nerves. The population of unmyelinated fibers (*C, group IV*) comprises nearly 50% of all primary afferents in most nerves. Their population histogram therefore had to be interrupted. (Data compiled from various authors by H. Handwerker in Schmidt 1995)

Table 4. Fiber composition of the lateral gastrocnemius-soleus muscle nerve in the cat

Type of nerve fiber	n	%
Myelinated[a]	1200	
Motor		
Aα, skeletomotor	382	53
Aβ, skeleto- and fusimotor	14	2
Aγ, fusimotor	324	45
Total	*720*	*60*
Sensory		
I		
Spindle primary endings (Ia)	144	30
Tendon organs (Ib)	72	15
II		
Spindle secondary endings	144	30
Spray (Ruffini) endings	5	<1
Lamellated (paciniform) endings	5	<1
III ("free nerve endings")	110	23
Total	*480*	*40*
Unmyelinated[b]	3000	
C (vasomotor)	1700	57
IV (sensory)	1300	43

[a] Data from Boyd and Davey (1968).
[b] Data from M. von Düring and R.F. Schmidt in Mitchell and Schmidt (1983).

50% of all afferent nerve fibers in cutaneous nerves. Their fiber diameter usually does not exceed 2 μm. As outlined in this chapter, their sensory endings are in part thermoreceptors, others serve as low-threshold mechanoreceptors and as nociceptors.

Cutaneous nerves also contain a great number of autonomic efferent nerve fibers, mostly unmyelinated postganglionic sympathetic fibers. It is usually assumed that about 50% of the C fibers belong to this fiber type, i.e., they are about as frequent as the unmyelinated cutaneous afferent nerve fibers. In the glabrous skin, the postganglionic sympathetic efferents innervate the smooth muscle of the blood vessels (vasoconstrictor fibers) and the sweat glands (sudomotor fibers). In the hairy skin these types of efferent fibers are supplemented by the pilomotor fibers innervating hairs.

2.4.3
Composition of Muscle Nerves

Motor Nerve Fibers. In addition to somatic afferent and sympathetic efferent fibers, muscle nerves contain three types of efferent fibers (Table 4): (1) The fast-conducting large-diameter motor axons called Aα-fibers stem from the α-motoneurons in the spinal or brain stem and innervate the extrafusal muscle fibers; (2) the much thinner and hence slower motor axons, commonly referred to as the Aγ-system, innervate the intrafusal muscle fibers of the muscle spindles; (3) a much less frequent type of motor axon with an intermediate diameter, the Aγ-motor axon, gives collaterals to intra- and extrafusal muscle fibers (Barker 1974).

Autonomic Nerve Fibers. The postganglionic sympathetic efferent fibers of muscle nerves innervate the smooth muscle of the blood vessel walls in the skeletal musculature. These fibers are C fibers. As in the skin, most, if not all, are vasoconstrictor in function. About 10% are thought to be cholinergic vasodilator neurons; however, species differences are likely (Horeyseck et al. 1976; Brody 1978; Rowell 1981).

Sensory Afferent Nerve Fibers. Each muscle nerve also contains four types of afferent fibers (Fig. 12B, Tables 1, 4). Groups I, II, and III have a myelin sheath; group IV is unmyelinated. Details of their sensory end structures are contained in Fig. 14B and in Table 1. As indicated in the figure, the labels Ia and Ib do not indicate a difference in conduction velocity but in the peripheral receptor organ of these fibers (primary muscle spindle ending versus Golgi organ, respectively), although, at least in some muscle nerves of the cat hind limb, the Ib afferents seem to be on average somewhat slower than the Ia units (Bradley and Eccles 1953).

Any muscle nerve thus contains a great variety of efferent and afferent nerve fibers. The exact composition of each muscle nerve can only be determined by careful histological examination with various denervation techniques. So far this has been done mainly with light-microscopic methods. Satisfactory fiber counts are therefore available for a wide variety of myelinated nerve fibers of muscle nerves in different species such as cat, dog, monkey, and human. The most extensive count is probably the one by Boyd and Davey (1968) on the hind limb nerves in the cat.

Unmyelinated Fibers in Muscle Nerves. In contrast, only sporadic attempts have been made to determine the number of unmyelinated efferent (C) and afferent (IV) nerve fibers in a muscle nerve. There is, however, general agreement that these fibers outnumber the myelinated ones by a factor of 2.5. For example, the combined nerve of the lateral gastrocnemius and soleus muscles of the cat contains some 1700 efferent (C) and some 1300 afferent (IV) unmyelinated fibers compared with 1200 myelinated ones (Table 4). This table also illustrates that because of the abundance of group IV units there are many more sensory than motor fibers in this and probably all muscle nerves. For instance, of the 1200 myelinated fibers, about 40%, i.e., some 480 fibers, are afferent, giving a total of 1480 afferent and 720 motor fibers. Of the afferent units, there are about 144 group Ia, 72 group Ib, 155 group II, 110 group III, and 1300 group IV units. Thus this muscle nerve contains nearly four times more fine muscle afferents than large ones (1410 vs. 370 fibers). In nerves to more distal muscles of the cat the number of myelinated afferents compared to the number of motor fibers is even higher. Overall both fiber types are equal in number (Boyd and Davey 1968). Similar relationships probably exist for muscle nerves of humans and other vertebrate species.

In the context of this chapter, it should be emphasized that any muscle nerve contains many more fine (groups III, IV) than large afferent units (groups I, II). The emphasis is needed because, mainly for technical reasons, until recently very little attention has been given to the fine muscle afferents in the literature.

2.4.4 Composition of Articular Nerves

General Considerations. Because of their small size and their limited accessibility, the composition of articular nerves has received much less attention than that of other sensory nerves. For instance, in their pioneering study on the composition of peripheral nerves, Boyd and Davey (1968), while quoting earlier literature on the innervation of the knee joint (Gardner 1944; Skoglund 1956), confined themselves to an analysis of the posterior articular nerve of that joint. Their study concentrated on myelinated afferents but, also in the studies just quoted as well as in that of Freeman and Wyke (1967), the light microscope did not provide sufficient resolution for proper counts of the unmyelinated afferent fibers (and for the postganglionic sympathetic ones, for that matter). However,

Table 5. Composition of the nerves innervating the knee joint of the cat (from Langford and Schmidt 1983; Heppelmann et al. 1988)

	Medial articular nerve (n)	Medial articular nerve (%)	Posterior articular nerve (n)	Posterior articular nerve (%)
Afferent fibers	**630**	**100**	**680**	**100**
Group I (Aα) fibers	0	0	27	4
Group II (Aβ) fibers	59	9	149	22
Group III (Aδ) fibers	131	21	94	14
Group IV (C) fibers	440	70	410	60
Sympathetic fibers	**500**		**515**	

as will be discussed in the next paragraph, the vast majority of joint afferents are of this fiber type (for further literature on joint nerves, see Sect. 2.5.3).

Innervation of the Knee Joint. The medial (MAN) and the posterior articular nerves (PAN) form the major innervation of the cat's knee joint (Freeman and Wyke 1967). Their afferent and efferent fiber content is shown in Table 5. Each nerve contains approximately 650 afferent fibers, mainly of groups III and IV, and another 500 unmyelinated sympathetic efferents (Langford and Schmidt 1983). In the group IV fiber range the maximum of the diameter distribution is between 0.3 and 0.4 μm for the afferent fibers whereas the maximum is around 0.8–0.9 m for the sympathetic efferent fibers (Heppelmann et al. 1988). The latter study and that of Langford and Schmidt (1983) also allowed determination of the respective numbers of the group I–III fibers (Table 5). The small contribution of the group I fibers in the PAN is presumably due to stray fibers joining in from muscle spindles of the popliteus muscle (McIntyre et al. 1978). Thus the majority of afferent fibers in both nerves belong to the fine afferents consisting of groups III and IV (MAN 91%, PAN 74%). As discussed earlier in this chapter, many of these fine afferents are mechanically insensitive under normal conditions (sleeping nociceptors) but become very active in the course of tissue damage (such as inflammation). Particularly under these circumstances, the sensitized nociceptors exert powerful actions on the autonomic nervous system (see, e.g., Sects. 4.1.2.4 and 4.5.2.1).

2.4.5 Composition of Mixed Nerves, Spinal and Dorsal Roots

The electric stimulation of either mixed nerves, spinal nerves or dorsal roots offers the experimental possibility to induce afferent volleys which in their composition will be quite different from those evoked by stimulation of pure cutaneous and muscles nerves, and this has often been of advantage in one or another experimental situation.

Composition of Mixed Nerves. Volleys evoked in mixed nerves, i.e., in nerves composed of cutaneous and muscle nerves, have frequently been used to study the impact of the somatosensory on the autonomic nervous system. Many examples of such work will be quoted and illustrated in this review. Particularly in the pioneering studies of this field, the sciatic nerve has been used as an easily accessible and powerful somatosensory input (see Sect. 3.1 for a discussion of these studies).

In mixed nerves electric stimulation of increasing intensity will, just as in muscle nerves, first activate α-motoaxons because these nerve fibers have the largest diameter and thus the fastest conduction velocity (see Sect. 2.4.7). The antidromic volley thereby induced will antidromically excite their motoneurons, and – via the motoaxon collaterals – the Renshaw cells. However, as far as is known, this activity does not influence the autonomic nervous system. Next, group I primary afferents from muscle will be activated, and thereafter the cutaneous as well as the muscle group II and III afferents. A further increase in stimulus strength will finally also activate the unmyelinated afferents of cutaneous and muscle origin. The complex composition of the afferent volleys thus produced, particularly when stimulating large mixed nerves (such as the sciatic nerve) at high to very high stimulus strength has to be appreciated when evaluating the central autonomic effects of such volleys.

Composition of Spinal Nerves. Spinal nerves have often been used as input in studies of the type reviewed here, mainly because they offer the unique advantage that all afferent activity evoked by their stimulation enters in one and the same segment. For instance, the segmental organization of the early and late somatosympathetic (see Fig. 18), and of the C reflexes (see Fig. 24) could be elucidated using spinal nerves as afferent input, and several other studies using spinal nerves when analyzing somatosympathetic relations will be discussed in later parts of this review. Another advantage of the stimulation of spinal nerves relative to the stimulation of mixed nerves is that because of the short conduction distance to the spinal cord the afferent activity in the slowly conducting high-threshold myelinated nerve fibers arrives at the spinal neurons almost simultaneously with the activity in the fast conducting low-threshold fibers (see also Sect. 2.4.7).

Composition of Dorsal Roots. Using dorsal roots as afferent input in the study of somatoautonomic interactions offers advantages similar to those of using spinal nerves. To stimulate all dorsal roots of a given segment may prove cumbersome, and should usually be avoided but the stimulation of one or a few rootlets offers the advantage of restricting the afferent input to certain parts of a spinal segment. Actually, with the introduction of the slice technique to the study of the spinal autonomic nervous system, the stimulation of the dorsal root filaments connected to the spinal slice has proved to be most valuable indeed (e.g., Yoshimura et al. 1986; Dun and Mo 1989).

2.4.6 Special Aspects of Cranial Nerves

Somatosensory Innervation of the Head. Of the cranial nerves (CN) the trigeminal nerve (CN V) carries the largest amount of somatosensory information from the tissues of the head to the CNS. To a lesser extent, three other CNs contribute, namely the facial nerve (CN VII), the glossopharyngeal nerve (CN IX) and the vagus nerve (CN X) as well as the upper three cervical nerves. All in all, the anatomical and physiological properties of the nerve fibers contained in this nerve are in no way different from those of the nerve fibers of the somatosensory supply of the body. Nevertheless, in the context of this review a few aspects deserve mentioning since somatosensory input from the face and the underlying structures, particularly nociceptive input, has powerful access to the autonomic nervous system (e.g., Fig. 58).

Composition of the Trigeminal Nerve. The trigeminal nerve, the largest of the cranial nerves, is a mixed somatic nerve consisting of a large sensory and small motor roots. The motor trunk supplies muscles of mastication, whereas the sensory root conducts sensation from the face and the anterior two-thirds of the head. Of the three major divisions of CN V, the ophthalmic (first and smallest division) and the maxillary (second division) are purely sensory (disregarding their autonomic fibers) whereas the third division, the mandibular nerve, is mixed. In the context of this review, it is worth noting that all three divisions, but particularly the maxillary nerve, innervate the vessels and the dura mater of the brain (i.e., they may perhaps be considered visceral afferents), thus being responsible for transmitting migraine and other headache sensations often inducing strong autonomic reactions (for literature and details on this innervation e.g., Andres et al. 1987a,b; Messlinger et al. 1993, 1995). The second point to mention is the special innervation of the tooth pulp which presumably consists exclusively or nearly exclusively of fine afferents with high (nociceptive) thresholds which only transmit pain sensations (for further discussion and references, see Jyväsjärvi and Kniffki 1989; Iwata et al. 1991; Ikeda et al. 1995). (As regards the somatosensory nerve fibers in the other cranial and upper cervical nerves mentioned above, the reader is referred to textbooks of anatomy, e.g., Romanes 1964; Clemente 1985).

2.4.7
Activation of Primary Afferent Nerve Fibers by Electrical Stimulation

2.4.7.1
Advantages and Disadvantages of Electrical Stimulation of Primary Afferents

Electrical stimulation, usually with short square pulses, is frequently used to activate in a precise way primary afferent fibers, both in animal and human experiments and for diagnostic and therapeutic reasons. When using such stimuli it has to be appreciated that the threshold for exciting axons with an electrical stimulus is an inverse function of axonal diameter. The larger the fiber, the lower is its threshold. In many experimental and clinical situations, advantage is taken of this property of axons to stimulate a selected population of axons by graded electrical pulses. The strength of stimulation is commonly expressed as a multiple of the largest axons of the nerve (e.g., 1.5T means 1.5 times the threshold of the most excitable axon in the nerve). To determine thresholds and the effects of increasing the strength of stimulation the afferent volley can either be monitored on the peripheral nerve "upstream" from the site of stimulation (see Fig. 23) or, under experimental conditions with the spinal cord exposed by a laminectomy, by monitoring the incoming volley at the dorsal root entry zone with a monopolar electrode (see Fig. 1D).

Activation of primary afferents by electrical stimulation, convenient and precise as it may be, has several drawbacks which have to be kept in mind when evaluating the effects resulting from such stimulation. Two are particularly worth mention. First of all, under most circumstances, electrical stimuli will coactivate afferent fibers from more than one class of receptor. Secondly, synchronous activation of the whole spectrum of afferent fibers is complicated by the fact that whereas all fibers are activated simultaneously the volley in the largest fibers will reach the spinal cord far in advance of the volley in the slowest fibers. The central neural circuits activated by the fastest fibers may cause an alternation in the effects that would otherwise have been produced by the slowest (i.e., smallest) afferent fibers.

2.4.7.2
Effects of Preferential Blocking of A and C Fibers

Since the pioneering work of Erlanger and Gasser in the late twenties, it has been known that low concentrations of anesthetics block impulse transmission in small diameter fibers before thick fibers are affected. Conversely, compression block suppresses transmission in myelinated fibers before C fibers are affected (for references, see Torebjörk and Hallin 1973). Using such experimental blocks in awake human subjects while recording from skin nerves with percutaneously inserted tungsten microelectrodes, Torebjörk and Hallin (1973) confirmed and extended the previous, more indirect observations, that within certain limits these two methods do indeed allow selective temporary blocking of one or the other fiber group. When activity was recorded in the superficial branch of the radial nerve from A fibers (under lidocaine block of the C fibers), weak electrical skin shocks were still felt as a tactile sensation. A strong stimulus was perceived as a short, sometimes sharp blow but the prolonged pain usually experienced from such shocks had disappeared. The preferential blocking of the A response by pressure was accompanied by an impaired discrimination of weak stimuli. Stronger stimuli evoked pain.

2.5 Spinal and Brain Stem Termination and Projection of Somatic Primary Afferents

2.5.1 Termination and Projection of Cutaneous Primary Afferents

In cats the primary afferent axons from SA I and SA II receptors project into the nucleus proprius of the spinal cord (Brown 1977; Brown et al. 1978). The SA I endings are distributed in a series of elliptical zones arranged rostrocaudally in a sagittal plane through laminae III, IV, and dorsal V. Many hair follicle afferent fibers project directly to the dorsal column nuclei by means of collaterals ascending in the dorsal columns. In addition, hair follicle afferent fibers synapse in the dorsal horn in "flame-shaped arbors," recurrent branches of afferent fibers that first descend into deeper layers of the dorsal horn and then turn dorsally to end in laminae III and IV (Brown et al. 1977). The terminals form a continuous rostrocaudal column. The primary afferent fibers of Pacinian corpuscles terminate in the medial part of the dorsal horn (Brown et al. 1980). The endings form a series of elliptical terminal zones arranged in a rostrocaudal plane in lamina III–VI (for a review, see Brown 1981).

Unmyelinated cutaneous primary afferents terminate quite differently, namely largely in the superficial dorsal horn, particularly in lamina II, i.e., the substantia gelatinosa. Figure 16 depicts the terminal arborizations of six single, functionally identified group IV afferents in the cervical and lumbar spinal cord (Sugiura et al. 1989). The area of termination of such fibers typically is a highly concentrated region of terminal enlargements and boutons which extends approximately 400–500 μm in the rostrocaudal direction and less than 200 μm in the medio-lateral dimension. There is a clear relationship between the central projection zone of somatic group IV afferents and the type of sensory unit. Nociceptors tend to terminate most superficially, in lamina I and lamina IIo, while the low-threshold mechanoreceptors have their circumscribed terminal plexus more deeply in lamina Iii. It may be added, as discussed in detail by Sugiura et al. (1989), that visceral afferents generally converge on the same spinal projection sites as the unmyelinated somatic afferents.

The spinal termination sites of fine myelinated afferents (group III) from nociceptors are similar to those of group IV afferents. They invade the interface between lamina I and lamina IIo. Similarly, the group III

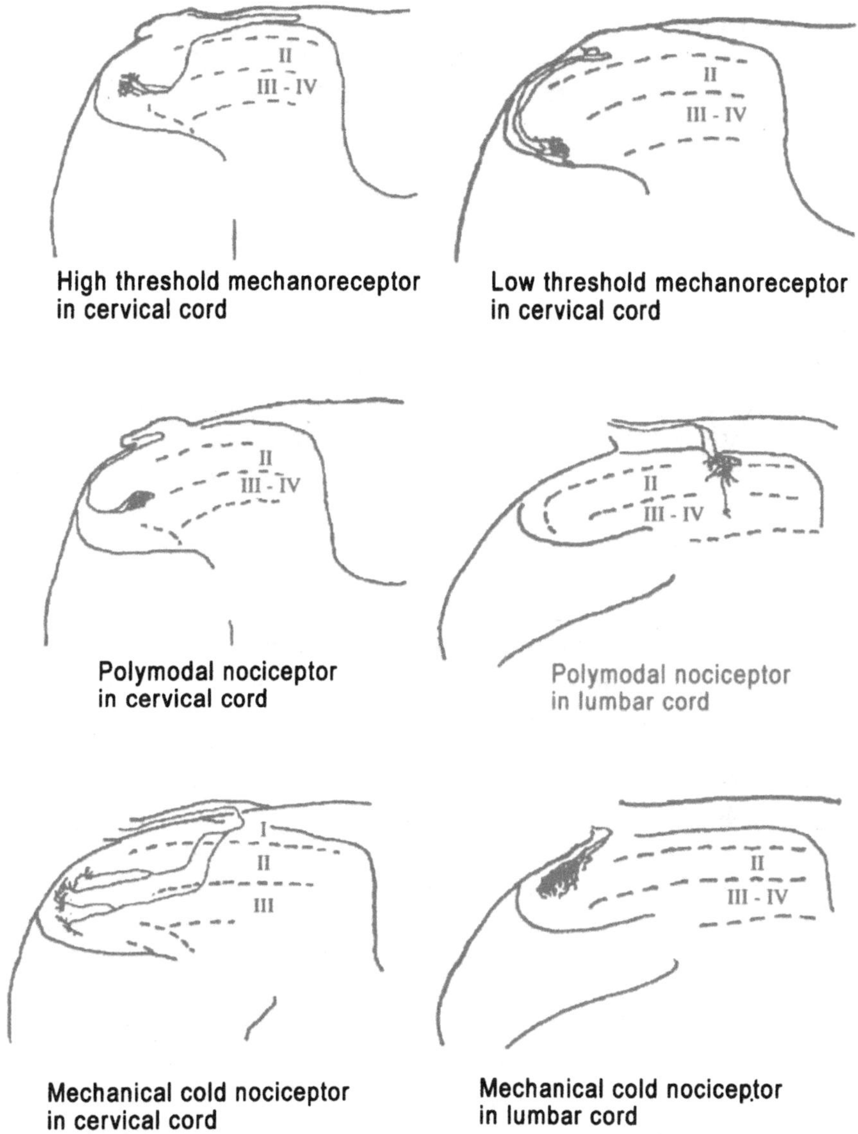

Fig. 16A–F. Spinal termination of cutaneous unmyelinated afferents. Locations of arborizations of functionally identified group IV cutaneous units. The transverse plane drawings were reconstructed from distributions observed in parasagittal sections of the cervical or the lumbar spinal cord. **A** High-threshold mechanoreceptor and **B** low-threshold mechanoreceptor in cervical cord. **C,D** Polymodal nociceptor in **C** cervical and **D** lumbar cord. **E,F** Mechanical cold nociceptor in **E** cervical cord and **F** lumbar cord.(From Sugiura et al. 1989, with permission)

afferents from low-threshold mechanoreceptors of the hairy skin (D-hair receptors) terminate at the interface between lamina III and IV, the nucleus proprius (for literature see Sugiura et al. 1989).

2.5.2
Termination and Projection of Muscle Primary Afferents

Large-diameter primary afferents from muscle (Ia and Ib) terminate in laminae IV–VI and in the interneuron and motoneuron pools of laminae VII and IX (Sugiura et al. 1989).

The spinal terminations of group II and III afferents from muscle and other deep tissue, i.e., thin myelinated afferents with conduction velocities of less than 40 m/s, were analyzed with single units which were identified as being either low-threshold mechanosensitive (LTM), high-threshold mechanosensitive (HTM, requiring noxious stimuli for activation), or muscle spindle secondary if they were activated in the typical manner by muscle stretch (Hoheisel et al. 1989).

Half of the fibers with HTM properties terminated exclusively in lamina I. The parent fibers ascended in Lissauer's tract or the lateral dorsal columns and issued collaterals that entered lamina I laterally. The other half of the HTM units terminated in both lamina I and deeper laminae of the dorsal horn, namely laminae IV–VI. Some of these units had their origin in gastrocnemius-soleus muscle, demonstrating that there is a projection of muscle afferents of the HTM type to lamina I.

The LTM units differed from the HTM ones in that they did not terminate in lamina I. The main projection areas were lamina IV–VI with some additional collaterals terminating in lamina II. In general, the LTM units were more heterogeneous in appearance than other types, i.e., the differences in the patterns of termination were greater.

The fibers from muscle spindle secondary endings exhibit a termination pattern identical to that described for spindle secondary endings of faster conduction velocities (Fyffe 1979, see also 1984). The fibers had no terminal arborizations in laminae dorsal to lamina IV, the main projection areas being laminae VI/VII and the ventral horn. A consistent finding in all these experiments was that lamina III was free of terminals. This lamina seems not to be an important relay for primary afferent information from muscle and other deep tissue. Thus lamina III can be considered a dorsal horn region which is reserved for the processing of primary afferent information from the skin.

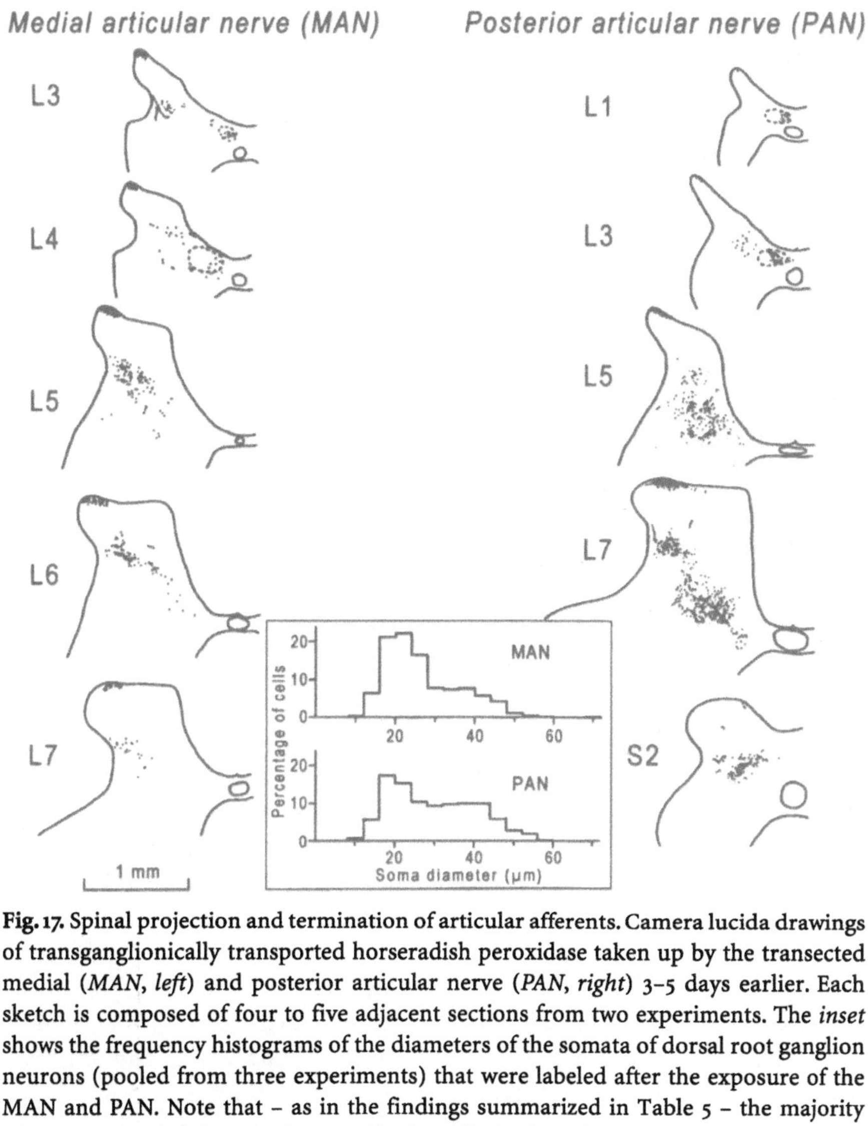

Fig. 17. Spinal projection and termination of articular afferents. Camera lucida drawings of transganglionically transported horseradish peroxidase taken up by the transected medial (*MAN*, *left*) and posterior articular nerve (*PAN*, *right*) 3–5 days earlier. Each sketch is composed of four to five adjacent sections from two experiments. The *inset* shows the frequency histograms of the diameters of the somata of dorsal root ganglion neurons (pooled from three experiments) that were labeled after the exposure of the MAN and PAN. Note that – as in the findings summarized in Table 5 – the majority of neurons labeled from both nerves had small-sized perikarya. (Modified from Craig et al. 1988, with permission)

2.5.3
Termination and Projection of Articular Primary Afferents

As demonstrated with electrophysiological and histological techniques, the primary afferents from the medial articular nerve (MAN) of the knee joint of the cat enter the spinal cord via the dorsal roots L5 and L6, whereas those in the posterior articular nerve (PAN) enter via L6, L7 and sometimes S1 (Lit in Craig et al. 1988; Schaible and Grubb 1993). After entering the spinal cord the fibers terminate mostly in the segments corresponding to the dorsal roots through which the afferents fibers coursed. However, as seen in Fig. 17, the articular afferent projection is not restricted to these segments but extends over several adjacent segments. A sparse projection was found caudally as far as S2, and rostrally up to L1, with a dense projection of the PAN into the medial portion of Clarke's column (Craig et al. 1988). In the monkey, labeling was identified in the dorsal roots L4-S1 after injection of HRP into the knee joint cavity (Wiberg and Widenfalk 1991) whilst in the rat injections of HRP into the elbow joints labeled neurons in dorsal root ganglia C4-T4 with a maximum in C7-T1 (Widenfalk et al. 1988). Injections of HRP into the temporomandibular joint in rats labeled the dorsal root ganglia at the level C2-C5 (Widenfalk and Wiberg 1990). It seems therefore that each large joint projects to several spinal segments (see Widenfalk and Wiberg 1989, for the rat knee joint). The cat's wrist joint is innervated by the carpal branch of the interosseus nerve (CBIN) which branches from the median nerve (Rausell and Avendano 1989). The CBIN seems to be a pure articular nerve, as has also been described for the human wrist (Romanes 1964). The primary afferents of CBIN, which is a very thin nerve with probably less than 50 myelinated afferents, seem to enter the spinal cord only at C8.

With regard to the termination sites of articular primary afferents in the various laminae of the spinal gray matter, most types of thick myelinated afferents of cat knee were found to terminate within the adjacent lumbosacral and lower thoracic spinal segments, although a few rapidly adapting knee joint afferents ascended in the dorsal columns (Clark 1972). The projection fields of MAN and PAN of the cat knee (these nerves contain mainly fine afferent nerve fibers, see Table 5 and the inset in Fig. 17) were identified in the cap of lamina I, in laminae V–VI and in the dorsal part of lamina VII in the segments adjacent to the entry roots. No terminations were found in laminae II–IV. In general, the patterns of intraspinal termination of MAN and PAN afferents were similar to those found with HRP labeling of muscle and visceral afferents in the cat (see Craig et al. 1988). By contrast, the brain stem terminations of axons originating in the temporomandibular joints of cats exhibited a different pattern since termination sites were observed in laminae I, II, and III of the medullary dorsal horn (Capra 1987). In the rat, Levine et al. (1984) compared the patterns of spinal termination of nerves in the ankle and knee joint. After injection of the knee no or only minimal labeling was found in the spinal cord but after injection of the ankle intense labeling was obtained in lamina I and the substantia gelatinosa of the fourth lumbar segment. None of the studies performed to date have correlated the spinal termination site and electrophysiologically identified types of joint afferents.

2.6 Summary and Conclusions

1. Mechanical stimulation of the glabrous skin of mammals will induce afferent impulses in various types of low-threshold mechanoreceptors with group II afferents. Vibratory stimuli are most effective in exciting the very rapidly adapting PC receptors (technically speaking "acceleration detectors," histological structure: Pacinian corpuscles). Touching and stroking the glabrous skin will, in addition, excite the rapidly adapting RA receptors ("velocity detectors," Meissner corpuscles). Pressing on and stretching of the skin are, respectively, the adequate stimuli for the slowly adapting SA I and SA II receptors ("pressure detectors," Merkel disks and Ruffini corpuscles, respectively).
2. Brushing the hairy skin of a mammal will induce impulses in subcutaeous PC receptors and in various types of rapidly adapting hair follicle receptors with group II and III afferents, such as type G, T, and D receptors. The composition of the afferent volley will not only depend on the intensity of stimulation but also on which mammal and which area of skin are being stimulated. Hairy skin also contains SA I receptors (dome-shaped Merkel touch corpuscles) and SA II receptors (Ruffini corpuscles) with thresholds and discharge characteristics similar to those in glabrous skin.
3. Specific cold receptors with fine myelinated afferents (group III) and specific warm receptors with unmyelinated afferents (group IV) innervate the glabrous and – in a usually less dense and more variable

way – the hairy skin of mammals. The glabrous skin of primates, particularly the face and the hands, contain especially high numbers of such specific thermoreceptors. Cold and warm receptors show dynamic responses to cooling and warming, respectively, in addition to static discharges to constant skin temperatures within their respective sensitivity ranges.

4. The skeletal muscles, including their tendons, contain low-threshold muscle spindle and tendon organ afferents (groups Ia, Ib, II) which seem to have little if any impact on the autonomic nervous system. In addition stretch-, pressure- and contraction-sensitive as well as chemosensitive units with uncorpuscular ("free") nerve terminals and fine afferents (groups III and IV) have been found which seem to signal the amount of work performed by a muscle. These units do evoke somato-sympathetic reflexes.
5. Articular afferents with low thresholds to local mechanical stimulation (touching, innocuous pressure) and to joint movement in the normal working range of the joint either have corpuscular endings of the Ruffini, Golgi, or Pacini type with group II afferents, or they have noncorpuscular ("free") nerve terminals stemming form group III or IV afferents.
6. Nociceptors with group III and IV afferents and noncorpuscular ("free") nerve terminals are ubiquitous in every somatic tissue. They all have high thresholds. Most nociceptors seem to be sensitive to more than one stimulus modality, i.e., they are polymodal, but some are more sensitive to noxious mechanical stimuli, others to noxious heat or cold, others to noxious chemical (algesic) stimuli. Within a given stimulus modality the threshold of nociceptors is not uniform. In normal tissue a certain and often considerable number of nociceptors have such high thresholds that they cannot be excited by even the most intense noxious stimuli (silent or "sleeping" nociceptors).
7. Tissue damage such as an inflammation sensitizes all types of nociceptors, including the "sleeping" ones which are "woken up." The percentage of spontaneously active nociceptors increases, as does the frequency of their discharges (these afferent impulses are the peripheral neural correlate of the pain at rest from an inflamed tissue such as a joint). Stimuli which normally are innocuous will now excite nociceptors and, therefore, induce pain (a phenomenon called allodynia), and noxious stimuli produce greatly enhanced afferent volleys and intense pain sensations (hyperalgesia). A further effect of the peripheral sen-

sitization of nociceptors is the sensitization of the secondary and later neurons in their central pathways. This central sensitization enhances the central actions of any incoming afferent volley far above what may be expected from the afferent volley alone.

8. The peripheral terminals of nociceptors (and possibly those of other "free" nerve endings, too) upon suprathreshold stimulation release neuropeptides, such as substance P (SP) or calcitonin gene-related peptide (CGRP), and possibly other substances into the surrounding tissue (the efferent function of fine afferents). The release takes place not only at the activated sites of the unit but, by spread of the afferent impulse into collaterals (termed the "axon reflex"), also from its other terminals. The released substances lead to vasodilatation (CGRP) and plasma extravasation (SP). This neurogenic inflammation, as it is called, possibly contributes to removal of noxious chemical substances from the site of injury.
9. Electrical stimulation of peripheral nerves, spinal nerves or dorsal roots will evoke afferent volleys, the composition of which depends entirely on the stimulus strength and the fiber spectrum, i.e., the threshold of the primary afferent fibers: the higher the stimulus strength, the more impulses in fine afferents will be included in the volley. Thus it has to be appreciated that modality-specific afferent volleys can only be evoked in a limited way by electric nerve stimulation5)

3 Central Processing (Pathways, Modes of Operation)

3.1 Introduction

Alexander (1946) seems to have been the first to record somatically evoked reflex discharges directly from sympathetic nerves. It was recognized that in anesthetized cats electrical stimuli applied to a somatic afferent nerve could provoke a reflex discharge in sympathetic nerves, with a short latency of less than one tenth of a second. Schaefer and his colleagues (Sell et al. 1958; Schaefer 1960; Weidinger et al. 1961) confirmed Alexander's results and, in addition, emphasized that somatic afferent inputs usually elicit massive generalized reflex discharges in cardiac and renal sympathetic efferent nerves. Sympathetic efferent fibers usually have spontaneous (or ongoing) activity with various rhythms (e.g., synchronous with heart rate and respiration) as discovered by Adrian et al. (1932). It was noted that the spontaneous discharges were inhibited after reflex discharges, and this was called postexcitatory depression (Schaefer 1960). In the experiments on somato-sympathetic reflexes cited above, usually peripheral limb afferent nerves were electrically stimulated. After spinal transection at the cervical level, the somato-sympathetic reflex discharges were abolished. Consequently, the reflexes were thought to be of medullary origin (Sell et al. 1958; Weidinger et al. 1961).

In contrast to the evidence outlined above and on the basis of recordings of arterial blood pressure in chronically spinalized animals, Sherrington (1906) and Brooks (1933) had predicted a somatically originating spinal reflex path to the visceral organs. In 1964, Beacham and Perl (1964a,b) were the first to record sympathetic reflex discharges of spinal origin. These were evoked from thoracic and lumbar white rami (i.e., preganglionic neurons) by single shocks to dorsal roots of spinal nerves in acutely spinalized cats.

In 1965, Sato and coworkers observed two kinds of sympathetic reflex discharges with different latencies in the lumbar sympathetic trunk following electrical stimulation of a sciatic nerve in anesthetized CNS-intact cats. These were the early spinal and late supraspinal reflexes. The resolution of these two reflex components required deafferentation of the baroreceptors in order to eliminate spontaneous discharges which were synchronous with the cardiac and respiratory cycles. The late reflex was found to be more sensitive to depression by baroreceptor afferent excitation (Kirchner et al. 1971). In 1966, Coote and Downman also observed early spinal and late supraspinal reflexes in cardiac and renal sympathetic nerves elicited by stimulation of thoracic spinal nerves. In 1966, Sato and Schmidt introduced an averaging recording technique for the study of somato-sympathetic reflexes. This technique eliminated irregular background activity and made it possible to measure the magnitude of reflex responses and quantitatively analyze the contributions of different somatic afferent fibers to the somato-sympathetic reflex. Using the averaging technique, Sato and Schmidt (1971) quantitatively analyzed the two different reflex components, the early spinal reflex and the late supraspinal reflex. The amplitude of early reflex was largest when the afferent volley entered the spinal cord at the same segment or at a segment adjacent to the white ramus from which the reflex was recorded. In contrast, the size of the late reflex was largely independent of the segmental level of afferent input.

In 1965, Fernandez de Molina and coworkers succeeded in making an intracellular recording of sympathetic preganglionic neuronal activity at the thoracic level in spinalized cats. In 1985, Seller and his colleagues (Dembowsky et al. 1985) were the first to record intracellularly the early spinal and late supraspinal (medullary) components of somato-sympathetic reflexes from the thoracic preganglionic sympathetic neurons in anesthetized CNS-intact cats. Yoshimura et al. (1986), and McKenna and Schramm (1983) initiated the use of the spinal cord slice preparation for studying the basic electrical properties of the sympathetic preganglionic neuron. The advantages of this technique are the stability of neuronal activity and the ability to study the neuron devoid of background synaptic input and in the absence of anesthetics. The neurotransmitters within the CNS which mediate the somato-sympathetic reflexes have not yet been conclusively identified. One candidate appears to be L-glutamate. Using the slice preparation technique, Dun and his colleagues demonstrated that the *N*-methyl-D-aspartate (NMDA) receptor is involved in somato-sympathetic spinal reflexes (Shen et al. 1990), suggesting that L-glutamate acts as a neurotransmitter in the spinal reflex pathway.

Wurster and his colleagues (Chung et al. 1979) demonstrated that the somatic afferent information to the brain which is responsible for producing sympathetic reflexes ascends the dorsolateral funiculus and dorsolateral sulcus of the spinal cord. This somatic afferent information is considered to be integrated within the supraspinal structures. Foreman and Wurster (1973) and Coote and Macleod (1975) demonstrated excitatory and inhibitory descending pathways in the spinal cord originating in the supraspinal structures and projecting to the spinal sympathetic preganglionic neurons. The involvement of the ventrolateral medullary neuronal groups in the medullary reflex pathway of the somato-sympathetic reflex was documented by Morrison and Reis (1989), Masuda et al. (1992) and Sun and Spyer (1991).

In addition to the early spinal and late supraspinal reflexes evoked by myelinated afferents, several groups of investigators demonstrated that sympathetic reflex discharges induced by unmyelinated group IV (or C) somatic afferent volleys (the group IV-sympathetic reflex or C reflex) had a longer latency than sympathetic reflex discharges induced by myelinated group II (or Aβ) and III (or Aδ) fibers (Fedina et al. 1966; Coote and Perez-Gonzalez 1970; Koizumi et al. 1970; Schmidt and Weller 1970; Sato 1973). The much longer latency of the C reflex, relative to reflexes induced by group II and III afferent stimulation, is almost entirely due to the low conduction velocity of the unmyelinated afferents. The C-reflex component could be selectively reduced by administration of opiates in anesthetized cats (Ito et al. 1983; Sato et al. 1986b) and dogs (Niv and Whitwam 1983; Wang et al. 1992).

On the other hand, in the anesthetized rat the C-reflex component was augmented by intravenous or intra-cisterna magna administration of morphine (Adachi et al. 1992b), and the μ-opioid receptors were proven to be the dominant modulators of this effect (Sato et al. 1995). Li et al. (1995) found that nitric oxide was involved in inhibition of the C reflex. Furthermore, in rats both NMDA and non-NMDA receptors were found to inhibit both A and C reflexes at the brain stem (Nagata et al. 1995). Both A and C reflexes were found to be influenced by chemoreceptor afferent excitation (Li et al. 1996).

Not only sympathetic, but also parasympathetic efferent nerve activity is reflexly influenced by somatic afferent stimulation. Iriuchijima and Kumada (1963) demonstrated that spontaneous activity of cardiac vagal efferents was depressed by somatic afferent stimuli in anesthetized dogs. Kimura et al. (1996a) recorded somato-vagal reflex discharges from a gastric vagal efferent nerve following electrical shock to a hindlimb afferent nerve in anesthetized rats. Using somatic afferent stimulation, Bradley and Teague (1968) elicited a reflex discharge in pelvic parasympathetic efferent nerves innervating the urinary bladder. A. Sato et al. (1983) recorded both reflex inhibition and excitation of the spontaneous efferent discharges in pelvic efferent branches to the bladder following a single electrical shock to hindlimb afferents in anesthetized cats. It was found that these reflex responses were facilitated when the bladder was expanded. Furthermore, A reflex and C-reflex components corresponding to excitation of myelinated (A) and unmyelinated (C) somatic afferents in a hindlimb nerve were identified (A. Sato et al. 1983). Morrison et al. (1996b) further demonstrated that pelvic efferent discharges in anesthetized rats are strongly inhibited by various somatic afferent stimuli, and also that there are A and C reflex excitatory components of spinal cord origin.

Each of these individual components of the somato-autonomic reflex is part of a complex physiological system. However, what at first must appear to be the confounding complexity of somato-autonomic reflexes in fact reveals the elegance of truly intergrated central processing mechanisms whose adaptive significance is now beginning to be appreciated.

3.2 Somato-sympathetic Reflexes

3.2.1 Sympathetic Reflex Responses to Stimulation of Myelinated (A) Somatic Afferents; the A Reflexes

3.2.1.1 Early Spinal, Late Supraspinal (or Medullary), and Very Late Suprapontine Components

In cats anesthetized with chloralose and whose baroreceptor afferents had been cut, single-shock electrical stimulation of the sciatic nerve produced two massive reflex discharges in the lumbar sympathetic trunk, with latencies of 25–50 and 80–120 ms, respectively. These were named the "early" and "late" reflex potentials (Sato et al. 1965). These reflexes were elicited by stimulation of somatic myelinated A afferent fibers and so they were also called A reflexes. These reflex responses could be elicited in a fairly reproducible fashion once discharges synchronous with the cardiac and respiratory cycles had been eliminated by baroreceptor deafferentation. Using a similar recording technique, the early and late reflex discharges were also recorded from the cardiac and renal sympathetic efferent nerves in anesthetized cats (Coote and Downman 1966). When electrical stimulation was coupled with transection of the brain stem, starting at the upper midbrain level down to the lowest level of the pons, the late as well as the early reflex could still be elicited. However, after spinal transection at the first cervical (C1) level, only the early reflex was preserved. These results indicate that the early responses are transmitted via spinal pathways, while the late response is transmitted via the medulla oblongata (Sato et al. 1965; Sato 1972a).

Sato and Schmidt (1971) quantitatively analyzed the two different reflex components recorded from the first lumbar (L1) white ramus. The reflex was elicited following a single electrical stimulation of high strength (50T) applied to spinal nerves from L1 to the first sacral (S1) nerve and to hindlimb nerves. The two components of the elicited reflex were: (1) the early spinal reflex with a short latency and (2) the late supraspinal (medullary) reflex with a long latency (Fig. 18). The size of the early reflex component recorded from the white ramus depended to a great extent on the proximity of the segment of afferent input to the segment of the white ramus from which the reflex was recorded. The amplitude was largest when the afferent volley entered the spinal cord at the same segment or at a segment adjacent to the white ramus. In contrast, the size of the late reflex was largely independent of the segmental level of afferent input (Koizumi et al. 1971; Koizumi and Brooks 1972; Sato and Schmidt 1971, 1973).

In mass recordings from lumbar white rami, the late reflex was evoked by low- and high-threshold myelinated afferents (group II (Aβ) and III (Aδ) afferents), whereas the early component appeared only when the high-threshold myelinated afferents (group III afferents) were included in the volley (Sato et al. 1969). Stimulation of the low-threshold myelinated somatic afferents (group II) could evoke the early reflex in chronic spinal cats, if the spinalized animals were kept for 3 months or longer after the operation (Sato 1973). This indicates that excitability of the spinal reflex is increased in the chronic spinal preparation. However, the late reflex component never reappeared (Koizumi et al. 1968). In regard to group I muscle afferents from primary endings of muscle spindles (group Ia fibers)

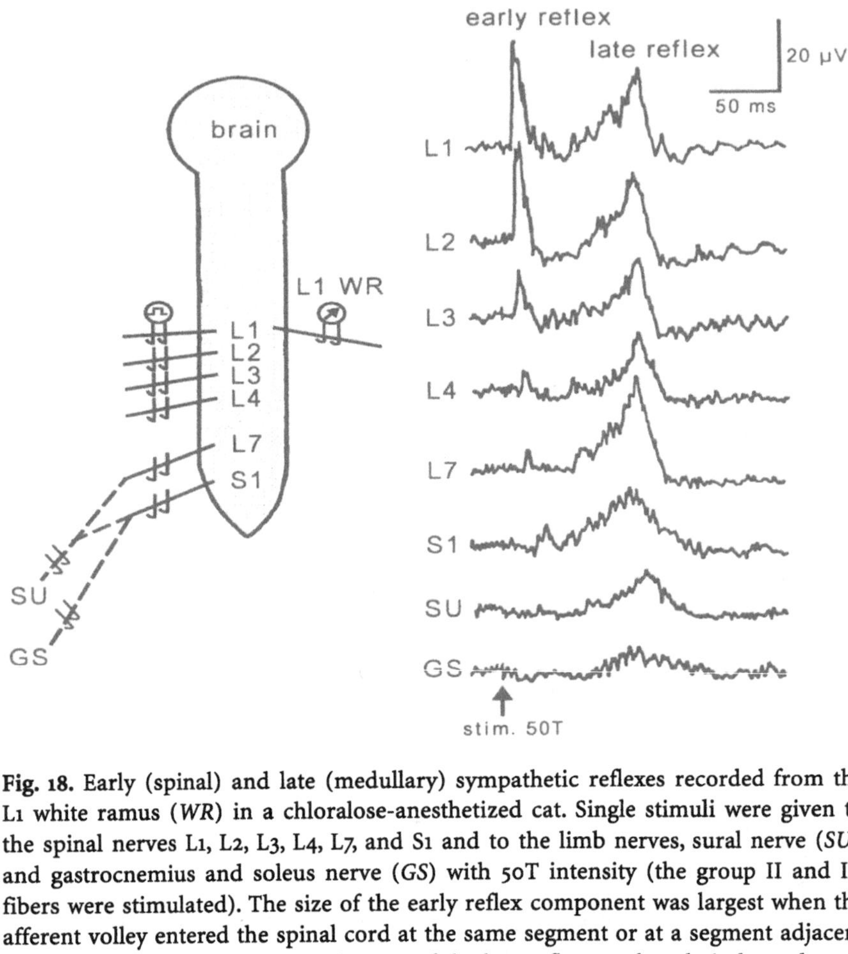

Fig. 18. Early (spinal) and late (medullary) sympathetic reflexes recorded from the L1 white ramus (*WR*) in a chloralose-anesthetized cat. Single stimuli were given to the spinal nerves L1, L2, L3, L4, L7, and S1 and to the limb nerves, sural nerve (*SU*), and gastrocnemius and soleus nerve (*GS*) with 50T intensity (the group II and III fibers were stimulated). The size of the early reflex component was largest when the afferent volley entered the spinal cord at the same segment or at a segment adjacent to the white ramus. In contrast, the size of the late reflex was largely independent of the segmental level of afferent input. (Modified from Sato and Schmidt 1971)

and from Golgi tendon organs (group Ib fibers), there is general agreement that these fibers usually have no excitatory or inhibitory influence on the sympathetic nervous system. Even after strychnine application or after prolonged tetanization of the afferent limbs, no reflex effects from group I fibers were seen (Sato et al. 1969).

Using cats, Fernandez de Molina et al. succeeded in 1965 in obtaining intracellular recordings of sympathetic preganglionic neuronal activity at the thoracic level. This technique has permitted resolution of the details of the action potential and naturally occurring, ongoing sympathetic activity (Fernandez de Molina et al. 1965; Coote and Westbury 1979; McLachlan and Hirst 1980). In 1985, using intracellular recording, Dembowsky et al. were able to demonstrate the early spinal and late supraspinal components of somato-sympathetic reflexes in thoracic preganglionic sympathetic neurons. Using chloralose-anesthetized, paralyzed and artificially ventilated cats, Dembowsky et al. (1985) obtained recordings from sympathetic preganglionic neurons of the third thoracic segment, which had been identified by antidromic stimulation of the white ramus. They reported that stimulation of segmental afferent fibers in the intercostal nerves evoked early excitatory postsynaptic potentials (EPSPs; amplitude, 3 mV; latency, 5–22.3 ms), and late, summation EPSPs (amplitude, up to 20 mV; latency, 27–55 ms). In most sympathetic preganglionic neurons, the intracellular recordings with the spinal cord intact or sectioned at C3 revealed distinct early EPSPs (Dembowsky et al. 1985). The early and late EPSPs which were evoked in sympathetic preganglionic neurons by stimulation of intercostal somatic afferent fibers correspond well to the early spinal and late supraspinal components of the somato-sympathetic reflex (see Koizumi et al. 1971; Koizumi and Brooks 1972; Sato and Schmidt 1971, 1973).

The results reviewed so far can be summarized by saying that in anesthetized cats, afferent volleys in myelinated somatic afferents have a twofold action on the sympathetic nervous system: a generalized action via the supraspinal sympathetic reflex centers, and a more circumscribed action on the preganglionic neurons at the segmental level. Similarly, in anesthetized CNS-intact rats, the early spinal reflex with a short latency and the late supraspinal reflex with a long latency were also elicited in the cardiac sympathetic efferent nerve by a single shock to either the third or fourth thoracic (T3-4) spinal afferent nerve (Kimura et al. 1996b; see Sect. 4.1.2.6).

In addition to the early spinal and late supraspinal reflex components induced by myelinated somatic afferents, Sato (1972a) reported the existence of a "very late" reflex discharge with a latency of 300–350 ms recorded from lumbar white rami (Fig. 19). This very late reflex was seen only in animals under light chloralose anesthesia, which may be the main reason why it had not been previously recognized. After surgical transection of the brain stem at the supra- or midpontine level, or after anemic decerebration, the very late reflex disappeared completely, although the other two reflex components, the early and the late one, remained (in Fig. 19,

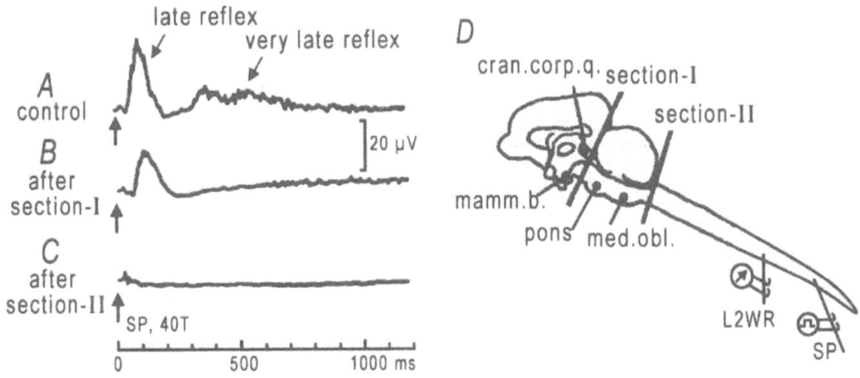

Fig. 19A–D. Effects of surgical decerebration on the late (medullary) and very late (suprapontine) sympathetic responses recorded from a whole L2 white ramus in chloralose-anesthetized cats. **A–C** Single stimuli were given to the superficial peroneal nerve (*SP*) with 40T intensity (the group II and III fibers were stimulated). After transection of the brain stem at the suprapontine level (*Section-I* in **D**), the very late reflex disappeared completely, although the late one was still observed (**B**). After spinal transection at the first cervical level (*Section-II* in **D**), the late reflex discharge also completely disappeared (**C**). (Modified from Sato 1972a)

there is no early reflex because of limb afferent nerve stimulation). These results suggested that the very late reflex discharge has a suprapontine reflex pathway.

3.2.1.1.1
The Spinal A Reflex and Descending Inhibitory Influences

Spinal reflex discharges in sympathetic nerves elicited by electrical stimulation of somatic afferent nerves were first demonstrated in acutely spinalized cats by Beacham and Perl (1964a,b). After measuring the latency of the reflex discharges, they concluded that the spinal component of the somato-sympathetic reflex took a polysynaptic, not monosynaptic pathway in the spinal cord.

In 1983, McKenna and Schramm recorded the extracellular activity of sympathetic preganglionic neurons using the isolated spinal cord of the neonatal rat, and later Yoshimura and his colleagues succeeded in making intracellular recordings from sympathetic preganglionic neurons using in vitro spinal cord slice preparations of adult cat. These studies provided fundamentally new insights into the electrophysiological properties of sympathetic preganglionic neurons. Intracellular recordings were obtained in the intermediolateral nucleus of the upper thoracic spinal cord from sympathetic preganglionic neurons which were identified by their antidromic responses to stimulation at various ipsilateral sites. Antidromic responses could be evoked by stimulation of the white ramus, the ventral root, the ventral root exit zone, the white matter between the latter and the outer edge of the tip of the ventral horn and the lateral edge of the ventral horn. Both EPSPs and inhibitory postsynaptic potentials (IPSPs) could be generated by focal stimulation (Yoshimura et al. 1986).

Dun and his colleagues (Shen et al. 1990) subsequently demonstrated that an NMDA receptor is involved in somato-spinal sympathetic reflexes. In this work, using transverse thoracolumbar spinal cord slices from neonatal rats (12- to 22-day-old), intracellular recordings were taken from sympathetic preganglionic neurons which had been identified by antidromic stimulation. The dorsal root-evoked EPSPs had a mean synaptic latency of 2.6 ms (range, 1.2–11 ms), suggesting a polysynaptic pathway. These EPSPs were suppressed by kynurenic acid, a nonselective glutamate receptor antagonist, and by the NMDA receptor antagonists D-2-amino-5-phosphonovaleric acid (APV) and ketamine. On the other hand, the EPSPs were unaffected by the non-NMDA receptor antagonist 6,7-dinitroquinoxaline-2,3-dione (DNQX; Fig. 20). Furthermore, the similarity of the electrophysiological characteristics of the EPSPs to those of NMDA-induced depolarization suggests that the sympathetic preganglionic neuron is endowed with typical NMDA receptors (Shen et al. 1990).

In the cat and monkey, Craig (1993) employed the immunofluorescent *Phaseolus vulgaris* leukoagglutinin technique to investigate anatomically the possibility that specific thermoreceptive and nociceptive influences on the sympathetic outflow are conveyed directly to spinal sympathetic regions by lamina I neurons. Iontophoretic injections made with physiological guidance were restricted to lamina I or to laminae I–II in the cervical (C6-8) or lumbar (L6-7) enlargement. Terminal arborizations were observed bilaterally but with, nonetheless, an ipsilateral predominance in the intermediolateral, intermediomedial and intervening regions of the thoracolumbar intermediate zone. In both the cat and the monkey, the labeling was most dense in the upper thoracic T2-4 spinal cord segments after cervical injection; however, labeling was also present in the T10-12 segments. On the other hand, after lumbar injection, labeling in the cat was restricted to the L4 segment while labeling in the monkey was present in both the T4-6 and T10-12 regions. The labeling obtained was of greater density in the monkey than in the cat. Since lamina I is known to be a

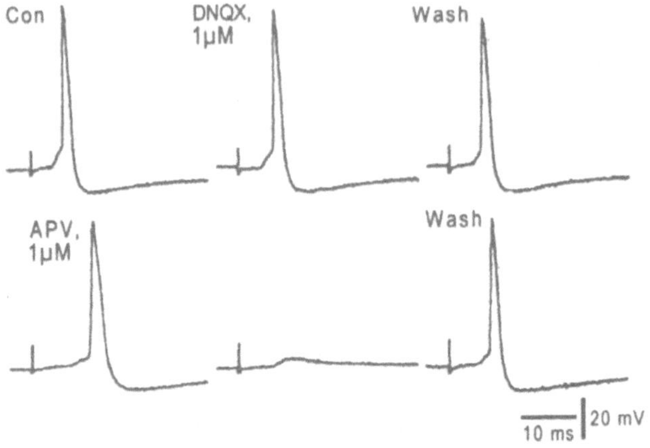

Fig. 20. Effects of glutamate receptor antagonists on synaptic responses evoked in a sympathetic preganglionic neuron in a transverse thoracolumbar spinal cord slice from a neonatal rat. Electrical stimulation (5–10 V) of dorsal root evoked in this neuron an action potential that was not blocked by *DNQX* (6,7-dinitroquinoxaline-2,3-dione, non-NMDA receptor antagonist) applied for 20 min. Superfusion of the slice with APV (D-2-amino-5-phosphonovalerate, NMDA receptor antagonist) for 10 min nearly blocked the synaptic response, and the effect was reversible after washing with Krebs solution. (From Shen et al. 1990)

primary site for integration of small-diameter input from a majority of tissues of the body, these observations reveal a spinal lamina I projection that could provide a direct pathway for the somato-sympathetic reflex effects of thermal and noxious stimuli.

Descending inhibitory influences on the spinal reflex component were first described by Sato et al. (1965). In chloralose-anesthetized cats, single pulse stimulation of the sciatic nerve produced two kinds of reflex potentials in the lumbar sympathetic trunk: a late reflex potential and a less pronounced early reflex. The early reflex component was augmented after spinal transection at the C1 or T8 level. From this augmentation of the early spinal reflex, the descending inhibitory effect on the spinal reflex was first suggested by these authors. Coote and Macleod (1975) noted that in chloralose-anesthetized cats electrical stimulation in the medulla caused an inhibition of the spinal reflex component elicited in thoracic white rami communicantes by stimulation of intercostal nerves, and the inhibitory effects were abolished by destruction of the ipsilateral funiculus and ventral quadrant of the spinal cord. Coote and Sato (1978) demonstrated

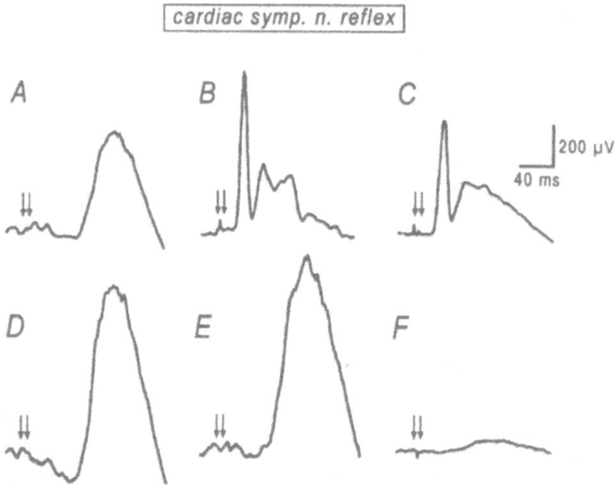

Fig. 21A–F. Reflex responses elicited in the inferior cardiac nerve following stimulation with two shocks, 200 Hz and 10 V (50–100T, group II and III fibers were stimulated) to **A–C** the third intercostal nerve and **D–F** the lateral popliteal nerve in a chloralose-anesthetized cat. **A,D** Following contralateral hemisection of the spinal cord and a small ipsilateral lesion in the dorsal part of the lateral funiculus at the C4 level. **B,E** 10 min after a further lesion at C4 in the ventral part of lateral funiculus and anterior funiculus, ipsilateral to recording site. **C,F** 10 min after complete spinal cord transection at the C4 level. In cats displaying only a long latency somato-cardiac reflex response, damage to the ventral quadrant of the ipsilateral cervical spinal cord (through which a bulbospinal inhibitory pathway runs) resulted in the appearance of shorter-latency reflexes to intercostal nerve stimulation. The early reflexes remained and the late reflex disappeared on subsequent complete transection of the spinal cord. (Modified from Coote and Sato 1978)

that in cats the spinal reflex component evoked in the cardiac sympathetic nerve by intercostal afferent nerve stimulation was tonically inhibited by fibers descending in the ventral quadrant of the ipsilateral spinal cord because the spinal reflex component was augmented after the destruction of this area of the spinal cord (Fig. 21).

Dembowsky et al. (1980) demonstrated in anesthetized cats that the amplitude of the early spinal reflex recorded from the T3 white ramus following stimulation of the intercostal and spinal nerves was increased and the onset latency of the spinal reflex was shortened, but supraspinal components were completely abolished following cooling the spinal cord between the second and third cervical segments, as well as during bilateral cold block of the dorsolateral funiculus alone. Hence, it appears that in the anesthetized cat the spinal component of the somato-sympathetic reflex is modulated by tonic inhibition descending via the dorsolateral funiculus of the spinal cord. Dembowsky et al. (1981) further showed that the increase in the amplitude of the spinal reflex induced by the cold block was reduced by the α_2-adrenoceptor agonist, clonidine. This effect was reversed by yohimbine, an α_2-adrenoceptor antagonist. These results suggest that these descending inhibitory influences are mediated by either adrenaline or noradrenaline, and originate in either the cranial part of area A1 and/or area A5.

Zanzinger et al. (1994) demonstrated in anesthetized cats that the amplitude of the spinal reflex recorded from the left T3 white ramus following stimulation of the ipsilateral T4 intercostal nerve was enhanced by cooling either the left or right rostral ventrolateral medulla (RVLM), although cooling of the left (ipsilateral) side was more effective. These results suggest that the descending inhibition of the spinal reflex is mediated bilaterally by the RVLM.

3.2.1.1.2 Supraspinal A Reflex and Medullary Regions Related to the Supraspinal A Reflex

Somatically evoked supraspinal reflexes have been recorded from various peripheral sites of the sympathetic nervous system, such as the renal, splanchnic, cardiac and long ciliary sympathetic nerves (Sell et al. 1958; Okada et al. 1960; Coote and Downman 1966; Katunsky and Khayutin 1968; Iwamura et al. 1969; Sato et al. 1969; Miyamoto and Alanis 1970; Kirchner et al. 1971). Activation of group I muscle afferents did not result in reflex discharges, whereas activation of group II and III afferent fibers of both muscle and cutaneous nerves did elicit distinct reflex discharges in lumbar white rami (Sato et al. 1969). When stimulating group II muscle (hamstring nerve), cutaneous (superficial radial nerve) and mixed (sciatic nerve) afferents in cats, Iwamura et al. (1969) recorded powerful reflex effects in renal sympathetic nerves.

The importance of the pressor area of the brain stem in maintaining both the tonic activity of sympathetic efferent nerves and the pressor reflex response following somatic afferent stimulation was established by Alexander's pioneering work (Alexander 1946) in anesthetized cats. It was confirmed that the brain stem was essential for the somatically elicited reflex discharges recorded from cardiac and renal sympathetic efferent nerves (Sell et al. 1958). The area of the brain stem responsible for eliciting somato-sympathetic reflexes has gradually become more circumscribed. In 1977, Ciriello and Calaresu suggested the lateral reticular nucleus as the most likely candidate for the reflex center. McCall and Harris (1987) determined that spontaneous sympathetic activity of the inferior cardiac nerve was increased, but reflex excitatory responses to electrical stimulation of sciatic nerve afferents were not effected by midline medullary lesions in anesthetized cats.

Because direct application of pentobarbitone sodium or glycine to the ventral surface of the medulla caused a profound fall in arterial pressure (Feldberg and Guertzenstein 1972; Guertzenstein and Silver 1974), the rostral ventrolateral medulla (RVLM) was initially considered to have a role in the regulation of sympathetic activity. This area has continued to be the subject of extensive investigations. A vast array of experimental studies has provided an enormous body of data about the function of the RVLM (see Calaresu and Yardley 1988; Loewy and Spyer 1990). A brief overview of the data relevant to our discussions on the involvement of RVLM in somatically induced regulation of sympathetic activity is presented below. The C1 area of the RVLM has been shown to play a critical role in the expression of the reflex pressor response to somatic afferent stimulation both in the cat (McAllen 1985) and in the rat (Stornetta et al. 1989). In chloralose-anesthetized, artificially ventilated cats, the bilateral topical application of glycine to the "glycine-sensitive area" of the RVLM reversibly abolished or severely attenuated both the pressor response and the renal nerve volley produced by electrical stimulation of the tibial nerve (McAllen 1985). The RVLM neurons involved in the cardiovascular regulation responded to somatic afferent nerve activation in cats (Arita et al. 1988; Blair 1991), rats (Morrison and Reis 1989; Sun and Spyer 1991; Ermirio et al. 1993; Ruggeri et al. 1995), and rabbits (Terui et al. 1987).

The RVLM neurons were also demonstrated by Morrison and Reis (1989) to comprise the medullary efferent arm of the somato-sympathetic reflex, or the medullary descending pathway of the supraspinal somato-sympathetic reflex. In their study, sympathoexcitatory neurons in the RVLM that project to the spinal intermediolateral nucleus (i.e., RVLM-spinal sympathoexcitatory neurons or RVLM-spinal-projecting neurons) were identified antidromically in urethane-anesthetized, paralyzed, ventilated rats. Single sciatic nerve stimuli evoked early (latency, 26 ms) and late (latency, 117 ms) excitation of RVLM-spinal sympathoexcitatory neurons. These discharges paralleled and preceded the biphasic increases in splanchnic nerve activity (peak latencies, 87 and 176 ms). The differences between the respective latencies of the RVLM unit responses and those recorded in the splanchnic nerve were accounted for by the conduction time in the sympatho-excitatory pathway from the RVLM to the splanchnic nerve. Both the threshold intensity and the relationship between response amplitude and stimulus intensity for the sympathetic response were comparable to those for the group III fiber component of the dorsal root response to sciatic nerve stimulation.

Sun and Spyer (1991) made recordings in anesthetized rats from 38 RVLM neurons with spinal-projecting axons. RVLM-spinal-projecting neurons that receive a powerful baroreflex inhibitory input are believed to play an important role in vasomotor control. With percutaneous electrical stimulation of the hindlimb, they elicited essentially the same response pattern regardless of whether the stimulation was applied to the ipsilateral or contralateral limb, although with contralateral stimulation there was a shorter latency period. In all 14 animals tested, hindlimb stimulation produced an early peak of excitation, and in 12 of 14 animals there was an early inhibition. There was also a later peak of excitation in two of 14 animals. Considering the latencies of responses from two sites of stimulation on the tail, they concluded that the early excitation and inhibition of RVLM-spinal projecting vasomotor neurons were elicited by excitation of peripheral pathways with conduction velocities characteristic of group III fibers. This research argues convincingly for somato-sympathetic reflex excitation of sympathetic preganglionic neurons leading to increases in sympathetic nerve activity and arterial pressure, with the efferent limb of the reflex originating in vasomotor neurons of the RVLM.

Zanzinger et al. (1994) demonstrated in anesthetized cats that the supraspinal reflex recorded from the left T3 white ramus following stimulation of the left T4 intercostal nerve was suppressed by cooling of the left (ipsilateral) RVLM. In contrast, cooling of the right (contralateral) RVLM had no significant effects. Similar results were obtained by ipsilateral and contralateral microinjection of the RVLM with kynurenic acid, a glutamate antagonist, and CNQX, a specific non-NMDA receptor antagonist. These findings suggest that the descending pathway of the supraspinal somato-sympathetic reflex component projecting to the intermediolateral cell column originates in the ipsilateral RVLM.

Spinal Tracts Involved in the Supraspinal A Reflex. Fibers traveling within the dorsolateral funiculus of the spinal cord are responsible for excitation of sympathetic nerves (Barman et al. 1976; Barman and Wurster 1975, 1978; Foreman and Wurster 1973; Gebber et al. 1973; Henry and Calaresu 1974a,b,c; Illert and Gabriel 1972; Kerr and Alexander 1964; Szulczyk 1976; Taylor and Brody 1976). In some studies, for example, stimulation of the dorsolateral funiculus led to activation of sympathetic nerves and concomitant increases in heart rate and blood pressure. In contrast, blood pressure and heart rate fell following bilateral lesioning or cooling of this area and, in the anesthetized cat, the supraspinal component of the somato-sympathetic reflex was eliminated (Foreman and Wurster 1975; Szulczyk 1976; Dembowsky et al. 1980, 1981).

However, since the dorsolateral sulcus and the dorsolateral funiculus also conduct afferent input to the supraspinal reflex (Chung and Wurster 1976; Chung et al. 1979; Coote and Downman 1966), it is unclear whether, for example, experimental block in this region is selectively interfering with either the afferent or the efferent limb of the supraspinal component of the somato-sympathetic reflex.

3.2.1.2 Inhibitory Somato-sympathetic Reflexes

3.2.1.2.1 Postexcitatory Depression

Often following mass somato-sympathetic reflexes, there is a period known as the "postexcitatory depression" or the "silent period," during which spontaneous activity is markedly reduced. In 1958, Schaefer and his colleagues (Sell et al. 1958) described such a postexcitatory depression or silent period after prior excitation of sympathetic renal and cardiac nerves by stimulation of somatic afferents. Subsequently, other researchers have confirmed this observation in both pre- and postganglionic fibers (Koizumi et al. 1968; Iwamura et al. 1969; Coote and Perez-Gonzalez 1970; Schmidt and Schönfuss 1970; Sato and Schmidt 1971).

By applying two stimuli successively to the same peripheral nerve fibers at different intervals, it was possible to determine whether a depression of reflex excitability accompanied the silent period. In recordings from the lumbar sympathetic trunk or from lumbar white rami, the late reflex appeared in response to the test shock stimulation at approximately 600 ms and gradually increased in amplitude until at last it recovered completely at about 2.2 s (Sato et al. 1967; Koizumi et al. 1968). In contrast to such a relatively early recovery of the late reflex, the early reflex in response to the test shock stimulation was first recognized after an interval of as long as approximately 1.3 s, and a complete recovery was not observed until after 2.5 s. The late reflex recovered somewhat faster than the early reflex. After high spinal transection the inhibitory time course of the early reflex shortened considerably. Dembowsky et al. (1985), using in vivo intracellular recording of sympathetic preganglionic neurons, observed a long-lasting membrane hyperpolarization and reduction of the on-going synaptic activity preceded by a subthreshold EPSP evoked in sympathetic preganglionic neurons in intact cats in response to stimulation of spinal pathways and segmental inputs. In spinal cats, a similar hyperpolarization was not recorded except in one sympathetic preganglionic neuron. These results indicate that most of the inhibition seen in the intact animal originates in the medulla or higher centers rather than in the spinal cord. Iwamura et al. (1969) showed that by lesioning the ventral medullary reticular formation the silent period could be abolished in cats, without affecting the excitatory period.

In anesthetized rats, electrical stimulation of the hindpaws, tail, or spinal cord evoked both a group III excitatory response and an inhibitory response in RVLM neurons with spinal-projecting axons. The group III afferent fibers responsible for eliciting the excitatory and inhibitory responses had similar conduction velocities, however, the inhibitory response usually followed the excitatory effect, suggesting a more complex central pathway for inhibition. Evidence was presented that the inhibitory response was mediated by a γ-aminobutyric acid (GABA)-like transmitter acting on $GABA_A$ receptors on the vasomotor-related neurons of the RVLM. Specifically, inhibition of vasomotor-related neuronal activity following hindpaw stimulation was blocked by ionophoresis of the $GABA_A$ receptor antagonist bicuculline, at a dose which did not completely suppress the baroreflex inhibition of these neurons (Sun and Spyer 1991). Thus GABAergic inhibition of the RVLM-spinal vasomotor neurons may be involved in nociceptive-evoked vasodepressor responses.

In urethane-anesthetized rabbits, stimulation of cutaneous myelinated afferents of the hindlimb sural nerve evoked a reflex response in the renal sympathetic nerve consisting of a small early excitatory component followed by a prolonged inhibition. When the RVLM, where sympatho-excitatory reticulospinal neurons originate, was injected bilaterally with the GABA receptor antagonist bicuculline, the somatically induced inhibitory response in the renal sympathetic nerve diminished. When the caudal ventrolateral medulla (CVLM) was injected bilaterally with the neurotoxic agent kainic acid, the inhibitory response was also reduced. Sympathoinhibitory neurons projecting to the RVLM are located in the CVLM. It therefore appears that stimulation of somatic afferents elicits a sympathoinhibitory response mediated by CVLM neurons projecting to the RVLM and by GABA receptors located in the RVLM (Masuda et al. 1992).

3.2.1.2.2
Early Depression

In addition to postexcitatory depression, inhibition of sympathetic activity may also occur without prior excitation. This has been reported by Coote and Perez-Gonzalez (1970) in renal sympathetic nerves, by Wyszogrodski and Polosa (1973) in sympathetic preganglionic neurons of the first thoracic segment, by Jänig and Schmidt (1970) in preganglionic sympathetic fibers of the cervical sympathetic trunk and by Koizumi and Sato (1972) in postganglionic sympathetic fibers in muscle nerves of anesthetized cats.

Supraspinal Inhibitory Components. The early depression of adrenal sympathetic efferent discharges appearing prior to the excitatory reflex has been observed in the anesthetized rat. Araki et al. (1981) demonstrated that stimulation of myelinated afferent nerve fibers of the T13 spinal nerve first depressed spontaneous ipsilateral efferent nerve discharges, with a latency of about 30 ms and a duration of about 70 ms, and then elicited a reflex discharge with a latency of about 90 ms and a duration of about 120 ms (Fig. 22A). In the spinalized animal, a single electrical stimulation of myelinated afferent fibers of the same T13 spinal nerve could elicit an excitatory A reflex discharge with a latency of about 25 ms without any depression (Fig. 22C). Thus there is a descending inhibitory influence on preganglionic sympathetic neurons to the adrenal nerves in CNS-intact rats, as has been shown for other sympathetic neurons in cats (Coote et al. 1969; Dembowsky et al. 1980).

*Spinal Inhibitory Components.*The inhibitory effect of sciatic and ulnar nerve afferent stimulation on the firing frequency of sympathetic preganglionic neurons of the T1 segment was studied in anesthetized cats, with intact spinal cords or spinalized at the C2 level (Wyszogrodski and Polosa 1973). In preparations with either the spinal cord intact or sectioned, ulnar or sciatic afferent nerve stimulation could depress the spontaneous firing and the firing evoked by antidromic stimulation, by iontophoretic glutamate, and by mechanical injury. The depression was not preceded by

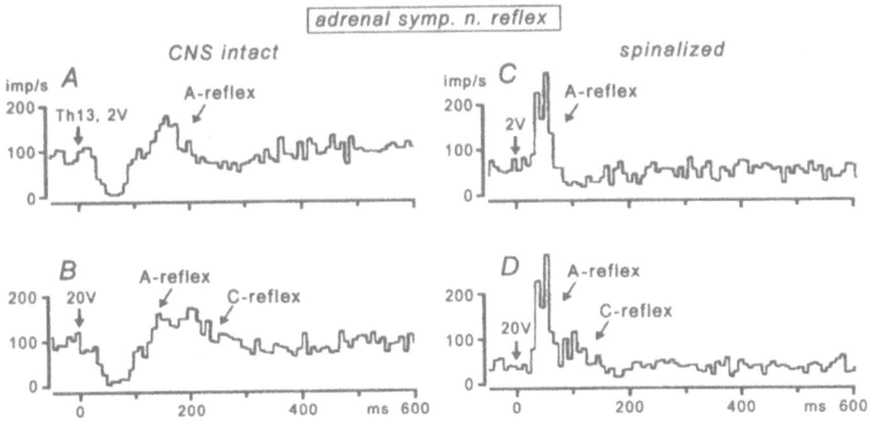

Fig. 22A–D. Poststimulus-time histograms of adrenal efferent nerve activity produced by a single electrical stimulation of the ipsilateral 13th thoracic spinal afferent nerve in chloralose/urethane-anesthetized rats. **A,B** In CNS-intact rats, the early depression of adrenal sympathetic efferent discharges appeared prior to the excitatory supraspinal A and C reflexes. **C,D** The spinal A- and C-reflex discharge components appeared after spinal transection at the C1–2 level. (From Araki et al. 1981)

excitation. The persistence of T1 sympathetic preganglionic neuron inhibition by both ulnar and sciatic nerve stimulation in the acute spinal cat makes it clear that entirely spinal inhibitory pathways exist, for both the segmental and the intersegmental effects (Wyszogrodski and Polosa 1973).

In vivo and in vitro studies have shown that stimulation of axons in the lateral funiculus (Dembowsky et al. 1985; Inokuchi et al. 1992) or focal stimulation of thoracic cord (Yoshimura et al. 1986) can elicit EPSPs and IPSPs in sympathetic preganglionic neurons.

In neonatal (12- to 22-day-old) rats, stimulation of dorsal root afferents in thoracolumbar spinal cord slices has been shown to induce IPSPs in antidromically identified sympathetic preganglionic neurons (Dun and Mo 1989). IPSPs were induced in some preganglionic sympathetic neurons by the application of NMDA, an agonist of the NMDA receptor of glutamate. Furthermore, these IPSPs were abolished by strychnine, a glycine antagonist, or by D-2-amino-5-phosphonovalerate, an NMDA receptor antagonist, and strychnine-sensitive unitary IPSPs were evoked in several preganglionic sympathetic neurons by the electrical stimulation of dorsal rootlets. IPSPs of 5–15 mV were elicited in 19 sympathetic preganglionic neurons by electrical stimulation of dorsal rootlets. A reversal potential of -60 to -75 mV could be induced in a low $[Cl^-]_o$ solution, but not in a low $[K^+]_o$ solution. The IPSPs were reversibly blocked by strychnine but essentially unaffected by bicuculline and picrotoxin, GABA receptor antagonists. Hence, the unitary and evoked IPSPs recorded in preganglionic sympathetic neurons are due primarily to an increase of Cl^- conductance by glycine (an inhibitory transmitter) or a glycine-like substance, released from interneurons, that can be activated by NMDA.

3.2.2 Sympathetic Reflex Responses to Stimulation of Unmyelinated (C) Somatic Afferents, the C Reflexes

Recording from renal sympathetic nerve in cats, Fedina et al. (1966) made the original observation that electrical stimulation of unmyelinated (C) somatic afferents resulted in a sympathetic reflex discharge with a longer latency than that arising from stimulation of myelinated somatic afferents. Many investigators have since confirmed this observation (Coote and Perez-Gonzalez 1970; Khayutin and Lukoshkova 1970; Koizumi et al. 1970; Schmidt and Weller 1970; Sato 1972b,c, 1973).

Using the averaging technique, Schmidt and Weller (1970) quantitatively analyzed the response recorded either from the cervical or lumbar sympathetic trunk, and called it the C reflex. They found that the C reflex was augmented or recruited by using stimulus frequencies higher than one per 8 s or by train pulse stimulation with a pulse interval of about 50 ms (Fig. 23).

In 1973, Sato resolved the spinal and supraspinal mechanisms of the C reflex:

1. Afferent inputs entering the spinal cord at the same level or at a segment adjacent to the level of the sympathetic outflow under observation evoke C reflexes (Fig. 24). These occur both in the anesthetized animal with an intact neuraxis and in acute and chronic spinal animals. The C reflexes are seen after single volleys and do not need any type of temporal facilitation. The pathway for these reflexes is complete at the spinal level.
2. Afferent inputs entering the spinal cord at segments different from those of the sympathetic outflow under observation evoke C reflexes in an anesthetized animal with an intact neuraxis. Temporal facilitation is needed to obtain maximum reflex effects. This reflex disappears after

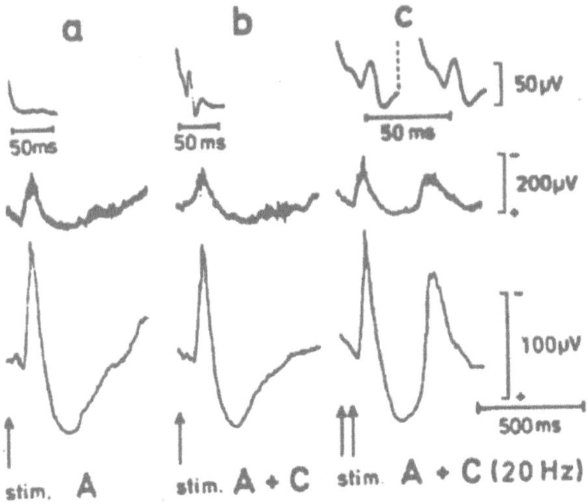

Fig. 23a–c. Somato-sympathetic C reflexes in a chloralose-anesthetized cat. In the *upper tracings*, the afferent volleys from the saphenous nerve are displayed at slow sweep speeds and at high amplification. The stimulus artifact of stimuli above 2T prevented the recording of the A volley. **a** Single stimulation with 4T (A fibers were stimulated). **b** Single stimulation with 200T (A and C fibers were stimulated). **c** Two stimuli at a 50-ms interval with 200T. The *middle tracings* show specimen photographs of the resulting sympathetic reflexes recorded from the lumbar sympathetic trunk caudal to the L5 ganglion. The *lower tracings* of the records show averaged responses. Single afferent A and C volleys did not evoke C reflexes of supraspinal origin in cats, and such reflexes only appeared when train pulse stimulation was used. (From Schmidt and Weller 1970)

spinal transection; its pathway is presumably complete at the medullary level.

3. In well-maintained chronic spinal cats, about 3 months after spinalization, afferent inputs entering the spinal cord at segments different from those of the sympathetic outflow under observation elicit C reflexes that are not seen in the intact anesthetized animal, and which do not need temporal facilitation for maximum reflex effects.

This C reflex has also been recorded from a dissected single preganglionic sympathetic nerve filament in the lumbar white rami (Sato 1972b), from a preganglionic sympathetic neuron, using intracellular techniques (Dembowsky et al. 1985), and from dissected single postganglionic sympathetic

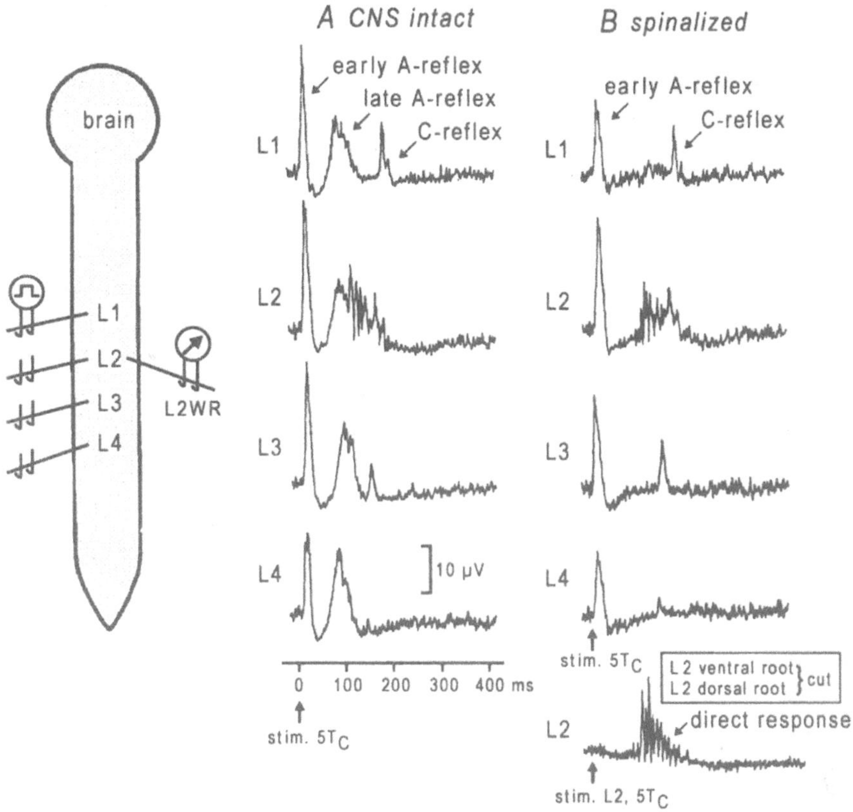

Fig. 24A,B. Effect of spinal transection on the sympathetic reflex discharges elicited by stimulation of the group IV afferents in various adjacent spinal nerves in a chloralose-anesthetized cat. Sympathetic activity was recorded from a whole L2 white ramus (*WR*) in **A** the CNS-intact condition and **B** about 1 h after spinal transection. The group IV afferent reflex (C reflex) appeared in a lumbar WR after adjacent spinal nerve stimulation in CNS-intact cats and remained even in acute spinal cats. In the *bottom recording* of B, L2 ventral and dorsal roots were totally severed. Some discharges were still present in the L2 WR after stimulation of the L2 spinal nerve. Therefore, discharges elicited at about 90 ms after stimulation of the L2 spinal nerve include not only reflex components, but also some action potentials of direct nerve fibers from the L2 spinal nerve to the L2 WR. (Modified from Sato 1973)

nerves, e.g., muscle vasomotor nerves, and cutaneous sudomotor and cutaneous vasomotor nerves (Jänig et al. 1972; Koizumi and Sato 1972).

In cats, single afferent A + C volleys rarely, if ever, evoked C reflexes of supraspinal origin, and such reflexes only appeared when some kind of tetanic potentiation was used (Schmidt and Weller 1970; Calaresu et al. 1978; Chung and Wurster 1978; Y. Sato et al. 1983). However, in rats, single A + C volleys induced regular and, as a rule, substantial C reflexes of supraspinal origin when recording from the rat's adrenal sympathetic nerve (Fig. 22B) (Araki et al. 1981; Isa et al. 1985; Fujino et al. 1987) or cardiac sympathetic nerve (Fig. 25) (Adachi et al. 1992b; Kimura et al. 1996b). Although C reflexes cannot be obtained from rat splanchnic efferents by single afferent A + C volleys of a hindlimb nerve (Nosaka et al. 1980; Morrison and Reis 1989), the evidence points to a better efficacy of C-afferent volleys in rats than in cats.

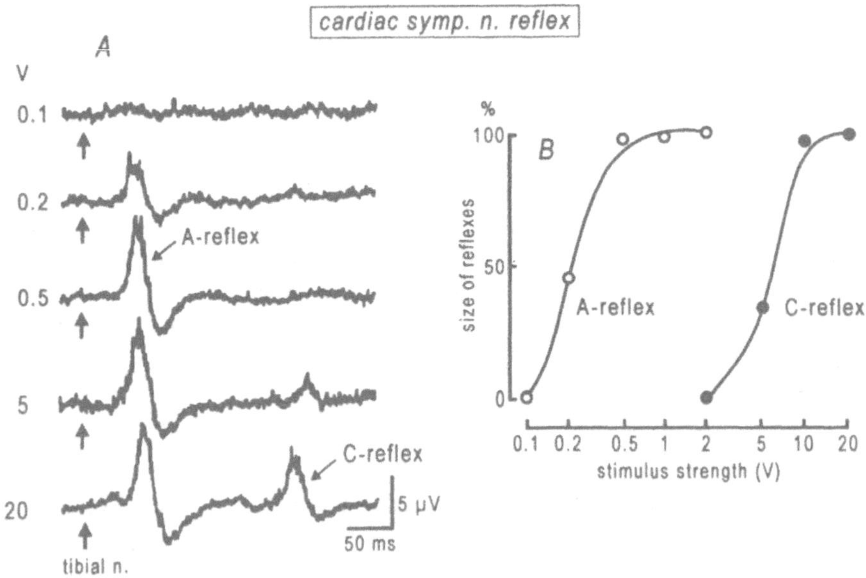

Fig. 25A,B. Somato-cardiac A and C sympathetic reflexes recorded from the inferior cardiac nerve following single electrical stimulation of the tibial nerve in a chloralose/urethane-anesthetized rat. With increasing stimulus strength first, the A reflex (latency, 45 ms) and then the C reflex (latency, 212 ms) appeared. (Modified from Adachi et al. 1992a)

3.2.2.1
Medullary Regions and Spinal Tract Involved in the Supraspinal C Reflexes

Cardiovascular neurons in the ventrolateral medulla (VLM) respond to activation of unmyelinated somatic afferent nerves in rabbits (Terui et al. 1987). In urethane-anesthetized rabbits, barosensory neurons were identified in the VLM. These neurons were spontaneously active and inhibited by stimulation of aortic nerve A fibers. Stimulation of group II (or Aβ) cutaneous afferents tended to inhibit barosensory VLM neurons, while stimulation of group III (or Aδ) and IV (or C) cutaneous afferents was excitatory. Stimulation of group III and IV muscle afferents generally, but not invariably, had an inhibitory effect on the spontaneous activity of the barosensory VLM neurons.

Short trains of electrical pulses of suprathreshold intensity have been used to stimulate afferent C fibers in the common peroneal nerve in anesthetized cats. Resulting C reflexes in the cervical sympathetic trunk were abolished by bilateral lesions at the T12 level of an ascending pathway in the dorsolateral sulcus area, suggesting that afferent input from C fibers ascends in the dorsolateral sulcus of the spinal cord (Chung et al. 1979).

Spinal C Reflexes. C reflexes were evoked in the cardiac sympathetic nerve of acutely spinalized cats (A. Sato et al. 1985) and rats (Kimura et al. 1996b) and in splanchnic (Nosaka et al. 1980) and adrenal (Araki et al. 1981; Isa et al. 1985) sympathetic nerve (Fig. 22D) of acutely spinalized rats by stimulation of adjacent spinal nerves.

In acutely spinalized cats, stimulation of somatic A and C afferents in the ipsilateral ulnar and upper thoracic intercostal nerves elicited sympathetic A and C reflexes of short latency (in the case of the ulnar nerve, approximately 30 ms and 140 ms, respectively) in the inferior cardiac sympathetic nerve (A. Sato et al. 1985). Spinal sympathetic C reflexes were elicited from fewer segmental levels than sympathetic A reflexes. Additionally, C reflexes were not evoked by stimulation on the contralateral side of the body.

3.2.3 Convergence of Various Somato-sympathetic Reflex Pathways on Single Sympathetic Neurons

3.2.3.1 Preganglionic Neurons

Sato (1972b) dissected individual preganglionic fibers in the lumbar white rami of anesthetized CNS-intact cats and determined to what extent the preganglionic neurons served or did not serve as a final common path for the different reflex pathways. A population of 76 units was tested for their excitation by the four major reflex pathways: (1) early spinal, (2) late supraspinal and (3) very late supramedullary A-reflex components elicited by myelinated afferent stimulation, and (4) C-reflex components (Fig. 26). All 16 possible combinations of these four responses were found and, surprisingly, there were just as many neurons showing a high degree of specificity (22 neurons were excited via one pathway only) as neurons having a high degree of convergence (six had convergence from all four pathways, 24 from three pathways). Similar findings were reported by Kaufman and Koizumi (1971) for the early and late A-reflex components evoked by myelinated afferents.

Intracellular recordings in anesthetized cats with an intact neuraxis revealed that most preganglionic sympathetic neurons showed distinct early, late and very late EPSPs in response to stimulation of segmental inputs (IC-T3 or T4) (Fig. 27) (Dembowsky et al. 1985). The early and late EPSPs represent the early spinal and late supraspinal components of the somato-sympathetic A reflex (Koizumi et al. 1971; Sato and Schmidt 1971).

Sympathetic preganglionic neurons of the third thoracic segment, identified by antidromic stimulation of the white ramus of T3, yielded two distinct subsets of very late EPSPs: those with latencies of 64–125 ms, and those with longer latencies of 173–240 ms. From earlier studies employing extracellular recordings, it is likely that the former EPSPs were evoked by stimulation of unmyelinated group IV afferents (Sato 1973), whereas the latter correspond to the suprapontine somato-sympathetic reflex component (Sato 1972a) (Dembowsky et al. 1985). The relative paucity of very late EPSPs of the second subset (latencies of 173–240 ms) probably reflects, on the one hand, the sensitivity of the suprapontine reflex to the conditions of anesthesia (Sato 1972a), and on the other, the sensitivity of the C reflex

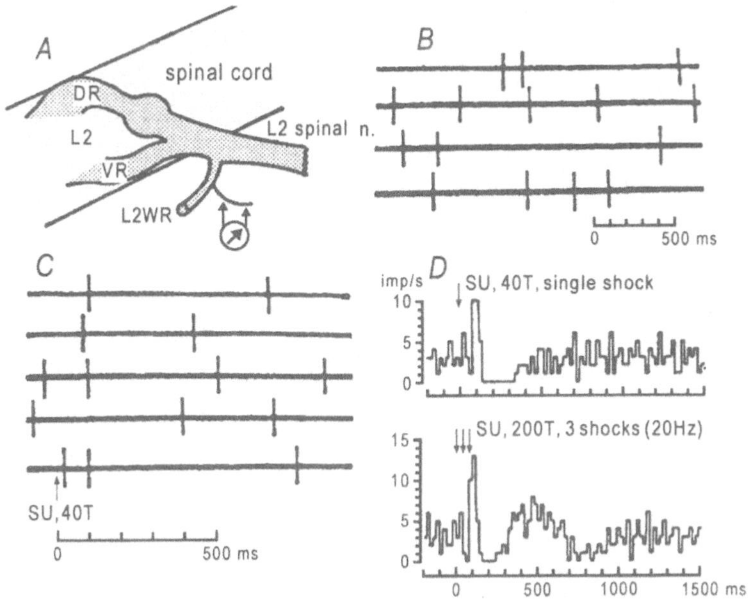

Fig. 26A–D. Modification of the spontaneous activity of a single preganglionic sympathetic fiber following somatic afferent stimulation in a chloralose-anesthetized cat. **A** The dissection and recording from a filament of the L2 white ramus (*WR*) are shown. *DR*, dorsal root; *VR*, ventral root. **B** Spontaneous discharges. **C** Single shocks were given to the sural nerve (*SU*). **D** The pulse-density histograms show the discharge probability evoked after stimulation of SU with a single shock of 40T intensity (supramaximal for the group II and III afferent fibers) or train pulses of 200T intensity (supramaximal for the group IV afferents). The A- and C-reflex discharges were evoked by stimulation of somatic afferents in a spontaneously active preganglionic sympathetic unit. (Modified from Sato 1972b)

to the conditions of stimulation employed (Sato 1973; Schmidt and Weller 1970).

3.2.3.2
Postganglionic Neurons

In postganglionic fibers of cutaneous sympathetic nerves of the hindlimb, stimulation of hindlimb afferents in anesthetized cats evoked the late medullary A reflex in the majority of units, whereas the C reflex was elicited in a much smaller percentage of units (Jänig et al. 1972). In muscle

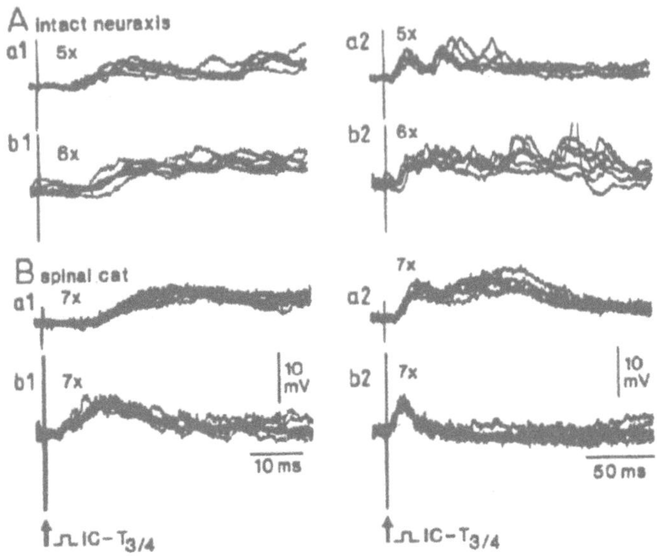

Fig. 27A,B. Early and late excitatory postsynaptic potentials (EPSPs) evoked in four different preganglionic neurons of **A** intact cats and **B** in cats spinalized at C3. Stimulation was to intercostal nerves of the third or fourth (IC-T3/4) segment (0.2–0.5 ms, 5–10 V, 1.9–2.9 s). The cats were anesthetized with chloralose. Preganglionic neurons were recorded from the third thoracic segment of the spinal cord. Superimposed sweeps on the left (*a1*, *b1*) and right side (*a2*, *b2*) are shown on different time scales. **A** Two neurons (*a* and *b*) recorded in different intact cats. *a1*, *a2* Early and late EPSPs are two clearly distinguishable components. *b1*, *b2* In this neuron, early and late components were fused to form one large summation EPSP which most probably also included very late EPSPs. The small size of the late EPSPs in *a* and *b* might be explained by damage to the T3 dorsal rootlets. **B** Two neurons (*a* and *b*) recorded in different spinal cats. *a1*, *a2* In spinal cats, half of the neurons showed both an early and late EPSP. *b1*, *b2* In the other half of the neurons in spinal cats, only an early EPSP was evoked. (From Dembowsky et al. 1985)

nerves of the hindlimb, the situation was reversed: more units showed C reflexes than the A reflexes following stimulation of myelinated afferents (Koizumi and Sato 1972). These efferents in muscle and cutaneous nerves were all thought to be vasoconstrictors, so that the differences in their responses must reflect different physiological functions.

Y. Sato et al. (1985) recorded unitary activity in filaments of the inferior cardiac nerve in cats anesthetized with chloralose and urethane. Responses in inferior cardiac nerve filaments were studied following electrical stimu-

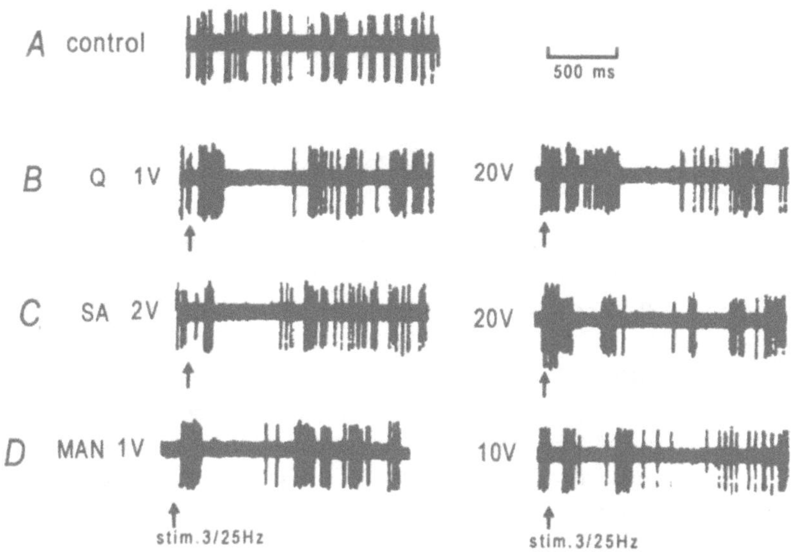

Fig. 28A–D. The A- and C-reflex responses of a sympathetic unit in a filament of the inferior cardiac nerve (ICN) to electrical stimulation of hindlimb afferent nerves in a chloralose/urethane-anesthetized cat. **A** Resting activity. **B–D** Effects of short repetitive stimulation (1–20 V, three stimuli at 40-ms interval; pulse duration, 0.5 ms) of the quadriceps (*Q*), saphenous (*SA*), and medial articular (*MAN*) nerves, respectively, at low (*left column*) and high stimulus strength (*right column*), as indicated. The ICN units were regularly excited by stimulation of A and C fibers in articular, cutaneous, and muscle nerves. (From Y. Sato et al. 1985)

lation of afferent A and C fibers in the articular, cutaneous and muscle nerves in the hindlimb. The inferior cardiac nerve units were regularly excited by electrically evoked single or short repetitive A volleys in articular, cutaneous and muscle nerves (Fig. 28). The excitation was followed by a silent period. Inclusion of C fibers in the afferent volleys gave a second, long-latency burst of impulses which was seen only with short repetitive stimulation.

3.2.4
Various Influences on Somato-sympathetic Reflexes

3.2.4.1
Modulation by Baroreceptor and Chemoreceptor Excitation

The spontaneous (or tonic) discharge activity of the sympathetic nervous system is well known (Adrian et al. 1932). This spontaneous discharge activity is reflexly inhibited by excitation of baroreceptor afferent nerves (Pitts et al. 1941; Gernandt et al. 1946), but reflexly facilitated by excitation of chemoreceptor afferent nerves (Downing and Siegel 1963; Lais et al. 1974; Trzebski et al. 1975; Kollai and Koizumi 1979; Trzebski and Kubin 1981; Dorward et al. 1987; Fukuda et al. 1989; Biesold et al. 1989b; Kimura et al. 1993). Both baroreceptor afferents (Thoren and Jones 1977) and chemoreceptor afferents (Fidone and Sato 1969) contain myelinated and unmyelinated fibers. However, it has not yet been determined whether it is the myelinated or unmyelinated afferents of these receptors which have reflex effects on sympathetic efferent discharges.

Activation of baroreceptor afferents depresses only the late medullary A reflex, and not the early spinal A reflex (Kirchner et al. 1971; Koizumi et al. 1971; Coote and Downman 1969; Baum and Shropshire 1979; Baum and Becker 1983). Activation of cardiopulmonary chemoreceptors by intravenous injection of veratrine also greatly attenuated the secondary (supraspinal) component but only slightly inhibited the amplitude of the initial (spinal) phase of the somato-sympathetic A reflex in anesthetized cats (Baum and Shropshire 1979). On the other hand, in anesthetized rats, systemic hypoxia enhanced somato-cardiac A and C reflexes (Li et al. 1996b). This effect of hypoxia was abolished after bilateral carotid sinus nerve section.

3.2.4.2
Involvement of Opioid Receptors

In unanesthetized animals, behavioral responses to noxious stimuli have been widely employed to gauge the analgesic effects of morphine. In addition, in both anesthetized and conscious animals, investigations of sensory neuronal processing mechanisms at the spinal and supraspinal levels have been conducted using noxious natural or electrical stimuli. In unanesthetized decerebrated and spinalized cats, electrical stimulation of unmyelinated somatic afferents evokes a somato-motor spinal C reflex which is depressed by morphine (Bell et al. 1980). Despite numerous reports on the spinal action of morphine, the question of the relative importance of spinal or supraspinal action of systemic opioids was not fully understood. The supraspinal sympathetic C reflex elicited by stimulation of unmyelinated afferent nerves may provide a useful experimental model for measuring the central effects of morphine.

It has been found that, in anesthetized cats, the C-reflex component can be selectively reduced by either systemic intravenous (i.v.) or intrathecal (i.t.) morphine administration (Ito et al. 1983; Sato et al. 1986b), whereas administration of morphine (Kato et al. 1992) or an enkephalin analogue (D-Met^2, Pro^5)-enkephalinamide (Sándor et al. 1985) to the brain stem via a vertebral artery slightly reduces both reflex components. These results have been taken to indicate that opioids suppress the C reflex mainly at the spinal level. Similarly, in anesthetized dogs, fentanyl applied i.v. or i.t. reduced the somato-sympathetic C-reflex component (Niv and Whitwam 1983; Wang et al. 1992). From these results, it was concluded that the somato-sympathetic C reflex may be a useful indicator to test the effects of morphine and other opioids in anesthetized animals.

Many studies on the analgesic and other CNS effects of opioids such as morphine have, however, been carried out not in cats and dogs, but in rats. These studies have extended the observations made on the effects of morphine on somato-cardiac sympathetic reflexes in cats (Ito et al. 1983; Sato et al. 1986b; Kato et al. 1992) to rats. In the anesthetized rat, morphine in a dose-dependent and naloxone-reversible manner either enhances or depresses these reflexes, depending on its route of application. Hence, application of morphine either into the femoral vein or into the subarachnoid space of the cisterna magna (i.c.m.) enhanced both the A and C reflexes in a dose-dependent manner, while application of morphine into the intrathecal space (i.t.) of the lumbar spinal cord selectively inhibited C reflexes (Adachi et al. 1992b).

No ready explanation is available for these apparent species differences between cats and rats in the modulation of the A and C reflexes by morphine applied i.v. and i.c.m. Obviously, in the rat a rather potent mechanism has to be involved, since the depressant action of morphine at the spinal level, which must take place after i.v. injection, is more than compensated for by the brain stem facilitation. The mechanism of this facilitation in rats remains unresolved. Inhibition of inhibitory neurons has been suggested as a common mechanism of opioid action (Nicoll et al.

1980). Thus one possible mechanism of the morphine-induced facilitation in rats may include inhibition of inhibitory neurons within or related to the somato-cardiac sympathetic reflex paths in the brain stem. On the other hand, the depressant action of morphine on the C reflex after i.t. application in rats is the same as that observed in cats (Sato et al. 1986b). Thus it can be confirmed in rats that morphine inhibits spinal synaptic transmission of C afferent volleys into the somato-cardiac sympathetic reflex pathways. I.t. application of the μ-opioid receptor agonist [D-Ala^2, *N*-methyl-Phe^4, Gly^5-ol]-enkephalin (DAMGO) selectively depressed the C reflex, while i.c.m. application of the same drug enhanced both the A and C reflex in anesthetized rats (Fig. 29) (Sato et al. 1995). Thus μ-opioid receptors are involved in both effects.

The modulatory effects of morphine in localized areas in the brain on both somato-cardiac sympathetic A and C reflexes have been examined using microinjection technique for application of morphine in anesthetized rats (Li et al. 1997). Morphine microinjected (0.2 μl/50 nl) into the nucleus tractus solitarius (NTS), usually produced a significant augmentation of the A and C reflexes, the augmentation of C reflexes being more dominant than that of A reflexes. Microinjection of the same dose of morphine into the RVLM produced a significant increase in the C reflex. Morphine microinjected into the other areas, such as CVLM, locus coeruleus, periaqueductal gray, and accumbens nucleus, had no significant effects on both reflexes. Augmentation of both A and C reflexes induced by microinjection of morphine into NTS may be caused by suppressing inhibitory baroreceptor information or by augmenting excitatory chemoreceptor information in the NTS. Augmentation of the C reflex induced by microinjection of morphine into the RVLM may be caused by suppressing inhibitory inputs there or by augmenting excitatory inputs there.

3.2.4.3
Involvement of Nitric Oxide

The role of nitric oxide (NO) in the two somato-sympathetic reflex arcs, i.e., the A and C reflexes, was examined using an NO-synthase (NOS) inhibitor in anesthetized rats (Li et al. 1995). The A- and C-reflex components were recorded from a cardiac sympathetic efferent nerve and elicited by stimulation of myelinated A and unmyelinated C afferent fibers in the left tibial nerve. N^G-nitro-L-arginine methyl ester (L-NAME), a NOS

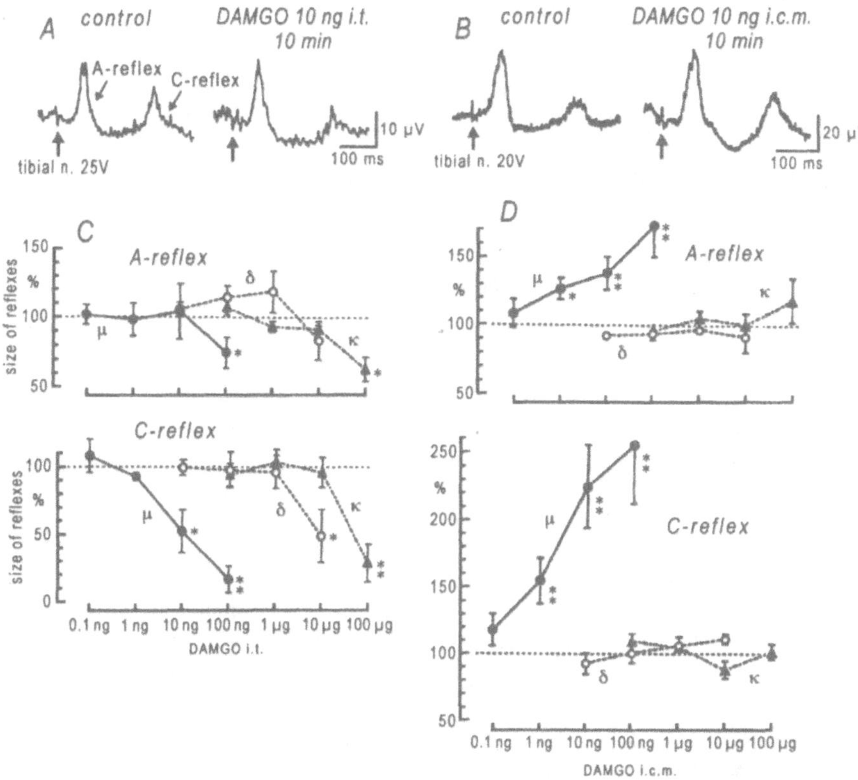

Fig. 29A–D. Effects of opioid receptor agonists injected **A,C** intrathecally (*i.t.*) or **B,D** into the cisterna magna (*i.c.m.*) on somato-cardiac sympathetic A and C reflexes elicited by stimulation of a tibial afferent nerve in chloralose/urethane-anesthetized rats. I.t. application of the μ-opioid receptor agonist (DAMGO, [D-Ala^2,*N*-methyl-Phe^4,Gly^5-ol]-enkephalin) selectively depressed the C reflex, while i.c.m. application of the same drug enhanced both the A and C reflex. Only at high doses did i.t. application of the δ-opioid receptor agonist (DPDPE, [D-Pen^{2,5}]-enkephalin) and κ-opioid receptor agonist (U-50,488H, trans-3,4-dichloro-*N*-methyl-*N*-[2-(1-pyrolidinyl)-cyclohexyl]-ben-zeneacetamide methanesulfonate) lead to a significant depression of the C reflex, and i.c.m. application of the same drugs did not affect the A and C reflexes. *$P<0.05$, **$P<0.01$; significantly different from the control reflexes. (Modified from Sato et al. 1995)

inhibitor, when administered either i.t. or i.c.m., augmented only the C reflex in a dose-dependent manner (Fig. 30). The effective i.t. dose of L-NAME to augment the C reflex was approximately 1000 times the i.c.m.

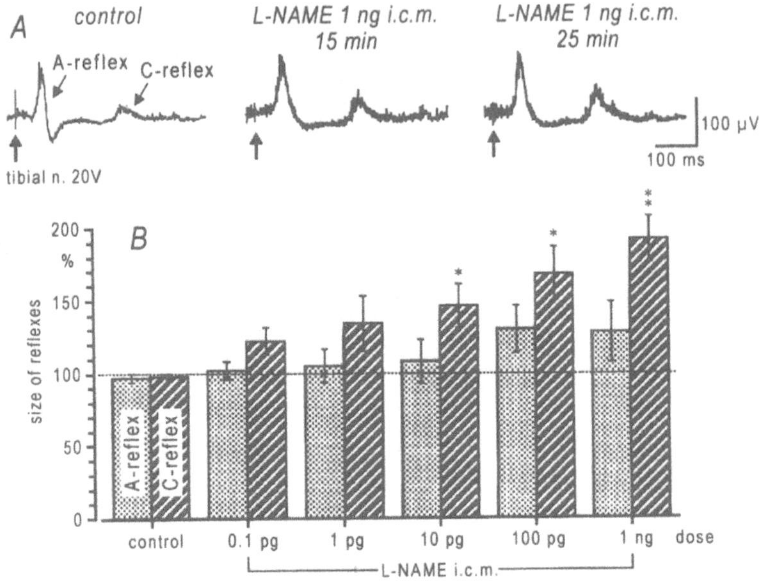

Fig. 30A,B. Effects of NO synthase inhibitor (L-NAME, N^G-nitro-L-arginine methyl ester), applied into the cisterna magna (i.c.m.), on somato-cardiac sympathetic A and C reflexes elicited by stimulation of a tibial afferent nerve in chloralose/urethane-anesthetized rats. L-NAME, applied i.c.m., augmented only the C reflex in a dose-dependent manner. *$P<0.05$, **$P<0.01$; significantly different from the control reflexes.(Modified from Li et al. 1995)

dose. N^G-nitro-D-arginine methyl ester (D-NAME), an isomer of L-NAME, had no effect on either A or C reflexes when administered i.c.m. Neither i.c.m. pretreatment nor posttreatment with L-arginine, a NOS substrate, influenced either A or C reflexes, but i.c.m. pretreatment with L-arginine abolished the facilitatory effect of L-NAME on the C reflex. These results suggest that NO, synthesized in the brain stem, plays an inhibitory role in the central modulation of the somato-cardiac sympathetic C reflex.

3.2.4.4
Involvement of Glutamate Receptors

The involvement of glutamate receptors in the central transmission of somato-sympathetic reflexes was studied by examining, in anesthetized rats, the effects of MK-801, an NMDA receptor antagonist, and CNQX, a non-NMDA receptor antagonist, on two reflex components, the A and C reflexes evoked in the left sympathetic renal nerve by a single shock to the left tibial nerve. The A reflex elicited by myelinated A fiber stimulation and the C reflex elicited by unmyelinated C fiber stimulation were depressed in a dose-dependent manner following i.c.m. administration of either MK-801 or CNQX (Fig. 31). I.t. administration of MK-801 did not have any effect on either A or C reflexes, while i.t. administration of CNQX had a slight effect on the A reflex (significantly on the A reflex only when treated with the highest dose, 100 ng; Nagata et al. 1995). Similarly, microinjection of the glutamate antagonists, kynurenic acid and CNQX to RVLM reduced the somato-sympathetic supraspinal A-reflex component in anesthetized cats (Zanzinger et al. 1994). These results indicate that both NMDA and non-NMDA receptors, stimulated by glutamate released possibly as a neurotransmitter, are involved in the central transmission pathways of somato-sympathetic reflexes at the level of the brain stem, but not the spinal cord.

3.2.4.5
Effects of Anesthesia

The effects of inspiring various concentrations of sevoflurane anesthetics on the sympathetic reflex responses evoked in the left inferior cardiac nerve branch following electrical stimulation to the ipsilateral superficial peroneal nerve were investigated in cats (Yanase et al. 1988). At an inspiratory concentration of 2% sevoflurane, two components of the somato-sympathetic reflexes with two different latencies, that is A and C reflexes, were recorded. Increasing the sevoflurane concentration from 2.0%–3.0% resulted in approximately a 50% attenuation of both the A and C reflexes. A further increase in the concentration of sevoflurane to 4.0% resulted in a further suppression of both reflexes.

The effects of nitrous oxide on the supraspinal somato-sympathetic A-reflex potentials induced in the lumbar sympathetic trunk by electrical stimulation of the femoral nerve were investigated in anesthetized (Ogawa et al. 1994). The A reflex was depressed by inhalation of 75% nitrous oxide in oxygen in CNS-intact cats. Midbrain decerebration itself caused marked potentiation of the A reflex, although subsequent inhalation of 75% nitrous oxide did not cause any depressive effect on the A reflex. From these results, the investigators concluded that anesthetic concentrations of ni-

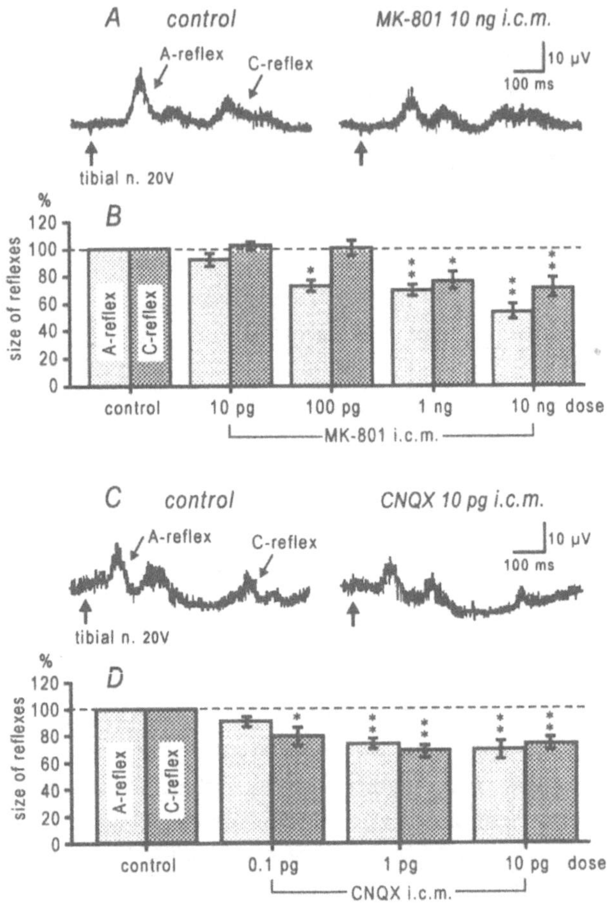

Fig. 31A–D. Effects of **A,B** an NMDA receptor antagonist (MK-801) and **C,D** a non-NMDA receptor antagonist (CNQX) applied into the cisterna magna (i.c.m.) on somato-renal sympathetic A and C reflexes elicited by stimulation of a tibial afferent nerve in chloralose/urethane-anesthetized rats. The A reflex and the C reflex were depressed, in a dose-dependent manner, following administration of either MK-801 or CNQX i.c.m. *$P<0.05$, **$P<0.01$; significantly different from the control reflexes. (Modified from Nagata et al. 1995)

trous oxide might activate a descending inhibitory system from higher areas in the CNS.

3.3 Somato-parasympathetic Reflexes

Relatively few studies have attempted to measure the reflex responses of parasympathetic fibers to stimulation of somatic afferents.

3.3.1 Cranial Parasympathetic Recording

Marguth et al. (1951) were the first to record activity from the cardiac efferent branches of the vagus nerve. However, this work and that of subsequent researchers (Green 1959; Okada et al. 1961) has been criticized (Calaresu and Pearce 1965) on the basis that these vagal branches also contain some sympathetic contribution from the stellate and inferior cervical ganglia (Cannon et al. 1926).

Iriuchijima and Kumada (1963, 1964) demonstrated in anesthetized dogs that the spontaneous unit activity of cardiac efferent branches of the vagus nerve was inhibited by stimulation of somatic afferents. Furthermore, a single pulse stimulation of the sinus nerve induced excitation, with a latency of 60 ms, of vagal cardiac efferents (Iriuchijima and Kumada 1963, 1964), and this method was used for identifying these efferent fibers.

In chloralose-anesthetized cats, Nakazato (1968) recorded pure parasympathetic efferent discharges of the vagus nerve after crushing the stellate and inferior cervical ganglia, and was able to record reflex responses in the cervical and thoracic, but not abdominal vagus nerve in response to sciatic nerve stimulation.

Koizumi et al. (1983) and Terui and Koizumi (1984) further demonstrated that cutaneous somatic afferent nerve stimulation initially inhibited cardiac vagus nerve activity and excited the cardiac sympathetic nerve. These initial responses were then followed by the opposite responses in the respective nerves, i.e., excitation of the vagus nerve and long-lasting inhibition of the sympathetic nerves. They concluded that the somato-vagal and somato-sympathetic reflex responses opposed each other. This pattern of opposing responses was elicited by excitation of group II afferents. When stimulus intensity was increased to recruit group III and group IV fibers, the pattern of response was essentially unchanged.

Kimura et al. (1996a) first recorded gastric vagal efferent reflex discharges caused by electrical stimulation of tibial afferent nerve in urethane-anesthetized rats. Group II and III afferent excitation caused excitatory reflex discharges with a latency of about 120 ms, and group IV afferent excitation caused additional reflex discharges with a latency of about 360 ms (see Sect. 4.2.5.2).

3.3.2
Sacral Parasympathetic Recording

The effects of somatic afferent nerve stimulation on pelvic parasympathetic efferent discharges innervating the bladder were first recorded by Bradley and Teague (1968) in anesthetized cats. They reported that stimulation of the pudendal nerve produced a supraspinal reflex potential with a latency of 150–200 ms. In anesthetized cats, De Groat and Ryall (1969) found that stimulation of hindlimb somatic afferents elicited a response in parasympathetic preganglionic neurons similar to that evoked by stimulation of the pelvic nerve. Both intracellular and extracellular recordings from the parasympathetic neurons showed short-latency IPSPs and inhibition of discharges in parasympathetic neurons followed by a supraspinal excitatory reflex. In chronic spinal cats, EPSPs were abolished but stimulation of somatic afferents continued to evoke short-latency IPSPs, as had been seen in CNS-intact animals.

In anesthetized cats, A. Sato et al. (1983) reported reflex discharges in pelvic efferent nerves to the bladder following hindlimb afferent volleys. Recordings were made both when the bladder was quiescent and when it was active during micturition contractions. Somatic afferent stimulations during micturition contractions (when pelvic efferent nerve activity is high) evoked the distinct parasympathetic reflex discharges, while the same stimulation did not produce the marked reflex discharges when the bladder was empty and quiet (when pelvic nerve activity is silent). The parasympathetic reflex responses were composed of excitation and inhibition of the efferent discharges in pelvic nerves. The excitatory pelvic parasympathetic reflexes were composed of spinal and supraspinal A-reflex discharge components with latencies of 90 ms and 320 ms, respectively, and a supraspinal C-reflex discharge component of 770 ms latency. The inhibition of pelvic parasympathetic discharges, elicited about 50 ms after the onset of stimulation, was preexcitatory inhibition, because this inhibition appeared before the reflex discharge was elicited. There was also postexcitatory depression of pelvic parasympathetic discharges, that is inhibition of spontaneous discharges immediately after the reflex discharges were elicited. After acute spinal transection these reflex potentials almost disappeared. In 2- to 19-month-old chronic spinal cats, in contrast to CNS-intact cats, reflex discharges were clearly elicited in the pelvic parasympathetic efferent nerves by volleys in hindlimb afferents independent of the volume of the bladder. There were A reflexes with a latency of 90 ms, and C reflexes with a latency of 340 ms. This suggests that in normal cats, the excitatory spinal reflex from the hindlimb somatic afferents is depressed by descending activity from supraspinal structures.

The effects of various somatic afferent volleys on the discharges in pelvic efferent nerves to the bladder were measured in anesthetized CNS-intact and acute spinal rats (Morrison et al. 1996a,b). Electrical stimulation of the tibial afferent nerve produced early reflex inhibition (I1) and excitation (E1), and late reflex inhibition (I2) and excitation (E2) of efferent pelvic discharges. The I1 and E1 responses represent A reflexes, while the I2 and E2 responses represent C reflexes. All of these reflex responses were easily demonstrable in the vesical branches of the pelvic nerve when pressure in the bladder was elevated nearly to the point at which vesical micturition contractions appeared, and the magnitudes of the reflex responses were dependent on bladder pressure. These reflexes disappeared after acute spinalization. Intravenous administration of L-NAME, an NOS inhibitor, had the following results: (a) a reduction in the level of resting discharge, (b) a reduction in the size of the first inhibitory component, (c) the disappearance of the second inhibitory component, and (d) the exaggeration of the late excitatory component. I.c.m. injection of L-NAME caused changes similar to those observed following i.v. injection. These results suggest that inhibitory components of the somato-pelvic parasympathetic reflex are mediated by pathways that utilize NO as a neurotransmitter or neuromodulator at the level of the brain stem (Morrison et al. 1996a). In contrast, stimulation of L6 and S1 segmental inputs from the perineo-femoral branch of the pudendal nerve produced short-latency excitatory A- and C-reflex discharge components of spinal cord origin in pelvic parasympathetic efferent nerve; both A and C pelvic spinal reflexes became larger and of increased duration following spinalization. This means that both components are under inhibitory influences from the supraspinal structures (Morrison et al. 1996b; see Sect. 4.3.6).

3.4 Visceral Regulation by Axon Reflex-Like Mechanisms Following Somatic Afferent Stimulation

The role of the axon reflex in vasodilatation in the skin was reported by Lewis and Marvin (1927) and the mechanism of this axon reflex-induced vasodilatation has been clarified (see Zimmermann 1984; Holzer 1988; Koltzenburg and Handwerker 1994). Excitation in cutaneous afferent fibers originating in a restricted area of skin can be conducted to branches of the same afferents innervating the cutaneous blood vessels. The excited afferent terminal releases vasodilative substances, such as substance P and calcitonin gene-related polypeptide (CGRP), which produce vasodilatation within a certain area around the stimulated spot. Interestingly, it has been suggested that some spinal afferent nerves have dichotomizing branchings to two separate nerves innervating two different tissues (Pierau et al. 1984; Dawson et al. 1992; Takahashi et al. 1993). There is the possibility that, by the axon reflex-like mechanism, these two branches act to produce a response in the collateral tissue after stimulation of the skin.

In anesthetized cats, in addition to the spinal and medullary reflex components, in 1973 Sato found unusual action potentials in the L2 white ramus following stimulation of the group IV afferent fibers in the ipsilateral L2 spinal nerve. These strange action potentials, with a latency of 90 ms, were not abolished even after complete transection of L2 ventral and dorsal roots close to the spinal cord, indicating that the action potentials were not reflex discharges through the CNS (Fig. 24B, bottom). A similar response was confirmed in the inferior splanchnic nerves (Bahr et al. 1981). Some of fibers in the inferior splanchnic nerves were activated by electrical stimulation of one of the lumbar white rami as well as by stimulation of the somatic nerve of the same spinal segment, with a stable latency. The responses of these fibers reliably followed high frequency stimulation of the nerves. The axons were unmyelinated, had no ongoing discharges, and were not excited by afferent stimuli via a spinal or supraspinal reflex pathway. These fibers were considered to be afferents (Bahr et al. 1981).

Laparotomy significantly attenuates ethanol-induced gastric mucosal lesions in the rat. It has been proposed that the laparotomy-induced protection is brought about by a somato-visceral or viscero-visceral axon reflex of afferent neurons that have collateral nerve endings both in the abdominal wall and in the gastric mucosa. Prostaglandins were suggested to be involved as mediators in the activation of these branches of afferent neurons in the abdominal wall following laparotomy (Yonei et al. 1990). Nerve blood flow in the sciatic nerve is suggested to be regulated via an axon reflex-like mechanism by unmyelinated afferent CGRP containing vasodilators with collaterals in the saphenous nerve (Hotta et al. 1996; see Sect. 4.1.3.6).

3.5 Comparison of Somato-autonomic and Somato-somatic Reflexes

One of the most important findings in the field of somato-autonomic reflex physiology has been that responses may be mediated at both the supraspinal and the spinal levels. This led, in turn, to the discovery of the strong segmental organization of the spinally mediated reflex component. Nonetheless, the existence of relatively localized somato-autonomic responses runs counter to Cannon's model of the generalized sympathetic reaction and, until recently, the weight of experimental evidence certainly argued against significant mediation of somato-autonomic reflexes at the spinal level. This, however, reflects a certain bias in experimental design.

Investigations of somato-autonomic reflexes were greatly influenced by parallel studies of spinal motor reflexes. These familiar reflexes, as described by Sherrington, rely on afferent fibers from somatic receptors synapsing on spinal motor neurons. Although not indifferent to descending influences from the brain, these reflexes show a very strong segmental organization, hence their utility in clinical practice. It will be noted, however, that the preponderance of investigations into spinal motor reflexes has involved stimulation of limb afferents (especially large, myelinated afferents), and it is the limb reflexes which enjoy the broadest application in clinical practice. When stimulation of limb afferents has been applied in order to elicit visceral responses, however, the evidence has pointed strongly to supraspinal mediation. Without an appreciation of the cytoarchitecture of the spinal cord, it was not apparent that the experimental model of limb stimulation was strongly prejudiced against revealing the spinal somato-autonomic reflex.

It can now be understood that the organization of the spinal cord acts to segregate segmental levels which display a bias towards either spinal motor or somato-visceral reflexes. Somatic afferent fibers from the limbs enter the spinal cord at the cervical and lumbar enlargements (see Fig. 95). These regions contain large tracts of somatic motor neurons, and so readily mediate somato-somatic reflexes in the limbs. These same regions, however, are essentially devoid of autonomic preganglionic neurons. Hence, stimulation of limb afferents may induce visceral responses, but these reflexes must involve supraspinal centers. Since such centers would be called upon to integrate input from throughout the body, the elicited responses would tend to be generalized rather than segmentally organized or target organ-specific. On the other hand, in humans the thoracic and upper lumbar cord, as well as the second to fourth sacral segments contain, respectively, sympathetic and parasympathetic preganglionic neurons. Somatic afferent fibers entering the spinal cord at these levels may synapse with local autonomic neurons as well as with projections to supraspinal somato-autonomic reflex centers. Such input could well elicit localized or target organ-specific responses. As has been repeatedly demonstrated experimentally, such responses are likely to be modulated by descending influences in the CNS-intact animal, but find full expression in the spinalized animal.

3.6 Conclusion

1. Since 1973, when Sato and Schmidt reviewed studies of the reflex pathways and afferent contributions to the somato-sympathetic reflexes, substantial findings have appeared concerning the spinal and supraspinal reflex components in the somato-sympathetic A and C reflexes in the various sympathetic and parasympathetic efferent nerves innervating different effector organs. It has been shown that the relative contributions of the spinal and supraspinal reflex components and also the relative importance of A and C reflexes are not uniform throughout the different autonomic functions, but rather are quite dependent on the different effector organs. It appears to be unwise to extend a conclusion obtained from one specific organ to other autonomic functions. Cannon's concept of a generalized sympathetic nerve reaction has been replaced by a more sophisticated concept of organ-specific autonomic responses. Furthermore, it is now apparent that, at least in part due to the cytoarchitectural organization of the spinal

cord, afferent input is treated differently depending upon whether it originates in the limbs or in segmental spinal nerves. Somato-autonomic reflexes elicited by stimulation of limb afferents appear to be mediated mainly at the supraspinal level, whereas stimulation of segmental afferents may elicit responses from both supraspinal and spinal reflex centers.

2. There have been many analyses of somato-autonomic reflex pathways using microelectrode and single fiber recording techniques, and many studies of neurotransmitters or modulators involved in somato-autonomic reflexes. Glutamate, opioids, catecholamines, serotonin, and NO have been proven to exist in the reflex pathways of the somato-autonomic reflexes. The relative importance of these transmitters or modulators as well as their receptors in the reflex pathways of different organ systems must be further elucidated.
3. There are still many unresolved details of the central pathways of the somato-autonomic reflexes. The polysynaptic complexity in the circuit is one confounding aspect. Nevertheless, major findings of the reflex characteristics thus far obtained seem applicable to the clinical treatment of malfunctions of the autonomic nervous system. Continued delineation of the reflex discharge characteristics using pathophysiological experimental models in animals is expected, since nowadays there are many such models of experimental animals with autonomic malfunctions.

4 Organ Reactions to Somatosensory Input ("Somato-autonomic Reflexes")

4.1 Somatosensory Modulation of the Cardiovascular System

4.1.1 Introduction

Since Carl Ludwig (1847) first developed the method of continuously recording arterial blood pressure using a mercury (Hg) manometer connected to a pen recorder, the cardiovascular system has been shown to be under the influence of neural, humoral and local metabolic factors. Four factors appear to be especially important for neural regulation: (1) mental or emotional, (2) baroreceptor, (3) chemoreceptor, and (4) somatosensory influences. Of these factors, the influence of mentation or emotion is beyond the scope of the present review. The latter three influences involve basically reflex mechanisms. The mechanisms and effects of baroreceptor and chemoreceptor influences on blood pressure are also excluded from this review because this subject has been reviewed repeatedly elsewhere.

During the 1860s and 1870s, Carl Ludwig and his colleagues made continuous recordings of responses of blood pressure to electrical stimulation of afferent nerves in the limbs of anesthetized animals, and studied the modifications of these responses that resulted from transections of the brain stem. From their results, they concluded that the pathways of somato-vasomotor reflexes ran from the spinal cord up to the medulla oblongata and back to the spinal vasomotor neurons.

Hunt (1895) observed that in anesthetized animals, depressor responses of arterial blood pressure were obtained when weak stimuli were delivered to the central cut end of a somatic nerve, but pressor effects occurred at higher stimulus strengths. He suggested that the peripheral nerves contained two types of afferents, "depressor and pressor fibers," characterized by differences in sensitivity to electrical stimuli, or to cooling of the somatic afferent nerves.

Bainbridge (1914) analyzed reflex acceleration of the heart beat induced by somatic afferent stimulation in anesthetized dogs. He described the reflex acceleration of the heart beat which followed stimulation of sensory fibers of the sciatic nerve, and concluded that this was brought about partly by reduction of vagal tone, partly by stimulation of the accelerator nerve, and partly by the entrance of adrenaline into the systemic circulation.

In 1916, Ranson and Billingsley introduced the important concepts of the "pressor and depressor centers" in the brain stem. Sherrington (1906) had noted that blood pressure was increased by somatic afferent stimulation in chronic spinalized animals. However, it was thought that the spinal cord played a minor role in central mediation of pressor responses because only small (Brooks 1933) or no pressor responses (Ranson and Billingsley 1916) were evoked by electrical stimulation of limb nerves in chronic spinal cats.

Systematic studies on the effectiveness of the various types of somatic afferents, as classified by fiber diameter, were reported by Laporte and Montastruc (1957) and in a more substantial work by Johansson (1962). The main findings and conclusions of Johansson's work can be summarized as follows. Repetitive electrical stimulation of muscle afferents produced a fall in blood pressure. These depressor effects could be ascribed mainly to the activity of group III afferents. They were most pronounced at low stimulus frequencies (5–20/s), but were also observed at stimulus frequencies as high as 100–400/s. Stimulation of muscle afferent nerves with intensities strong enough to activate group IV afferents produced pressor effects at high stimulus frequencies, whereas at low frequencies the effects of the low-threshold depressor afferents were often dominant. Repetitive stimulation of myelinated cutaneous afferents produced only small or insignificant depressor responses that were probably due to the activation of group III afferents. At stimulus intensities sufficient to excite the group IV afferents, pressor responses were always observed, irrespective of the impulse frequencies used.

Evidence for the existence of metabolic receptors (chemoreceptors) in skeletal muscles was presented by Coote et al. (1971) and McCloskey and Mitchell (1972). They observed that in anesthetized animals, tetanic contraction of the hindlimb muscles elicited by stimulation of the ventral roots of the spinal nerves caused a rise in arterial blood pressure which was usually accompanied by small increases in heart rate and pulmonary ventilation. The authors noted that the stimulus was chemical rather than mechanical, and they suggested that the "metabolic receptors" for this exercise reflex were the free endings of group III and IV sensory nerve fibers located around the skeletal muscle blood vessels.

Sato and his group (Sato et al. 1976, 1981, 1984; Kaufman et al. 1977; Sato and Schmidt 1987; Kimura et al. 1995) used quantitative electrical and qualitative natural stimulation of somatic afferents of the skin, muscles and joints, and recorded blood pressure and heart rate as well as activity from sympathetic and parasympathetic efferents to the heart and blood vessels. In a majority of anesthetized cats (Kaufman et al. 1977) and rats (Sato et al. 1976), a reflex increase in the heart rate was elicited after natural stimuli with different modalities applied to the skin at various segmental levels, as long as the core temperature was well maintained. The cardiac response was noted to be dependent on the body temperature (Sato et al. 1976) and depth of anesthesia (Kurosawa et al. 1989). This cutaneo-cardiac acceleration reflex has been proven to be produced mainly by an increase in cardiac sympathetic efferent nerve activity.

The majority of work concerning somatic reflex effects on cardiovascular responses highlights the importance of sympathetic nerves, not vagal nerves, as the efferent reflex limb. This conclusion sometimes contradicts the findings in conscious humans. In conscious humans, the contribution of the vagus nerve to the reflex arc cannot be neglected (Nishijo et al. 1991). It must be kept in mind that many anesthetics used in the study of somato-autonomic reflexes, such as urethane, chloralose, pentobarbital, halothane and sevoflurane appear to maintain cardiovascular sympathetic efferent activity well, but strongly suppress cardiac vagal efferent activity. Clearly, drugs that facilitate cardiac vagal activity could not be used routinely as anesthetics, because of the danger associated with cardiac arrest during surgery.

In studies of CNS-intact animals, the responses of heart rate and blood pressure were generally elicited from multiple segmental skin areas, while in acutely spinalized animals, the responses were elicited only by cutaneous stimulation of limited spinal segments in rats (Sato et al. 1976; Kimura et al. 1995) and cats (Kaufman et al. 1977). Much larger heart rate and blood pressure responses were elicited from specific segmental areas in acutely spinalized rats than in CNS-intact rats. These results suggested that the somatically induced cardiovascular reflex responses were a consequence of cardiovascular sympathetic responses, and these sympathetic responses had a strong segmental spinal reflex component and a generalized supraspinal reflex component. The central reflex pathways of the somatically induced cardiovascular sympathetic reflexes were clarified by analyzing reflex discharge potentials from cardiac sympathetic efferent nerves following electrical single stimulation of somatic afferent nerves. A- and C-reflex discharges evoked by stimulation of myelinated A and unmyelinated C somatic afferent nerves were proven to exist in rats (Kimura et al. 1996b). In CNS-intact rats, spinal and supraspinal reflex components of both A and C reflexes were observed only when stimulation was delivered to a thoracic spinal afferent nerve. On the other hand, only the supraspinal component of both A and C reflexes was elicited when stimulation was delivered to a hindlimb afferent nerve. The spinal reflex pathway is segmentally organized, because the spinal reflex is evoked only when stimulation is delivered to the afferent nerves close to the cardiac sympathetic outflow segments. When the CNS is intact, the spinal reflex component is depressed by descending inhibitory pathways which originate in the brain.

4.1.2 Somatosensory Modulation of Heart Rate and Blood Pressure

4.1.2.1 Cutaneous Stimulation

4.1.2.1.1 Mechanical Stimulation

*Innocuous Stimulation.*There have not been many reports on the effects of innocuous mechanical stimulation of the skin on cardiovascular responses in anesthetized animals. This seems to be due either to the lack of a response, or to an extremely small response. In anesthetized cats, a small increase in heart rate following rubbing of the skin of the perineum was reported, but the response was reported to be much less than the response following noxious stimulation of the same skin by pinching (Kaufman et al. 1977). Innocuous mechanical stimulation of any cutaneous area of the face, forelimb, back or hindlimb does not produce any significant changes in blood pressure in anesthetized rats (Adachi et al. 1990) or rabbits (Terui et al. 1981).

Noxious Stimulation. In contrast, there have been many reports on the effects of noxious mechanical stimulation of the skin on cardiovascular responses. In anesthetized cats (Kaufman et al. 1977), rats (Sato et al. 1976; Kurosawa et al. 1989; Gibbs et al. 1989; Sun and Spyer 1991; Adachi et al. 1990; Kimura et al. 1995) and rabbits (Dorward et al. 1987; Terui et al. 1981; Fukunaga et al. 1990), noxious mechanical stimulation of skin by pinching caused reflex increases or decreases in the heart rate and blood pressure. The depth of anesthesia, the body temperature or the site of stimulus can profoundly and qualitatively affect the cardiovascular reflex responses to noxious mechanical stimulation of skin (Sato et al. 1976; Kurosawa et al. 1989; Gibbs et al. 1989; Fukunaga et al. 1990).

In rats, at a normal depth of anesthesia and with normal body temperature, reflex increases in the heart rate and blood pressure are elicited after noxious mechanical stimulation applied to any skin area (Sato et al. 1976; Kimura et al. 1995; Adachi et al. 1990). The responses following fore- and hindpaw stimulation are larger than the responses following stimulation of other skin areas. Sometimes depressor responses are observed by pinching of the abdomen (Sato et al. 1975a) or chest skin (Sato et al. 1976), depending on body temperature and depth of anesthesia.

Sato et al. (1976) systematically studied the neural mechanisms of the cardiac reflex response to cutaneous pinching in anesthetized rats. When rectal temperature was 38.0°–38.9°C, noxious cutaneous stimulation produced a monophasic cardiac acceleration reflex response in about 70% of all tested cases. In the other 30% there were either biphasic responses (first deceleration and then acceleration) or no response. Figure 32 is a representative example of the reflex increases in heart rate produced by noxious stimulation of 20 s duration delivered to the chest skin area in a CNS-intact rat. The latencies of the responses were about 1–2 s. Maximal increases were 20 beats/min. The heart rate responses mentioned here are centrally mediated, as demonstrated by the fact that they are completely abolished after destroying the thoracic spinal cord segments by passing a small wire cable back and forth several times in the vertebral canal. The cutaneously induced heart rate response was observed even after bilateral removal of the adrenal glands. This indicates that hormones secreted from the adrenal medulla are not primarily responsible for the cutaneo-cardiac response. The contributions of the cardiac nerves to the cutaneo-cardiac responses were examined by performing denervation of cardiac sympathetic or vagal nerves. Bilateral vagal severance decreased the reflex increases in some cases but not in other cases (Fig. 32A). How-

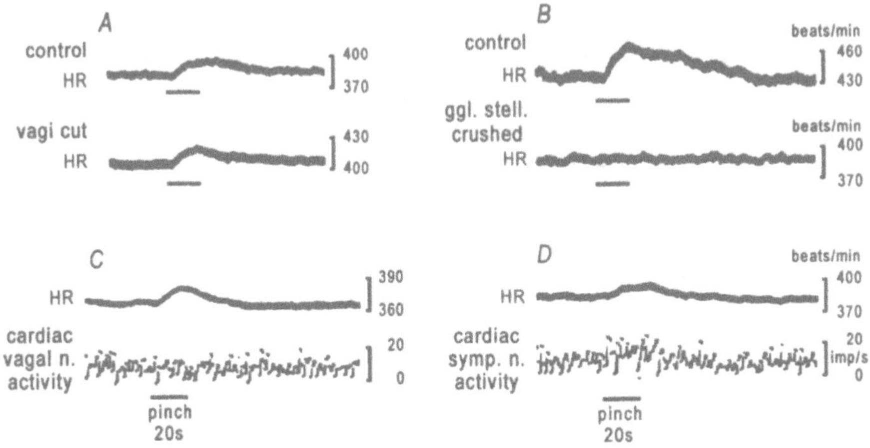

Fig. 32A–D. Pinching the chest skin increased the heart rate (*HR*) and cardiac sympathetic efferent nerve activity in chloralose- or chloralose/urethane-anesthetized rats. **A,B** Both cardiac sympathetic and parasympathetic nerves were kept intact for control recording (*upper trace*), and then either **A** the vagi or **B** the ganglia stellatum and the neighboring two thoracic paravertebral ganglia were cut or crushed bilaterally. The heart rate responses disappeared after crushing the sympathetic ganglia. **D** Cutaneous stimulation increased the efferent mass discharge activity of the cardiac sympathetic nerve and **C** did not influence that of the cardiac vagal parasympathetic nerve. (Modified from Sato et al. 1976)

ever, crushing both the stellate ganglia and the two proximal thoracic paravertebral ganglia abolished the responses in all cases (Fig. 32B). The efferent activities of the cardiac sympathetic nerve and the parasympathetic nerve were recorded in conjunction with cardiac responses, and it was found that cardiac sympathetic activity increased in parallel with the reflex cardiac acceleration (Fig. 32D). However, similar cutaneous stimulation did not influence right cardiac vagus activity during the reflex cardiac acceleration (Fig. 32C). From these results, it was concluded that in rats the reflex increase in the efferent discharge activity of the cardiac sympathetic nerve after cutaneous stimulation is responsible for the cutaneo-cardiac acceleration, while the efferent discharge activity of the cardiac parasympathetic (vagus) nerve does not seem to contribute to the reflex changes in heart rate. However, in dogs and cats, the cardiac parasympathetic vagal nerves do make a small contribution to heart rate changes following stimulation of somatic nerves (Kumagai et al. 1975; Norman and Whitwam 1973a,b; Kaufman et al. 1977; Rosenblueth and Freeman 1931; Sato et al. 1981).

Sato et al. (1976) examined the effects of noxious stimulation of skin on the heart rate in anesthetized rats, at different body temperatures. At normal body temperature, the response was usually an increase in heart rate as mentioned above, whereas the response at a rectal temperature of 36.0°–36.9°C was most often a decrease. Cardiac sympathetic nerve activity decreased in parallel with the cardiac deceleration. The cardiac vagus was not found to play a significant role in this cardiac deceleration.

Kurosawa et al. (1989) examined the effects of different inspiratory concentrations of sevoflurane (fluoromethyl-1,1,1,3,3,3-hexafluoro-2-propylether) on blood pressure and heart rate responses to pinching of a hindpaw in rats. Blood pressure and heart rate were increased by pinching at a concentration of 2.1% sevoflurane. The responses were attenuated at a concentration of 3.1% sevoflurane and almost disappeared at a concentration of 4.2%. These responses were noted to be closely related to the cardiac and renal sympathetic nerve responses. Gibbs et al. (1989) reported that in rats, increasing the end-tidal halothane concentration from sub-MAC (MAC is the minimum alveolar concentration, i.e., anesthetic potency) to supra-MAC levels was frequently associated with a reversal of the blood pressure response to noxious stimulation (to tail, hindfoot or forefoot), changing from a pressor response to a depressor response. The depressor responses were often, but not always, accompanied by decreases in heart rate. The depressor responses observed in deeply anesthetized rats were not influenced by vagotomy or muscarinic cholinergic blockade, and were associated with concurrent decreases in both cardiac output and systemic vascular resistance (Gibbs et al. 1989). Thus the hemodynamic changes associated with the depressor responses were consistent with a centrally mediated withdrawal of sympathetic tone (Gibbs et al. 1989).

Responses of blood pressure to cutaneous noxious stimulation are closely associated with responses in renal sympathetic nerve activity (Terui et al. 1981; Dorward et al. 1987). Terui et al. (1981) showed that noxious stimulation of the facial skin induced hypotension and attenuation of renal sympathetic nerve activity in anesthetized rabbits. Dorward et al. (1987) found that in anesthetized rabbits, stimulation of cutaneous nociceptors usually produced weak excitation of renal postganglionic neurons, accompanied by a small rise in arterial pressure.

The contributions of spinal and supraspinal reflex components to the cutaneo-cardiac and blood pressure responses have been clarified. In CNS-intact animals, the responses of heart rate and blood pressure were generally elicited from multiple segmental skin areas, while in acutely spinalized animals, the responses were elicited only by cutaneous stimulation of limited spinal segments in rats (Sato et al. 1976; Kimura et al. 1995) and cats (Kaufman et al. 1977). Kimura et al. (1995) systematically examined the effects on heart rate and blood pressure of pinching the skin at various segmental levels. They also studied responses of cardiac and renal sympathetic nerve activities in anesthetized rats with the CNS-intact or acutely spinalized at the cervical level. The stimuli were delivered to 12 different segmental areas of the body on the left or the right side: inputs to the brain (cheek), upper cervical (neck), lower cervical and thoracic (forepaw, medial side of arm, scapula, chest, upper back, abdomen), lumbar (rump, lateral side of thigh, hindpaw), and sacral (perineum) spinal cord. In CNS-intact rats, pinching of all 12 different segmental areas, both on the left and right, always produced an increase in the heart rate (Fig. 33A,C), blood pressure (Fig. 35A,C) and sympathetic nerve activities (Figs. 34A,B,E,

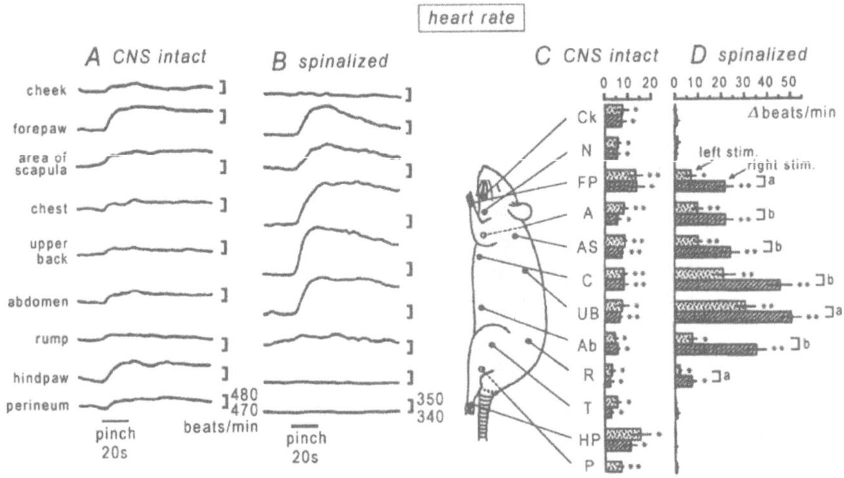

Fig. 33A–D. The responses of heart rate in urethane-anesthetized rats **A,C** with the CNS intact and **B,D** acutely spinalized at the C2 level to pinching of various skin areas. **A,B** Sample recordings of heart rate responses. **C,D** The maximum changes in heart rate in response to left-sided (*stippled column*) and right-sided (*hatched column*) stimulation. *Ck*, cheek; *N*, neck; *FP*, forepaw; *A*, arm (medial side); *AS*, area of scapula; *C*, chest; *UB*, upper back; *Ab*, abdomen; *R*, rump; *T*, thigh (lateral side); *HP*, hindpaw; *P*, perineum. In CNS-intact rats, pinching of various segmental areas produced an increase in the heart rate. In spinalized rats, only stimulation of restricted segmental skin areas was effective. $^{*}P<0.05$, $^{**}P<0.01$; significantly different from prestimulus control values, $^{a}P<0.05$, $^{b}P<0.01$; significantly different between responses to left and right side stimulation. (Modified from Kimura et al. 1995)

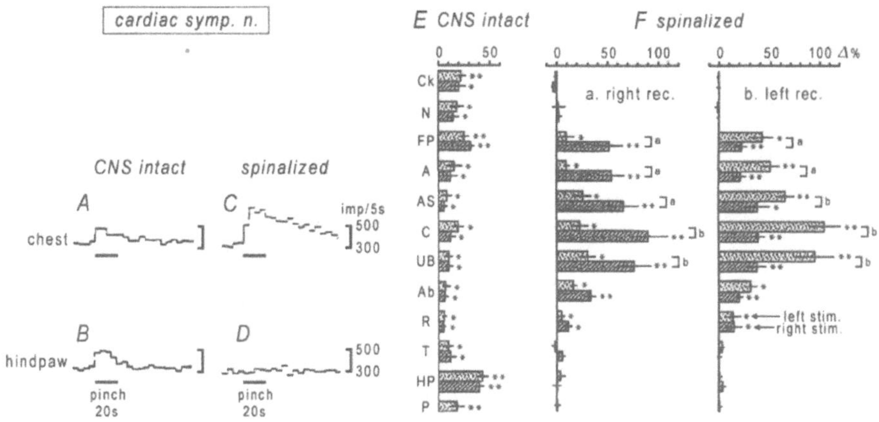

Fig. 34A–F. The responses of cardiac sympathetic nerve in urethane-anesthetized rats **A,B,E** with CNS intact and **C,D,F** acutely spinalized to pinching of various skin areas. **A–D** Sample recordings following pinching of chest and hindpaw. **E,F** The changes in right or left cardiac sympathetic nerve activity. Other details are the same as in Fig. 33. In CNS-intact rats, pinching of various segmental areas produced an increase in nerve activity, but in spinalized rats, only stimulation of restricted segmental skin areas was effective. The response of the cardiac sympathetic nerve displayed some laterality in spinalized rats. (Modified from Kimura et al. 1995)

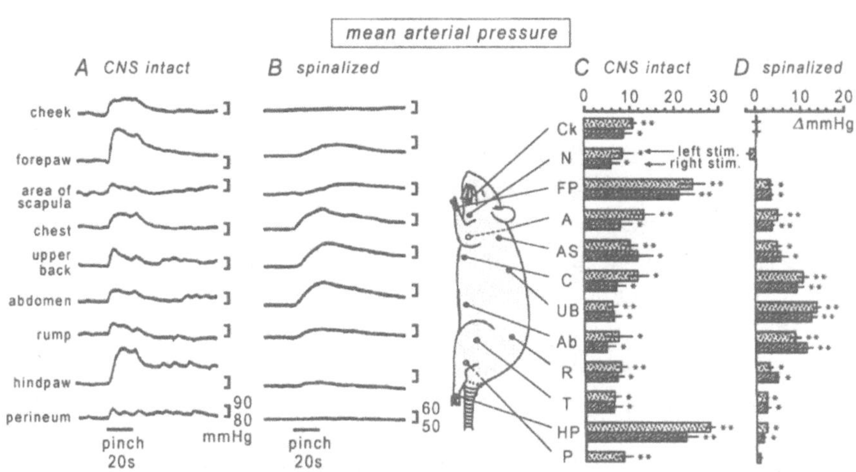

Fig. 35A–D. The responses of mean arterial pressure (MAP) in urethane-anesthetized rats **A,C** with CNS intact and **B,D** acutely spinalized to pinching of various skin areas. Other details are the same as in Fig. 33. (Modified from Kimura et al. 1995)

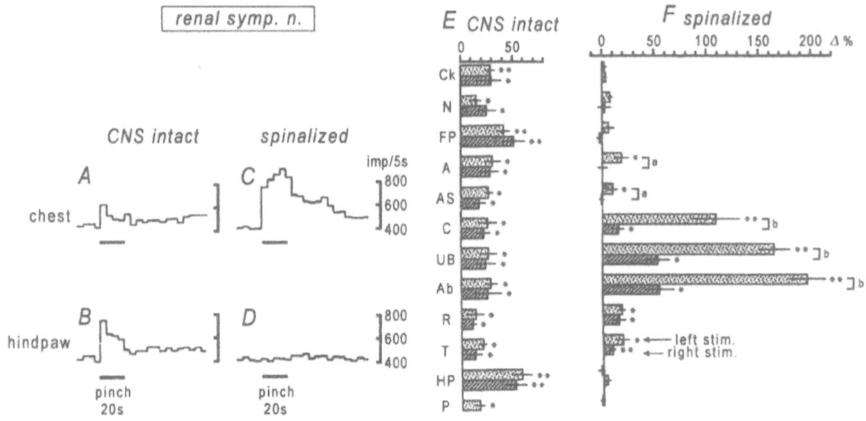

Fig. 36A–F. The responses of renal sympathetic nerve in urethane-anesthetized rats **A,B,E** with CNS intact and **C,D,F** acutely spinalized to pinching of various skin areas. Left renal sympathetic nerve activity was recorded. Other details are the same as in Figs. 33 and 34. (Modified from Kimura et al. 1995)

36A,B,E). There were no differences between the responses to stimulation of the left and right sides of each area. The responses following fore- and hindpaw stimulation were larger than the responses following stimulation of other skin areas. In spinalized rats, only stimulation of restricted segmental skin areas was effective in producing increases in heart rate (Fig. 33B,D), cardiac sympathetic nerve activity (Fig. 34C,D,F), blood pressure (Fig. 35B,D) and renal sympathetic nerve activity (Fig. 36C,D,F). Pinching the chest, abdomen and back of the body produced large increases, while neck, hindlimb and perineal stimulation induced only small increases or no increase in heart rate, blood pressure and sympathetic nerve activity. These results suggest the existence of two types of reflex responses, supraspinal and propriospinal, in the somato-cardiovascular reflex. The supraspinal component represents diffuse reflex organization, while the propriospinal reflex displays strong segmental organization.

It is important to emphasize that the cardiovascular response of spinal origin was proven to exist in acutely spinalized animals. It is particularly interesting to note that heart rate and blood pressure responses elicited by narrow segmental areas in acutely spinalized rats were much larger than those in CNS-intact rats. This indicates that in CNS-intact rats, these spinal reflex pathways are tonically inhibited by the brain, probably through descending inhibitory pathways (see Sect. 3.2.1.1). After spinal transection, these tonically active, inhibitory descending pathways are eliminated and thus unmask the segmental spinal reflex responses and result in large reflex effects. Tonically active, inhibitory descending effects on the spinal reflex pathway to the cardiac sympathetic efferent nerves have been demonstrated in anesthetized cats by Coote and Sato (1978). They transected the ventral quadrant of the spinal cord, and observed that spinal reflex discharges were greatly augmented.

It is of note that in spinalized rats the heart rate response was much larger when the stimulus was delivered to the skin on the right side of body. The laterality of the heart rate responses may depend on the strong influence of the cardiac sympathetic nerve on the right side of the heart, particularly on the sinus node (Levy et al. 1966; Irisawa et al. 1971; Kamosinska et al. 1989). Laterality, as in the heart rate responses, was not observed in the systemic blood pressure responses of spinalized rats. Blood pressure responses seem to be more general, with bilateral integration of the right and left visceral vasculatures. In spinalized rats, ipsilateral stimulation produced larger responses in cardiac and renal sympathetic nerve activities than did contralateral stimulation. The lack of laterality in heart rate and blood pressure responses in the CNS-intact preparation means that somatic afferent information ascends to the brain stem, and then bilaterally integrated information descends equally to both left and right sympathetic efferent preganglionic neurons, resulting in equal left and right cardiac and renal sympathetic responses.

Neurons in the rostral ventrolateral medulla (RVLM) appear to be involved in the supraspinal reflex pathway of somato-cardiovascular reflexes. Sun and Spyer (1991) recorded the responses of spinal projecting vasomotor neurons of the RVLM following mechanical stimulation in anesthetized rats. Noxious mechanical stimulation of either hindpaw led to an increase in the firing rate of the RVLM neurons, and a subsequent increase in mean arterial pressure. These results suggest that vasomotor neurons of the RVLM respond to noxious cutaneous input and contribute to the efferent or descending excitatory limb of the cutaneo-cardiac sympathetic reflex.

Thermal Stimulation. In anesthetized cats, a reflex increase in heart rate occurred after stimulation of small areas of skin on the perineum with a thermoprobe set at various temperatures (Kaufman et al. 1977). The threshold temperature for evoking cardiac acceleration was between 13° and 19°C for cold stimulation, and around 40°C for warm stimulation. Either

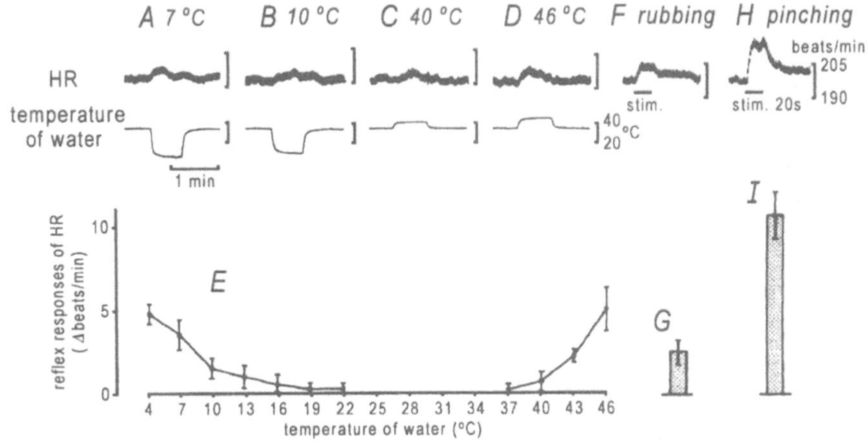

Fig. 37A–I. The heart rate (*HR*) response to thermal and mechanical stimulation of the perineal skin in chloralose/urethane-anesthetized cats. **A–D** Sample recordings of the heart rate. **E** Graph of the relationship between water temperature in the thermoprobe (*abscissa*) and the maximal magnitude of the heart rate reflex responses (*ordinate*). The water temperature in the thermoprobe was changed from the control temperature of 34°C to 4–46°C. Either innocuous warm (<45°C) or cool (>10°C) stimulation causes a small increase; noxious heat and cold stimulation produce a larger increase in heart rate. **F–I** Responses of the heart rate **F,G** after rubbing or **H,I** after pinching the skin. (From Kaufman et al. 1977)

innocuous warm (<45°C) or cool (>10°C) stimulation caused a small reflex increase in heart rate (Fig. 37). Thermal stimulation of the neck, chest and abdomen produced similar results, although the responses were usually somewhat smaller than those produced by thermal stimulation of the perineum. Stimulation by heat and cold in the noxious ranges (>45°C or <10°C) produced larger increases in the heart rate, suggesting that activation of cutaneous thermal nociceptors could also produce additional cardiac responses (Kaufman et al. 1977).

The responses of heart rate and blood pressure to noxious radiant heat were studied in anesthetized cats (Abram et al. 1983). The skin temperature of the hind footpad was raised to 53°C for 20 s using radiant heat and systemic blood pressure, heart rate and afferent activity recorded from the tibial nerve were monitored. The averaged tibial afferent nerve activity increased markedly as skin temperature approached 52°C. Within 2–3 s of the onset of increased tibial afferent nerve activity, systolic blood pressure and heart rate increased.

In anesthetized rats, the cardiovascular response to immersing the tail in 53°C water was examined after intrathecal injections of opioid agonists (Nagasaka and Yaksh 1995). Intrathecal injection of μ- and δ-, but not κ-agonists produced a dose-dependent blocking of the cardiovascular responses to noxious stimulation, and these effects were readily reversed by the opioid antagonist naloxone. Theses data indicate that the agonist occupancy of spinal μ- and δ-, but not κ-receptor sites can profoundly modulate the cardiovascular response evoked by thermal stimuli.

Vasopressor- and vasodepressor-related neurons of the RVLM integrate afferent input from various sensory receptors. Sun and Spyer (1991) investigated the effects of thermal stimulation on spinal-projecting vasomotor-related neurons of the RVLM in anesthetized rats. Innocuous thermal stimulation (38°C) or moderate pressure or brushing of the hindpaws did not significantly affect arterial pressure or the behavior of RVLM neurons. However, noxious thermal stimulation (52°C) of the hindpaws caused increases in both variables.

4.1.2.2 Muscle Stimulation

4.1.2.2.1 Induced Muscle Contraction

It is well known that exercise produces various cardiovascular responses. There have been many reports on the effects of skeletal muscle contraction on the cardiovascular system (see reviews by Mitchell and Schmidt 1983; Mitchell 1990). Two main theories were proposed to explain exercise-induced cardiovascular responses. One theory is that the initiation of motor activity originating in the brain is coupled with autonomic nervous control in the brain, and this information is transmitted to the cardiovascular system via autonomic efferent nerves. Thus, during exercise, central information in the autonomic nervous system or central commands from the brain are the main cause for exercise-induced cardiovascular responses. This theory can be called the "central command theory." The other theory is that during exercise, skeletal afferent nerves carry information about skeletal muscle contractions to the CNS. This muscle afferent information is integrated in the CNS, and integrated information is then transmitted to the cardiovascular system through the autonomic nervous system. This latter theory can be called the "reflex theory." In this case, the skeletal muscle afferents, autonomic efferents and the CNS are important reflex components. The central command theory is beyond the scope of this text, and readers are referred to the review by Mitchell (1990). However, the second theory will be discussed further.

In order to study reflex control of the cardiovascular system originating in skeletal muscle contractions, skeletal motor nerves in the spinal ventral root (Coote et al. 1971; McCloskey and Mitchell 1972; Fisher and Nutter 1974; Clement 1976; Coote and Dodds 1976; Mitchell et al. 1977; Streatfeild et al. 1977; Crayton et al. 1979; Aung-Din et al. 1981; Longhurst et al. 1981; Perez-Gonzalez 1981; Iwamoto et al. 1982, 1984a,b, 1985; Iwamoto and Botterman 1985; Waldrop and Mitchell 1985; Victor et al. 1989) or in the peripheral nerves (Fisher and Nutter 1974; Tibes 1977; Tallarida et al. 1981, 1985, 1990; Gelsema et al. 1983; Beaty 1985a,b) or skeletal muscle itself (Clement et al. 1973; Clement and Shepherd 1974; Clement 1976) were electrically stimulated and muscle contractions were elicited. It is possible to obtain either "tetanic" (or "static") contractions with high frequency stimulation, or "rhythmic" contractions using low-frequency stimulation.

In anesthetized dogs, Tallarida and his colleagues (1985) studied cardiorespiratory reflex responses during the initial phase of rhythmic and static contractions of hindlimb muscles. Muscle contractions were elicited by stimulating the femoral and gastrocnemius motor nerves for 20 s at 3 and 100 Hz with an intensity of 2.0–2.5 times the motor-evoking threshold. Rhythmic contractions (produced by stimulating at 3 Hz) caused a decrease in arterial pressure and heart rate, and increased pulmonary ventilation by increasing frequency without significantly changing the tidal volume of respiration (Fig. 38B). Tetanic contractions (produced by stimulating at 100 Hz) provoked an increase in arterial pressure and heart rate, and hyperpnea resulting from a rise in both frequency of respiration and tidal volume (Fig. 38A). Both patterns of cardiovascular and respiratory responses were reflexes initiated by activation of muscle receptors, as verified by interrupting the afferents from the contracting muscles.

There have been many reports supporting the hypothesis that the increases in blood pressure and heart rate during tetanic contraction of skeletal muscles are reflex responses whose afferent pathways involve muscle afferents (Coote et al. 1971; McCloskey and Mitchell 1972; Mitchell et al. 1977; Crayton et al. 1979). However, in some studies, the depressor effects induced by rhythmic contraction of skeletal muscles were explained as the result of metabolically induced vasodilation, because they remained after sectioning of the somatic afferents (Kaufman et al. 1984; Perez-Gonzalez 1981). The efferent pathway of these cardiovascular reflex responses

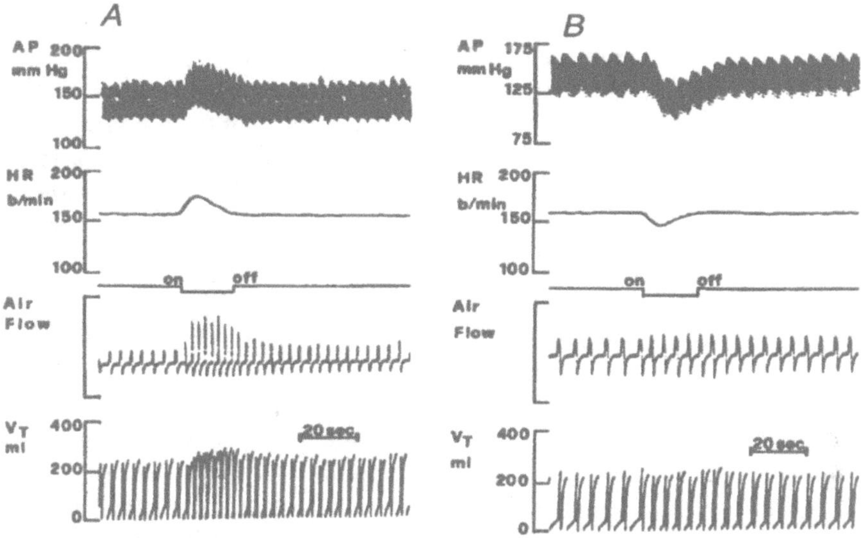

Fig. 38A,B. Cardiorespiratory responses to **A** induced tetanic contractions and **B** rythmic contractions of hindlimb muscles in chloralose-anesthetized dogs. **A** Response to induced tetanic hindlimb contractions (femoral nerve stimulated at 100 Hz and 2T; 2T intensity activated group I and II fibers). Note increase in arterial pressure (*AP*) and heart rate (*HR*) and rise in both frequency of ventilation and tidal volume (*VT*). **B** Response to induced rhythmic hindlimb contractions (3 Hz, 2T). Note decrease in AP and HR and increase in frequency of ventilation. (From Tallarida et al. 1985)

was confirmed in a direct way by recording the sympathetic nerve activity (Clement 1976). The effects of tetanic hindlimb muscle contraction on renal nerve activity were confirmed in anesthetized cats by Matsukawa et al. (1990) and Victor et al. (1989).

In anesthetized cats, the intrathecal administration (Kaufman et al. 1985, 1986) or microinjection into the dorsal horn region (Wilson et al. 1992) of an antagonist or an antibody to substance P (SP) blunted the pressor response to static contraction of the triceps surae muscle. Furthermore, Wilson et al. (1993) showed that static skeletal muscle contraction elicited by electrical stimulation of the L7-S1 ventral root in anesthetized cats caused the release of SP in the L7-dorsal horn region of the spinal cord, and this increase in SP was greatly attenuated after cutting the L7 and S1 dorsal roots, or completely abolished by muscle paralysis. It has been suggested that SP in the spinal cord plays a role in mediating the cardiovascular responses evoked during muscle contraction.

Vasopressin has also been implicated in the mediation of cardiovascular reflex responses to muscle contraction. Stebbins et al. (1992) examined the pressor, myocardial contractile (*dP/dt*), and heart rate responses in cats following electrically induced static contraction of the hindlimb muscles. Static contraction was accompanied by increases in both mean arterial pressure and *dP/dt*, which were further augmented by lumbar intrathecal injection of a vasopressin 1 (V1) receptor antagonist. Hence, V1 receptors in the lumbar spinal cord may have a role in the transmission of sensory input from hindlimb muscles, and vasopressin may thereby normally act to attenuate cardiovascular reflex responses to sustained muscle contraction.

Iwamoto et al. (1982, 1984b, 1985) analyzed the CNS structures mediating the pressor reflex evoked by hindlimb skeletal muscle contraction. They suggested that the caudal ventrolateral medulla may be a key integration site for the cardiovascular reflex response evoked by muscular contraction induced by ventral root stimulation. In anesthetized or decerebrated cats, increases in heart rate and blood pressure in response to sustained muscle contraction were augmented somewhat by transection of the brain at the midcollicular level. Transection of the lower medulla 5 mm rostral to the obex slightly attenuated the response, whereas transection of the spinal cord at the C1 level almost completely abolished the cardiovascular response (Iwamoto et al. 1985). Bilateral lesioning of an area of the ventrolateral medulla (VLM) which includes the lateral reticular nucleus also abolishes cardiovascular reflex responses to static muscle contraction (Iwamoto et al. 1982). As demonstrated by radioactive glucose studies, the metabolic rate of this area of the medulla increases during the reflex increase in blood pressure associated with muscular contraction (Iwamoto et al. 1984b; see Mitchell 1990).

Ascending spinal pathways were investigated by Kozelka and Wurster (1985) in anesthetized dogs. It was observed that the reflex increases in blood pressure and heart rate induced by muscle contraction were abolished by bilateral lesions of the dorsolateral sulcus in the L1-L3 region of the spinal cord. They concluded that ascending spinal pathways mediating somato-cardiovascular reflexes are located in the lateral funiculus, extending from the dorsal root entry zone to a position somewhat ventral to the dentate ligament.

4.1.2.2.2
Mechanical Stimulation

Innocuous Stimulation. In anesthetized or decerebrated cats, moderate stretching of forelimb and hindlimb muscles within physiologically innocuous ranges characteristically elicits a biphasic response consisting of an initial decrease in systemic blood pressure, followed by a subsequent pressor response (Skoglund 1960). In anesthetized rabbits, innocuous stretching of the triceps surae and quadriceps muscles, employing tensions of up to 500 g, did not significantly affect mean arterial pressure, heart rate or perfusion pressure of the hindlimb (Tallarida et al. 1981).

In anesthetized cats, tension on the triceps surae of 0.3 kg induced a small depressor response, while that in the range of 0.5–8.0 kg induced a pressor response (Stebbins et al. 1988). Similarly, pressures in the range of 125–300 mmHg generated by a cuff applied to the muscle led to a small increase in mean arterial pressure but not heart rate. The cardiovascular reflex responses to both passive stretching and external pressure were abolished by sectioning the dorsal roots from L5 to S1.

In anesthetized dogs (Tibes 1977) and rabbits (Tallarida et al. 1981), passive rhythmic movement of the knee joints elicited small depressor responses similar to those produced by active rhythmic movement. The reflex nature of these responses was established by transection of the local somatic afferents, which abolished the cardiovascular effects of joint movement. However, the most important afferent input to the reflex appears to arise from the muscles rather than the joint per se. Local anesthetic applied to the knee joint and adjacent articular tissues did not significantly affect the reflex response, but neuromuscular paralysis abolished the response (Tallarida et al. 1981).

Noxious Stimulation. In anesthetized rabbits, noxious mechanical stimulation of the triceps surae or quadriceps muscles, either by static or rhythmic pressure, produced an increase in arterial blood pressure (Tallarida et al. 1981).

4.1.2.2.3
Chemical Stimulation

Innocuous Stimulation. The injection of succinylcholine, which selectively activates group I and group II muscle afferent fibers, into a muscle artery in anesthetized cats caused no change in blood pressure (Sato et al. 1982) (Fig. 39E). Similarly, intra-arterial injection of nicotine, adenosine, adeno-

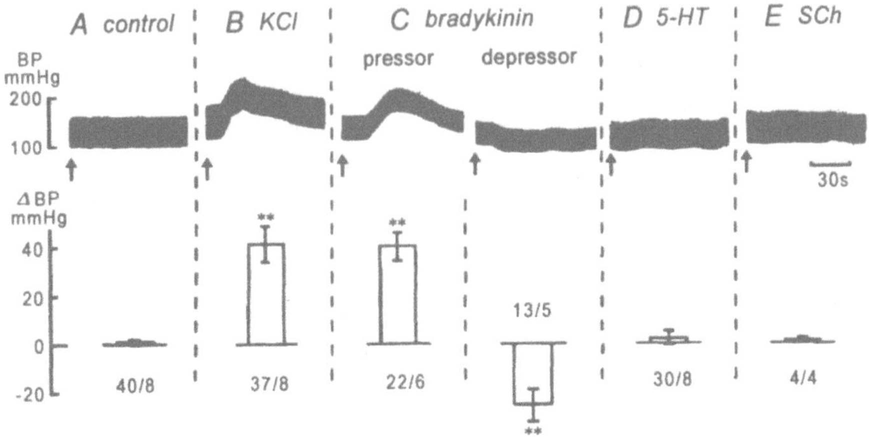

Fig. 39A–E. Blood pressure changes induced by intra-arterial injection of algesic substances into hindlimb muscles in chloralose/urethane-anesthetized cats. Each column shows a representative specimen record and the mean of the peak blood pressure changes upon the indicated injections of the saphenous artery. **A** Tyrode solution (control). **B** 0.3–0.5 ml twice isotonic KCl solution. **C** 26–44 μg bradykinin. **D** 45–225 μg serotonin (5-HT). **E** 164 μg succinylcholine. The KCl injection induced a pressor response, and the bradykinin injection induced pressor or depressor responses, but serotonin and succinylcholine injections were ineffective. $^{**}P<0.05$; significantly different from prestimulus control values. (From Sato et al. 1982)

sine triphosphate, adrenaline, noradrenaline, angiotensin, vasopressin and oxytocin also failed to elicit significant cardiovascular reflex responses (Tallarida et al. 1979).

Noxious Stimulation. Intra-arterial injection of various chemical algesic agents into skeletal muscle elicits blood pressure and heart rate responses in anesthetized animals. In anesthetized cats in which vagus and carotid sinus nerves were cut, intra-arterial injection of algesic, twice isotonic potassium chloride (KCl) into muscles of a hindlimb regularly induced acceleration of the heart rate and an increase in blood pressure (Fig. 39B) (Sato et al. 1982). With bradykinin, both accelerations and decelerations were observed (with or without accompanying pressor or depressor responses, respectively; Fig. 39C). In anesthetized rabbits, only hypotension and bradycardia were observed with intra-arterial injections of algesic bradykinin or potassium ions (Tallarida et al. 1979). The responses induced in both rabbits and cats by intra-arterial injection of bradykinin into muscle were abolished by sectioning the somatic nerves of the injected limbs

(Tallarida et al. 1979; Stebbins and Longhurst 1985), indicating that these responses are due to excitation of muscle afferent nerves.

Another algesic substance, serotonin (5-HT) alone has no such effects (Fig. 39D), however subsequent injection of either hypertonic KCl or bradykinin causes enhanced tachycardia and increases in blood pressure (Sato et al. 1982). Algesic prostaglandin E_2 (PGE_2) potentiates the cardiovascular response to bradykinin stimulation of skeletal muscle afferents (Stebbins and Longhurst 1985). In anesthetized cats, the cardiovascular response to bradykinin was reduced following inhibition of prostaglandin synthesis with indomethacin. The cardiovascular response to bradykinin after inhibition of prostaglandin synthesis can be restored by injection of PGE_2.

Injection of capsaicin into the hindlimb in anesthetized cats (Longhurst et al. 1980) and dogs (Crayton et al. 1981; Hussain et al. 1991) produced increases in mean arterial pressure and heart rate. After transection of the afferent neural connections to the stimulated hindlimb, the responses to the injection of capsaicin were abolished (Longhurst et al. 1980; Crayton et al. 1981).

Longhurst and Zelis (1979) found that regional hindlimb hypoxemia equivalent to that seen during severe exercise was associated with significant increases in heart rate and mean systemic blood pressure. However, regional hindlimb acidosis and hypercapnia failed to induce similar cardiovascular responses. In anesthetized animals, injection of hypertonic solutions of sodium chloride (NaCl) and glucose normally leads to increases in heart rate and blood pressure (Tallarida et al. 1979).

4.1.2.3
Acupuncture-Like Stimulation

Clinical studies have demonstrated that acupuncture will attenuate hypertension (Tam and Yiu 1975; Williams et al. 1991). This cardiovascular effect of acupuncture or acupuncture-like stimulation was also observed in conscious spontaneously hypertensive rats (SHR) (Yao et al. 1982) and hypertensive dogs (Li et al. 1983). Low-frequency electrical stimulation of a sciatic nerve for 30 min in conscious SHRs induced a poststimulatory depressor response with a parallel reduction in splanchnic sympathetic nerve activity (Yao et al. 1982). The poststimulatory depressor response in SHRs was naloxone reversible (Yao et al. 1982). In the hypertensive dog, electro-acupuncture at the Tsu-San-Li point reduced blood pressure significantly (Li et al. 1983). However, these depressor responses observed in conscious hypertensive animals could not be reproduced in anesthetized animals (Yao et al. 1982; Li et al. 1983).

In anesthetized normotensive rats and rabbits, arterial blood pressure decreased during electro-acupuncture stimulation of Tsu-San-Li (Chiu and Cheng 1974; Kline et al. 1978). The hypotensive effect of electrical stimulation of the hindleg muscles was suggested to be a consequence of somato-sympathetic reflexes where the afferent path was a hindleg nerve, because the response disappeared after cutting the nerve proximal to the site of stimulation (Kline et al. 1978).

The effects of acupuncture-like stimulation of a hindlimb on renal nerve activity and mean arterial blood pressure were studied in anesthetized normotensive Wistar rats (Ohsawa et al. 1995). A 160-μm diameter, stainless steel acupuncture needle was inserted into muscles through the skin of a hindlimb to a depth of about 5 mm. This area corresponds approximately to the Tsu-San-Li point in humans. The needle was twisted right and left at a frequency of about 1 Hz for 60 s. The responses were dependent upon the depth of anesthesia. In rats anesthetized intraperitoneally (i.p.) with a mixture of urethane 500 mg/kg and α-chloralose 50 mg/kg, acupuncture-like stimulation of a hindlimb for 60 s did not produce any consistent responses in renal nerve activity and mean arterial pressure (Fig. 40A–C). However, in about 70% of trials with rats under deeper anesthesia (1000 mg urethane/kg and 100 mg α-chloralose /kg, i.p.), the same stimulation induced a decrease in mean arterial pressure which was accompanied by a decrease in renal nerve activity (Fig. 40D,E). In these deeply anesthetized rats, acupuncture-like stimulation applied to the muscles alone induced inhibition of renal nerve activity and mean arterial pressure, but stimulation to the skin alone was ineffective (Fig. 41). Transection of the ipsilateral sciatic and femoral nerves completely abolished the responses of renal nerve activity and mean arterial pressure. Acupuncture-like stimulation of the hindlimb excited the femoral and common peroneal afferent nerves. These results indicate that the decrease in mean arterial pressure induced by acupuncture-like stimulation of the hindlimb was a reflex response. The afferent pathway is composed of hindlimb muscle afferents, while the efferent pathway is composed of sympathetic vasoconstrictors, including the renal nerves. The responses were identical to stimulation throughout an area of about 1 cm^2. It is reasonable to assume that the delivery of stimuli within an area of about 1 cm^2 would excite the muscle afferent nerve innervating the corresponding muscles which contain mechanical sensory receptors.

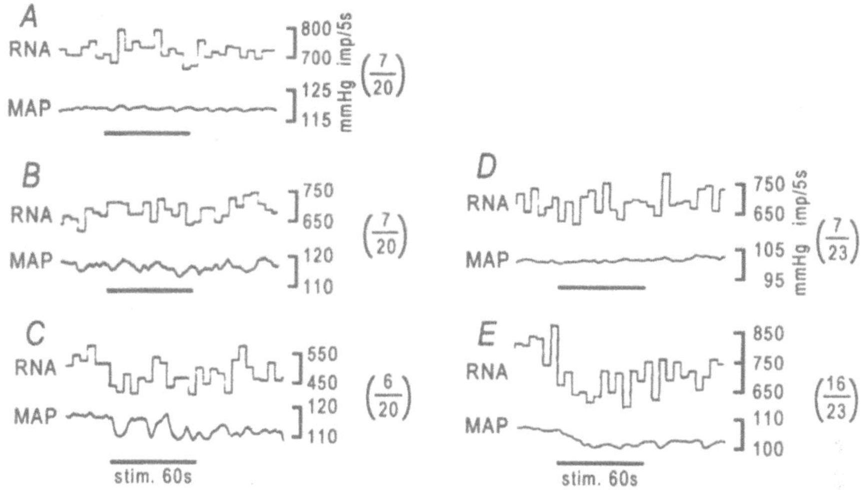

Fig. 40A–E. Effect of acupuncture-like stimulation to a hindlimb on renal sympathetic nerve activity (*RNA*) and mean arterial pressure (*MAP*) under different conditions of anesthesia in rats. **A–C** Sample recordings of three different response patterns of RNA and MAP following 500 mg urethane/kg and 50 mg α-chloralose/kg. Acupuncture-like stimulation of a hindlimb did not produce any consistent responses in RNA and MAP. **D–E** Sample recordings of two different response patterns following 1000 mg urethane/kg and 100 mg α-chloralose /kg. RNA and MAP decreased with acupuncture-like stimulation in 70% of trials. (From Ohsawa et al. 1995)

It has been proven that acupuncture-like stimulation produces an increase in endogenous opioids in the CNS (Sjölund et al. 1977). However, endogenous opioids may not be strongly involved in the reflex responses of renal sympathetic activity and blood pressure, since intravenous injection of naloxone, an opioid receptor antagonist, did not influence the reflex (Ohsawa et al. 1995).

4.1.2.4 Joint Stimulation

4.1.2.4.1 Mechanical Stimulation of the Knee Joint

Innocuous Stimulation. In anesthetized cats, passive movements of knee joints, such as rhythmic flexions and extensions, static outward rotations, and rhythmic inward and outward rotations of the knee joint within its

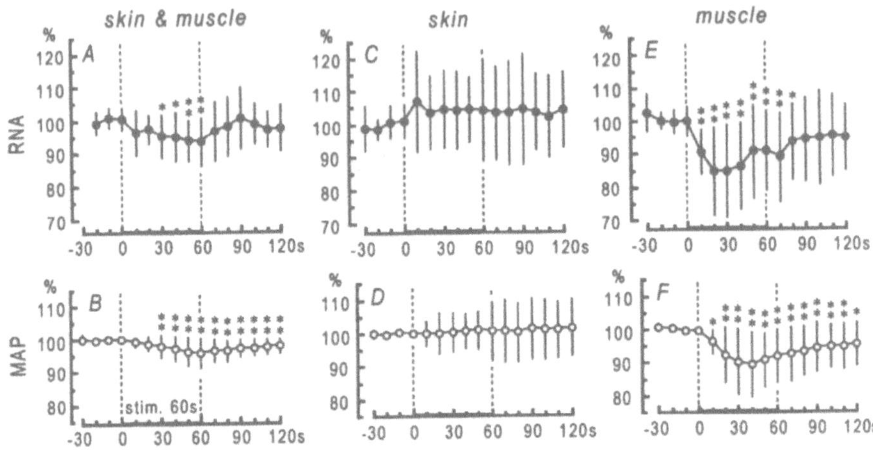

Fig. 41A–F. Acupuncture-like stimulation of **A,B** hindlimb skin and muscle and **E,F** underlying muscles alone decreased renal sympathetic nerve activity (*RNA*) and mean arterial pressure (*MAP*), but **C,D** acupuncture-like stimulation of skin alone did not change MAP and RNA in rats anesthetized with 1 g urethane/kg and 100 mg α-chloralose /kg. *P<0.05, **P<0.01; significantly different from prestimulus control values. (Modified from Ohsawa et al. 1995)

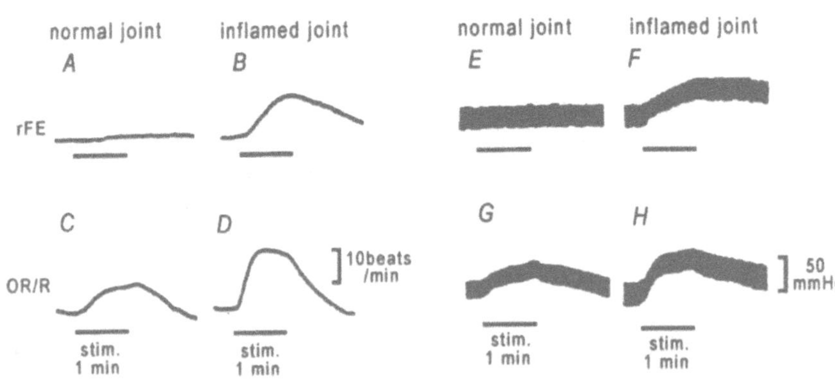

Fig. 42. A–D Heart rate and **E–H** blood pressure changes induced by movements of **A,C,E,G** normal and **B,D,F,H** inflamed knee joints in a halothane-anesthetized cat. Rhythmic flexions and extensions (*rFE*, innocuous movement) of a normal knee joint did not change heart rate and blood pressure, and noxious outward rotations (*OR/R*) of the normal knee joint increased these variables. The same movements of an inflamed knee joint caused definite increases in heart rate and blood pressure. (Modified from Sato et al. 1984)

physiological working range do not have any significant influence on blood pressure or heart rate (Fig. 42A,E) (Sato et al. 1984).

Noxious Stimulation. In nonanesthetized decerebrated cats (Barron and Coote 1973) and in baroreceptor denervated anesthetized cats (Sato et al. 1984), noxious movement of the knee joint induced increases in heart rate and blood pressure. Movement of a normal knee joint beyond the normal working range induced heart rate and blood pressure increases of the same order of magnitude as those elicited by pinching of a hindpaw (Fig. 42C,G) (Sato et al. 1984).

Cardiovascular responses are particularly pronounced when the joint receptors are sensitized by inflammation. Rhythmic flexions and extensions as well as rhythmic inward and outward rotations of an inflamed knee joint, even within its normal physiological working range, caused definite increases in blood pressure and heart rate (Fig. 42B,F). Static outward rotations in the noxious range significantly increased blood pressure and heart rate (Fig. 42D,H). The increases in blood pressure and heart rate induced by noxious outward rotation of the inflamed joint regularly exceeded those elicited by noxious squeezing of the hindpaw. Since the response is abolished after severance of medial and posterior articular nerves, it is evident that the response is a reflex that is evoked by excitation of knee joint afferents (Sato et al. 1984).

The effects of passive movements of normal and inflamed knee joints on unitary efferent activity in filaments of the inferior cardiac sympathetic nerve (ICN) were examined in anesthetized cats (Y. Sato et al. 1985). Passive movements in the normal working range of the joint did not influence the efferent activity of ICN units. However, noxious joint movements led to pronounced excitation of ICN units accompanied by increases in blood pressure. Most of these effects could still be seen after all nerves to the hindlimbs, except the medial articular nerve, were cut. The medial articular nerve contains the majority of afferent fibers originating in the knee joint in the cat (Langford and Schmidt 1983; Schaible and Schmidt 1983a,b). Therefore, the response of the cardiac sympathetic nerve during knee joint stimulation is elicited by knee joint afferent excitation. This articulo-cardiac sympathetic reflex is undoubtedly involved in the articulo-heart rate response. The reflex responses of the cardiovascular system mentioned above were abolished after surgical spinal transection at the cervical level. The articulo-cardiovascular reflex originating in the knee joint was therefore thought to be a supraspinal reflex.

4.1.2.4.2
Mechanical Stimulation of the Spine

The effects of mechanical stimulation of the spine on blood pressure, heart rate and the activity of renal sympathetic nerves were examined in anesthetized rats (Sato and Swenson 1984). Spinal segments from T10-13 or from L2-5 were isolated from surrounding muscle, and the upper and lower segments of the four segment units were fixed by means of spinal clamps. Forces of 0.5–3.0 kg were applied to the lateral aspect of the two mobile segments. Stimulation of the thoracic or the lumbar region produced large decreases in blood pressure and small decreases in heart rate (see Fig. 84B). Additionally, large and immediate decreases were observed in renal nerve activity. While responses attenuated during the course of stimulation, they generally outlasted the stimulus. Following acute spinalization at the C1-C2 level, mechanical stimulation of the spinal column produced small increases in blood pressure and increases in adrenal nerve activity and renal nerve activity. Thus the decreases in blood pressure and renal nerve activity during manipulation of the spine are thought to be supraspinal reflexes.

In order to confirm that the observed responses were mediated by spinal joint afferents, the response to lumbar spine stimulation was examined while cutting successive dorsal (sensory) roots. Destruction of dorsal roots T10-L2 abolished the response to stimulation of the L5 segment. This was somewhat surprising, since destruction of the cauda equina below L3 did not have any effect on the response. Additionally, lower lumbar spinal column stimulation produced increases in activity in dorsal root filaments from T10 to L2. However, this observation is compatible with the hypothesis that spinal afferents ascend the spine somewhat before entering the dorsal root. One possible pathway involves the lumbar sympathetic chain, especially in consideration of the connections which have been described between spinal afferents and sympathetic nerves (Kimmel 1961; Pederson et al. 1956). The observed responses were found to be due not to spinal cord compression, but to afferent fiber-mediated reflexes (Sato and Swenson 1984).

4.1.2.5
Electrical Stimulation

4.1.2.5.1
Stimulation of Cutaneous Afferent Nerves

Heart rate and blood pressure can be changed by electrically induced repetitive activity in cutaneous afferent fibers of the hindlimb in anesthetized cats (Sato et al. 1981), decerebrated cats (Gelsema et al. 1985) and anesthetized dogs (Kissin and Green 1984; Pitetti et al. 1989), and by repetitive stimulation of the infraorbital afferent nerve in anesthetized rabbits (Terui et al. 1981).

In a few cases involving anesthetized cats with the vagi intact, stimulating the superficial peroneal nerve at strengths sufficient to excite all group III and some or all of the unmyelinated group IV fibers gave rise to a brief bradycardia followed after some seconds by a more pronounced tachycardia (Sato et al. 1981) (Fig. 43A). The bradycardia was not only briefer but also of smaller magnitude than the subsequent tachycardia. The latter was already maximal at a stimulation strength sufficient to fully activate only group III fibers. Cutting both vagi resulted in the complete disappearance of the bradycardia and a continuous increase in the tachycardia as more and more group IV fibers were excited (Fig. 43B). Similarly, the blood pressure changes became more uniform after vagotomy (Fig. 43C,D). The bradycardia response was thus due to activation of vagal efferents to the heart. It is noteworthy that the vagally related bradycardia reflex response is very rare in anesthetized animals. This may indicate that anesthetics strongly depress cardiac vagal activity, but not cardiovascular sympathetic activity. In the majority of cases of anesthetized cats with vagi intact, the sole reaction was tachycardia in association with a pressor response.

In anesthetized vagotomized cats, the nature and magnitude of the heart rate changes induced by stimulation of hindlimb afferents have been shown to depend to a large extent on the fiber type being stimulated: stimulation of group II fibers from skin was almost ineffective; with group III fiber activation, the heart rate either increased in about 70% and decreased in about 15% of all trials; and with group IV volleys, there was a consistent increase (Fig. 44A–D) (Sato et al. 1981). These results have been confirmed in anesthetized dogs (Kissin and Green 1984; Pitetti et al. 1989) and decerebrated cats (Gelsema et al. 1985).

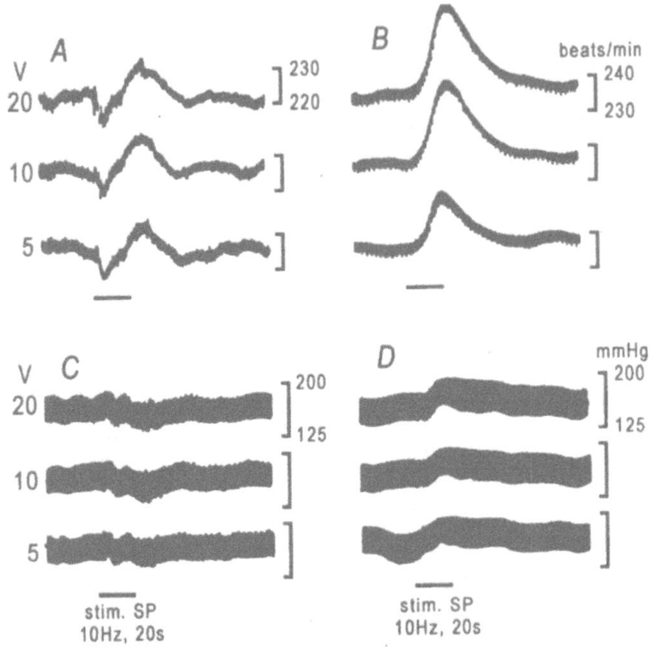

Fig. 43. A,B Changes in heart rate and **C,D** blood pressure responses to electrical stimulation of cutaneous nerve (superficial peroneal nerve, *SP*) **A,C** before and **B,D** after cutting the vagus nerves in a chloralose/urethane-anesthetized cat. SP stimulation caused a biphasic response (an initial decrease followed by an increase) in the vagi-intact condition. After the vagi were cut, SP stimulation caused a continuous increased response. (From Sato et al. 1981)

Fig. 44A–E. Increases in heart rate induced by volleys in cutaneous and muscle afferents in chloralose/urethane-anesthetized cats. **A,B** Specimen records of heart rate changes and corresponding blood pressure changes induced by volleys of the indicated stimulus strength in the superficial radial (*SR*) nerve. **C** The specimen of group IV afferent volleys recorded in SR nerve. **D** The maximum changes in heart rate at each stimulus strength are shown for two cutaneous (*SR*, superficial radial nerve; *SP*, superficial peroneal nerve) and two muscle nerves (*HA*, hamstring muscle nerve; *GS*, gastrocnemius and soleus muscle nerve). **E** The group IV cutaneous afferent volleys of SP nerve increased the discharge activity of the cardiac sympathetic nerve (*top trace*), heart rate (*second trace*), and blood pressure (*third trace*). (Modified from Sato et al. 1981)

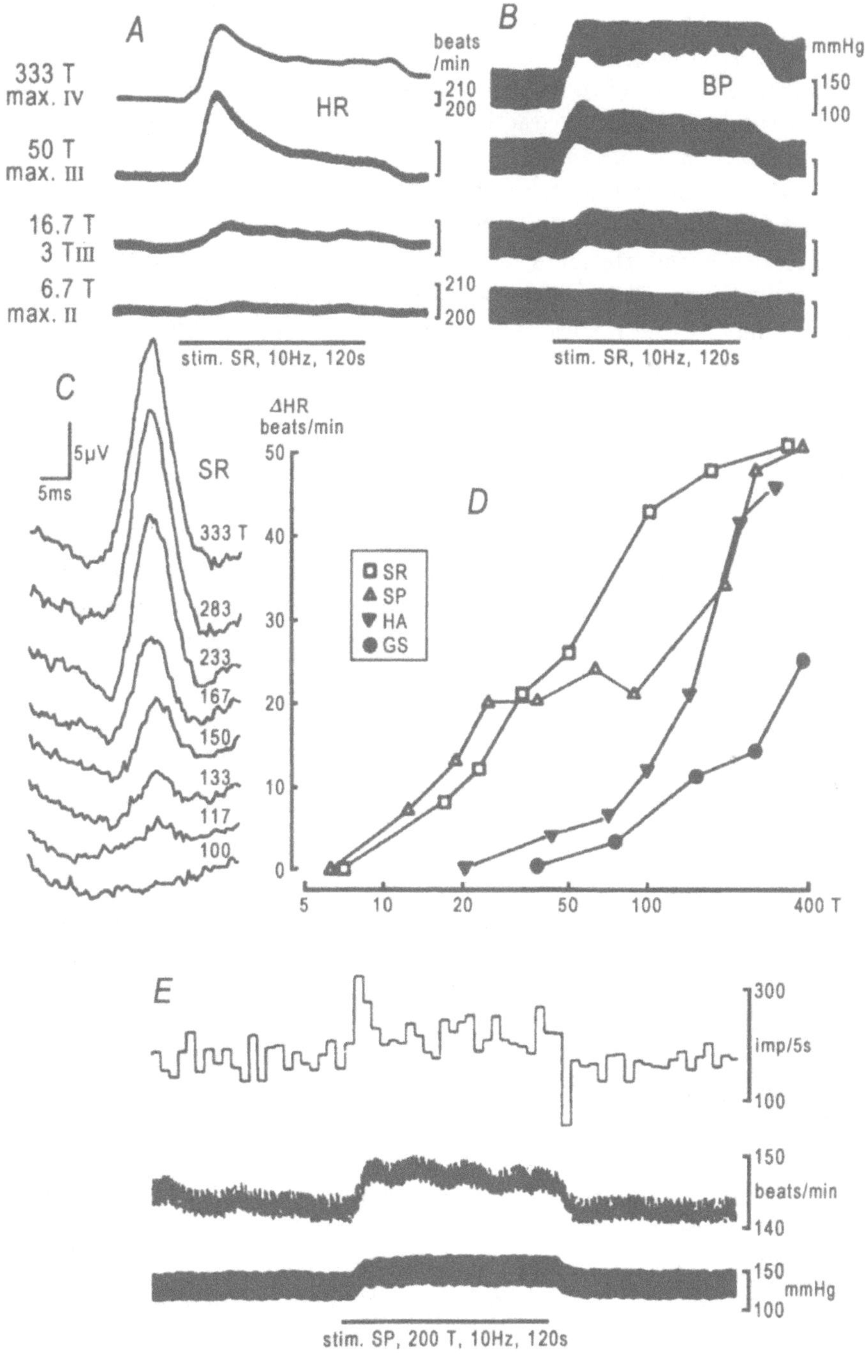
A
B
beats
/min
210
200
HR
BP
mmHg
150
100
333 T
max. IV
50 T
max. III
16.7 T
3 TIII
6.7 T
max. II
stim. SR, 10Hz, 120s
stim. SR, 10Hz, 120s
C
5μV
5ms
SR
333 T
283
233
167
150
133
117
100
ΔHR
beats/min
D
SR
SP
HA
GS
400 T
E
300
imp/5s
100
150
beats/min
140
150
mmHg
100
stim. SP, 200 T, 10Hz, 120s

Since elimination of the adrenal glands did not modify the responses, it has to be assumed that the heart rate changes were produced by reflexly changing the tonic outflow in the cardiac sympathetic nerves. Such an increase in the efferent nerve activity to the heart has indeed been observed (Fig. 44E) (Sato et al. 1981).

No simple and straight forward relationship between heart rate responses and blood pressure changes has been seen. Although at first sight, increases in heart rate seem to be coupled to pressor responses and decreases in heart rate to depressor responses, in quite a few cases heart rate changes were observed without any modification of the blood pressure. In other cases, definite increases in heart rate were accompanied by hypotensive or mixed responses (Sato et al. 1981).

In anesthetized and immobilized rabbits, electrical stimulation of group III and IV fibers of the infraorbital nerve always resulted in decreases in arterial pressure and renal sympathetic nerve activity (Terui et al. 1981). It is interesting to note that cutaneously induced cardiovascular sympathetic reflex responses are not always the same in different species of animals, i.e., pressor responses are more frequently seen in anesthetized rats and cats, and depressor responses are more frequently seen in anesthetized rabbits.

4.1.2.5.2
Stimulation of Muscle Afferent Nerves

It has been widely reported that in anesthetized animals, heart rate and blood pressure are largely unresponsive to electrical stimulation of group I and II muscle afferents (McCloskey and Mitchell 1972; Sato and Schmidt 1973; Coote 1975; Tibes 1977; Sato et al. 1981). However, Khayutin et al. (1986) reported that in anesthetized cats, electrical stimulation of group II tibial nerve afferents was associated with a depressor response in about 50% of the trials conducted.

Stimulation of group III afferents can either increase or decrease the heart rate in anesthetized cats (Sato et al. 1981; Khayutin et al. 1986) (Fig. 45), and can increase the heart rate in decerebrated cats (Gelsema et al. 1985) and anesthetized dogs (Pitetti et al. 1989). Qualitatively, the response of blood pressure to stimulation of group III afferents varies according to the stimulation frequency or strength. At low frequencies, stimulation of group III afferents leads to decreases in blood pressure, whereas stimulation of the same fibers at higher frequencies can lead to increases in blood pressure (Laporte et al. 1960). Electrical stimulation of

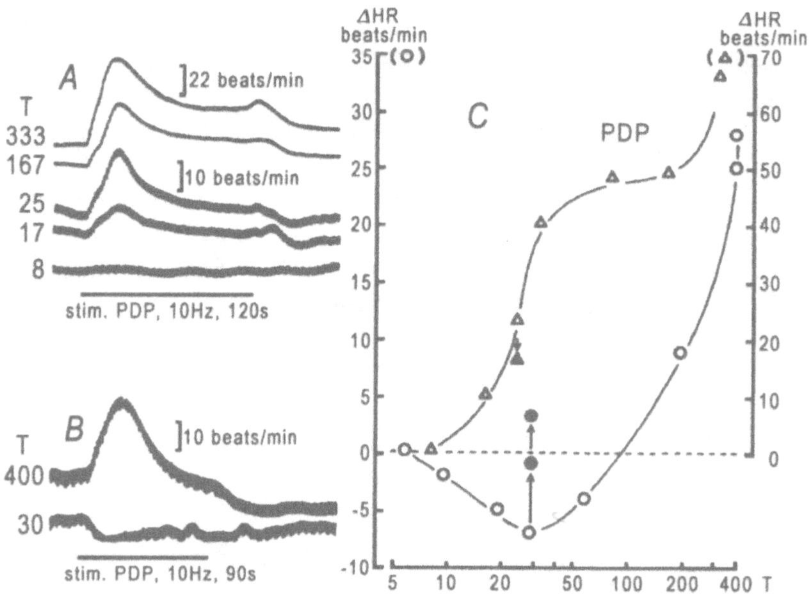

Fig. 45A–C. Changes in heart rate induced by muscle afferent volleys (muscle branches of the peroneal and deep peroneal nerves, *PDP*) in chloralose/urethane-anesthetized cats. **A,C** (*open triangles*) Group I and II volleys (<8T) were totally ineffective; group III volleys (17–30T) induced increases in heart rate. The *single filled triangle* illustrates, at this stimulus strength, the effect of a supplementary dose of 20 mg urethane/kg and 10 mg chloralose/kg. **B,C** (*open circles*) In this experiment, group III volleys induced bradycardia at 16.00 h. This effect disappeared at 20.00 h (*lower filled circle*, –1 beat/min) and was reversed to a heart rate increase at 11.00 h the next morning (*upper filled circle*, + 3 beats/min). (Modified from Sato et al. 1981)

low-threshold group III afferent fibers gave depressor responses, whereas that of high-threshold group III afferent fibers elicited pressor responses (Coote and Perez-Gonzalez 1970). A variety of receptors in muscle are innervated by group III afferents, and Sato et al. (1981) have suggested that this may account for the considerable variation reported regarding blood pressure responses to group III afferent stimulation. Hence, the activation of one or another population of afferents serving a specific set of receptors may dictate the nature of any reflex change in arterial blood pressure.

Cardiovascular function is, often to various degrees, coupled with respiratory function. Respiratory responses can be elicited by muscle afferent stimulation (Mizumura and Kumazawa 1976; Kumazawa et al. 1980). However, changes in respiratory movement are not implicated in the cardiac acceleration elicited by stimulation of group III muscle afferents, since the cardiac response is essentially identical before and after paralysis with a neuromuscular blockade (Gelsema et al. 1985).

In contrast to the variability seen with group III afferents, stimulation of group IV muscle afferents is consistently reported to induce increases in arterial blood pressure and heart rate (Laporte et al. 1960; Johansson 1962; Mitchell et al. 1968; Coote and Perez-Gonzalez 1970; Tibes 1977; Sato et al. 1981; Mitchell and Schmidt 1983).

4.1.2.5.3
Effects of Anesthesia and Levels of Decerebration

The responses of heart rate to stimulation of group III muscle afferent fibers were the most diversified responses seen in anesthetized cats, with increases in about 30% of all trials and decreases in about 40% (Sato et al. 1981). One of the causes of these variabilities is depth of anesthesia. For example, the very marked tachycardia induced by excitation of group III fibers was reduced slightly by the i.v. injection of a small additional dose of α-chloralose and urethane (10 mg and 20 mg/kg, respectively). With larger doses of anesthetic (e.g., 35 mg α-chloralose/kg plus 70 mg urethane/kg), previously marked tachycardiac responses were changed to mild bradycardia. In another experiment, stimulation of a particular nerve with certain stimulus parameters initially evoked bradycardia but after 19 h had passed, only a tachycardiac response to group III volleys was obtained (Fig. 45).

In dogs anesthetized with halothane (at approximately 1.1 vol.% = 1 MAC), noxious electrical stimulation of the superficial peroneal nerve produced a reflex increase in heart rate; however, the magnitude of the increase varied inversely as the concentration of halothane delivered to the animals. At an end-tidal halothane concentration of 2.2 vol.%, the reflex increase in heart rate previously seen in response to somatic nerve stimulation was completely abolished (Kissin and Green 1984).

In chloralose-urethane anesthetized rats, pressor responses evoked by stimulation of the sciatic nerve and depressor responses evoked by stimulation of the tibial nerve were both attenuated in the presence of 1.0% halothane (Samso et al. 1994).

In anesthetized cats, activation of low-threshold subgroups of group III afferents in tibial nerve elicited depressor responses exclusively (Khayutin et al. 1986). In contrast, activation of high-threshold group III afferents elicited depressor or pressor responses depending upon the level of anesthesia. Similarly, in unanesthetized decerebrated cats, the response to stimulation of group III afferents depended upon the level of decerebration. With high-mesencephalic transection, stimulation of high-threshold group III afferents invariably produced a pressor response, whereas stimulation of low-threshold subgroups of group III afferents produced both pressor and depressor responses. In prebulbar cats, only depressor responses were elicited. These qualitative variations in reflex blood pressure changes depending on the level of decerebration and anesthesia are inconsistent with the classical concept of somatic depressor afferents.

4.1.2.5.4 Medullary Reflex Center and Ascending and Descending Spinal Pathways

There is now a great deal of evidence that the pressor reflex response to somatic afferent stimulation is mediated by neurons of the reticular nucleus of the rostral ventrolateral medulla (RVLM). Selective lesions of the RVLM abolished the somatic pressor responses induced by activation of hindlimb afferents, but the reflex was preserved after transection of the anterior pons (Sato and Schmidt 1973; McAllen 1985; Lebedev et al. 1986; Stornetta et al. 1989). In anesthetized paralyzed rats, electrical stimulation of the sciatic or sural cutaneous afferent nerves could be used to evoke a pressor reflex. This response was preserved with midpontine transection but abolished by hemisection of the contralateral lumbar spinal cord, or by electrolytic or kainic acid lesioning of the contralateral RVLM. Corresponding lesions of the ipsilateral lumbar spinal cord or ipsilateral RVLM do not effect the reflex response. Similarly, the reflex was unaffected by injection of kainic acid into the lateral reticular nucleus or by electrolytic lesions of the parabrachial nucleus, the nucleus tractus solitarii, the A5 region, or the inferior cerebellar peduncle. These results strongly suggest that in rats, somatic afferent input ascending contralateral spinoreticular fibers enters the C1 adrenergic area of the RVLM, which in turn mediates the pressor response (Stornetta et al. 1989).

In cats, the afferent limb of supraspinal somato-cardiovascular reflexes elicited by the activation of myelinated fibers involves pathways ascending bilaterally in the spinal dorsolateral funiculus. Input to cardiovascular reflexes from unmyelinated somatic afferents ascended principally in a small region surrounding the dorsolateral sulcus (Chung and Wurster 1976;

Chung et al. 1979). In the dog, this same area of the dorsolateral sulcus conducted input from somatic afferents to somato-sympathetic cardiovascular reflex responses (Kozelka et al. 1981).

The efferent component of somato-cardiovascular reflexes was conducted through the VLM (Ciriello and Calaresu 1977) to descend in the dorsolateral spinal cord (Foreman and Wurster 1973; Gebber et al. 1973; Dembowsky et al. 1980). This was essentially the same descending pathway as that taken by the excitatory component of the arterial chemoreflex (Sato and Schmidt 1973; Szulczyk 1976).

4.1.2.6
Central Processing of Somato-cardiac Reflexes

As mentioned before, it is thought that analysis of reflex discharge potentials from autonomic efferent nerves following electrical single or train stimulation of somatic afferent nerves is the most useful method for studying central mechanisms, especially the central reflex pathways, of the somato-autonomic reflexes. In order to eliminate the time lags between autonomic nerve transmission and effector organ response, cardiac sympathetic and renal sympathetic nerves have been used for the monitoring of neural regulation of cardiac and cardiovascular responses.

Regarding autonomic cardiac regulation, there have been few reports on vagal efferent action potentials following somatic afferent stimulation. There seems to be very little or no response in the vagal efferent nerves to the heart following somatic afferent stimulation, probably due to strong suppressive effects of anesthesia as mentioned before.

In CNS-intact anesthetized cats, single shock of a hindlimb nerve was first noted by Hans Scheafer and his group (Sell et al. 1958) to elicit reflex discharges in the cardiac and renal sympathetic nerves. These discharges persist after decerebration rostral to the medulla oblongata, but are abolished after spinal transection at the cervical level. It was concluded that the reflex involves a supraspinal pathway. This reflex was the A reflex that is elicited by stimulation of myelinated A fibers. The C reflex, elicited by stimulation of unmyelinated C fibers in the hindlimb nerves, was discovered later (Schmidt and Weller 1970; Fedina et al. 1966; Koizumi et al. 1970). The C reflex evoked by hindlimb afferent nerve stimulation also takes a supraspinal pathway, because it is abolished after spinal transection (Sato 1973).

In acute spinal cats, Coote and Downman (1966) found renal sympathetic reflex discharges of short latency (11 ms) following electrical stimulation of the thoracic spinal afferent nerve. In retrospect, this can be explained as the spinal A reflex evoked by myelinated fibers in the spinal nerve.

It was noted that the spinal reflex is inhibited from the brain in CNS-intact anesthetized cats. Coote and Sato (1978) proved that the spinal component was greatly enhanced by section of the ventral guadrant, which is thought to contain descending pathways which inhibit the spinal somato-cardiac sympathetic reflex pathways.

Recently, in anesthetized rats, somato-cardiac sympathetic reflexes were investigated from the viewpoint of the somatic myelinated and unmyelinated components, and spinal and supraspinal reflex pathways (Kimura

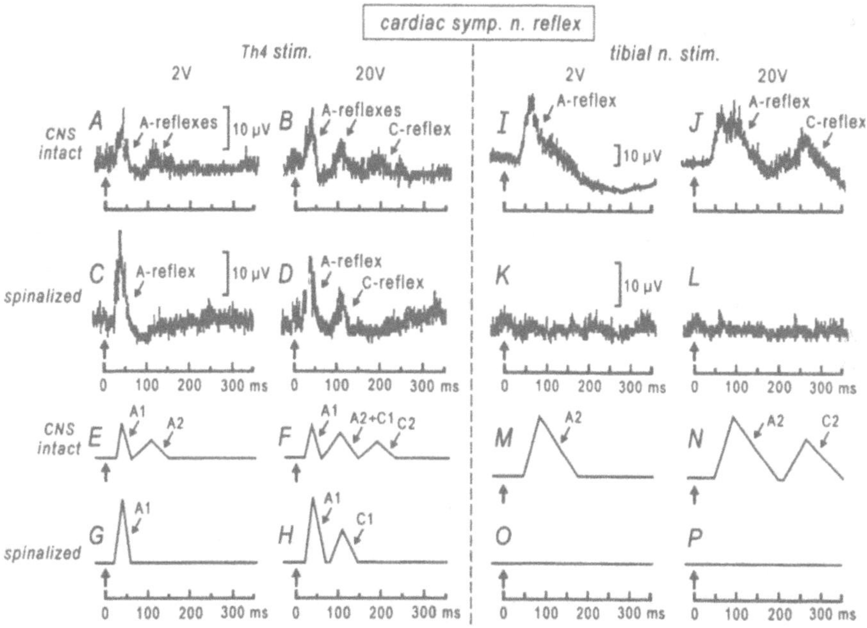

Fig. 46A–P. Somato-cardiac sympathetic reflexes evoked by stimulation of **A–H** a T4 spinal afferent nerve and **I–P** a tibial afferent nerve in urethane-anesthetized rats **A,B,E,F,I,J,M,N** with CNS-intact and **C,D,G,H,K,L,O,P** acutely spinalized (at C1 level). **A–D, I–L** Sample records of somato-cardiac sympathetic reflexes recorded from a inferior cardiac nerve following single shock with 2 V and 20 V. **E–H, M–P** Somato-cardiac A and C reflexes. Stimulation of a T4 spinal afferent nerve produced spinal and supraspinal A and C reflexes, while that of a tibial afferent nerve produced only supraspinal A and C reflexes. (Modified from Kimura et al. 1996b)

et al. 1996b). In CNS-intact anesthetized rats, single shock to a T3 or T4 spinal afferent nerve produced early (A1) and late A-reflex discharges (A2) with latencies of 20 ms and 62 ms, respectively, and a C reflex (C2) with a latency of 136 ms in a cardiac sympathetic efferent nerve (Fig. 46A,B,E,F). After spinalization at the C1 level, stimulation of the same spinal afferent nerve produced an A reflex (A1) with the same latency as an early A reflex in CNS-intact rats, and a C reflex (C1) with a latency of 86 $\pm$ 3 ms (Fig. 46C,D,G,H). Spinal transection augmented the reflex discharges in the cardiac sympathetic efferent nerves evoked by stimulation of a spinal afferent nerve whose segment corresponded to the sympathetic outflow segments. On the other hand, single shock to a tibial afferent nerve evoked an A-reflex discharge (A2) with a latency of 41 ms and a C-reflex discharge (C2) with a latency of 210 ms in CNS-intact rats (Fig. 46I,J,M,N). These A and C reflexes elicited by stimulation of a tibial afferent nerve were not observed after spinalization (Fig. 46 K,L,O,P). It was concluded that A- and C-reflex discharges evoked by stimulation of a segmental spinal afferent nerve in CNS-intact rats are of spinal and supraspinal origin, and those evoked by tibial nerve stimulation are of supraspinal origin. The spinal reflex pathway is segmentally organized, because the spinal reflex is evoked only when stimulation is delivered to the afferent nerves close to the cardiac sympathetic outflow segments. With the CNS intact, the spinal reflex component is depressed by descending inhibitory pathways originating in the brain.

4.1.3 Somatosensory Modulation of Blood Flow

Blood flow in various organs changes in response to somatic sensory stimulation. In this chapter, we deal with blood flow responses within the vessels of the skin, muscles, genital organs, brain and peripheral nerves.

4.1.3.1 Skin Blood Flow

It has been established that thermosensitive structures in the brain, spinal cord, and skin can each give rise to vasoconstriction or vasodilatation in the skin vessels; the constriction being reflexly produced by cooling and the dilatation by warming (see review by Hellon 1983). Peripheral cooling or heating caused antagonistic changes of skin and bowel temperatures in humans (Grayson 1949). In anesthetized animals, antagonistic changes of cutaneous and visceral sympathetic activity and, correspondingly, of regional blood flow occurred during thermal stimulation of skin (Riedel et al. 1972; Engel et al. 1992), spinal cord (Walther et al. 1970), and hypothalamus (Iriki et al. 1971). Cooling at any of the three sites resulted in an increase in cutaneous sympathetic activity, while visceral sympathetic activity decreased. Conversely, heating at any of the three sites induced a decrease in cutaneous sympathetic activity and simultaneously an increase in visceral sympathetic activity.

The putative influence of the thermoregulatory state on skin blood flow responses following various sensory stimuli was studied in healthy human subjects exposed to different ambient temperatures (Oberle et al. 1988). Skin blood flow was monitored by laser Doppler flowmetry and photoelectrical pulse plethysmography. Intraneural electrical sensory stimulation and mental stress were accompanied by virtually identical changes in skin blood flow; warm subjects responded with cutaneous vasoconstriction, whereas cold subjects responded with vasodilatation. The data indicate that the thermoregulatory state profoundly influences the nature and extent of various cutaneous vasomotor reflex responses. Furthermore, there were differences between responses in hands and feet. The intraneural electrical stimulation-evoked vasoconstriction in warm subjects was significantly more pronounced in the hand than in the foot, suggesting a spatial organization of vasomotor control.

Sympathetic activity can be directly measured from limb skin nerves of human subjects with microelectrode methods (Hagbarth et al. 1972). The effect of vibration applied to the hand upon skin sympathetic activity (SSA) in the lower limbs was studied in healthy subjects (Sakakibara et al. 1990). SSA was monitored in the right tibial nerve at the popliteal fossa, while a plethysmogram from the right second toe and perspiration from the sole of the right foot were measured as vibration of 100 m/s^2 at 60 Hz was applied to the left palm for 1 min. The SSA response to vibration differed between subjects, but every subject showed an increase in SSA from the tibial nerve when vibration was applied to the hand. A decrease in the amplitude of the plethysmogram from the toe was also found in all subjects. One subject displayed a remarkable increase in perspiration on the sole of the foot together with a great increase in SSA. These findings indicate that even vibration exposure to the hand triggers sympathetic activity in the tibial nerve innervating the foot, and causes vasoconstriction of the toe and perspiration on the sole of the foot.

Parasympathetic reflex vasodilatation in the lower lips can be elicited by activation of trigeminal and nasal afferents in anesthetized cats (Izumi and Karita 1992). Electrical stimulation of the infraorbital nerve and the maxillary buccal gingiva caused an increase in ipsilateral lip blood flow. The reflex vasodilator response was unaffected by transection of the ipsilateral cervical sympathetic trunk and facial nerve root, but was completely abolished by ipsilateral transection of the glossopharyngeal nerve root. These results suggest that there is a somato-autonomic reflex vasodilator system mediated via parasympathetic efferent vasodilator fibers which emerge from the brain stem with the glossopharyngeal nerve in the cat mandibular lip.

In addition to the vasodilation via parasympathetic efferent vasodilator fibers, there is also vasodilatation in the skin due to the axon reflex (Lewis and Marvin 1927; Celanderr and Folkow 1953). In the skin, the mechanism of axon reflex-induced vasodilatation has been well clarified (see Zimmermann 1984; Koltzenburg and Handwerker 1994). Excitation in cutaneous afferent fibers originating in a restricted area of skin can be conducted to the other branches of the same afferents innervating the cutaneous blood vessels. The excited afferent terminal releases vasodilative substances such as calcitonin gene-related polypeptide (CGRP) and substance P, which produce vasodilatation and increased permeability of blood vessels in a certain area around the stimulated spot.

4.1.3.2
Muscle Blood Flow

It has long been known that blood vessels in mammalian skeletal muscle are generally innervated by postganglionic sympathetic efferent vasoconstrictor fibers. They are considered to be spontaneously or tonically active, since vasodilatation is seen after sympathetic denervation. Although the evidence is somewhat indirect, it has been concluded from measurement of blood flow in skeletal muscle that the spontaneous discharge of efferent vasoconstrictor fibers is modified reflexly by stimulation of somatic afferent nerves (Johansson 1962; Clement et al. 1973).

In anesthetized cats, repetitive electrical stimulation of group III somatic afferent nerves produced a reflex fall in blood pressure (Johansson 1962). Initially, the muscle blood flow increased transiently but was soon reduced to less than its original value. In order to find out whether this latter reduction of muscle blood flow was merely a passive phenomenon, reduced to less than its original value. In order to find out whether this latter reduction of muscle blood flow was merely a passive phenomenon, caused by reduced perfusion pressure, the stimulation was repeated while the arterial inflow pressure was kept constant by adjusting a clamp around the abdominal aorta. The stimulation was then seen to produce a more clear-cut increase in muscle blood flow. In contrast, stimulation of group IV somatic afferents produced only a temporary reduction in the skeletal muscle blood flow, instead of the pressor response. It was proposed that the reflex vasodilatation was due to an inhibition of the tonic activity in sympathetic efferent vasoconstrictor fibers, and the reflex vasoconstriction was the result of increased tonic activity in these fibers (Johansson 1962). Crayton et al. (1981) have shown that injection of capsaicin into the arterial supply of skeletal muscle in anesthetized dogs causes a reflex increase in mean arterial pressure, but skeletal muscle blood flow remains near control values, indicating active vasoconstriction in the hindlimb muscle caused by increased tone of sympathetic vasoconstrictor fibers.

The existence of a cholinergic sympathetic vasodilator nerve supply to skeletal muscle is also recognized (Bülbring and Burn 1935; Folkow and Uvnäs 1948, 1950). These nerves are thought to be silent when the animal is at rest and to not be involved in depressor reflexes, at least in anesthetized animals (Folkow et al. 1950, 1965; Uvnäs 1960). They can, however, be activated by electrical stimulation of certain areas of the cortex, hypothalamus, midbrain tegmentum and medulla (Eliasson et al. 1951; Lindgren 1955; Abrahams et al. 1960). In anesthetized cats, Koizumi and Sato (1972) showed that 90% of all sympathetic fibers dissected from muscle nerves were spontaneously active and reflexly affected by stimulation of somatic afferents. Repetitive stimulation of group II and III somatic afferents at rates of 5–10/s produced a long depression, whereas repetitive stimulation of the group IV afferents of the same nerve at the same frequency resulted in an excitation. These fibers are most likely vasoconstrictors and can, of course, produce both constriction and dilatation of blood vessels in skeletal muscle by changes in their discharge frequencies. Another 10% of sympathetic fibers to muscle were silent and were not involved in reflex reactions evoked by somatic afferent excitation. They may belong to the group of cholinergic vasodilators.

Sympathetic nerve activity to muscle can be recorded as postganglionic sympathetic nerve activity using microneurography in human peripheral nerves (Vallbo et al. 1979). Muscle sympathetic nerve activity remained unchanged during the initial 0- to 30-s period of the cold pressor test and increased remarkably during the later 30- to 90-s period of the test (Yamamoto et al. 1992). Muscle sympathetic nerve activity increased transiently during rotation of the acupuncture needle applied to the Tsu-San-Li acupuncture point of the same limb (Sugiyama et al. 1995).

4.1.3.3 Cutaneous and Muscle Sympathetic Nerve Activity

In anesthetized cats, somato-sympathetic reflexes in postganglionic neurons to hairy skin and to muscle were studied following mechanical stimulation of skin (Horeyseck and Jänig 1974a,b; Jänig 1975). In 60% of the cutaneous postganglionic neurons, innocuous stimulation of hairs induced excitation followed by a slight depression of the spontaneous activity. In another 30% of the neurons, the spontaneous activity was depressed to a greater or lesser degree by these stimuli. In most muscle postganglionic neurons, the spontaneous activity was depressed by innocuous stimulation of hairs. In both types of neurons, the reflexes were produced by activity in hair follicle receptors with group II afferents. Reflexes in postganglionic neurons could be elicited by stimulation of hairy skin anywhere on the body surface. In most cutaneous units, the spontaneous activity was depressed during noxious stimulation of skin. The depression of the spontaneous activity was maximal when stimulation was applied to the skin area which was innervated by the cutaneous postganglionic neurons, and much weaker or not elicitable from other skin areas. Most muscle units were excited during noxious stimulation of skin. This excitation could be elicited from anywhere on the body surface. The cutaneous afferent fibers which are involved in these reflexes are the group III axons, which are excited by noxious mechanical stimuli, and the group IV axons, which are excited by noxious mechanical and thermal stimuli. These investigations reveal that the somato-sympathetic reflexes have opposing organizations in cutaneous and muscle postganglionic fibers. In anesthetized cats, noxious heat stimulation of the skin above 45°C increased muscle vasomotor activity and decreased cutaneous vasomotor activity. These site-specific responses were in accord with the former findings by Riedel et al. (1972) of cutaneous vasomotor and visceral vasomotor reactions to temperature. The increase in muscle vasomotor activity persisted even in the spinalized cat, but the decrease in cutaneous vasomotor activity reversed to an increase in both medullary and spinalized animals, indicating that the decrease in vasomotor activity during noxious heating of the skin was generated in the CNS above the medulla oblongata, probably in the hypothalamus (Jänig 1975).

Jänig and Szulczyk (1980) found that in anesthetized cats, 27% of lumbar preganglionic neurons projecting through L2 and L3 white rami to lumbar ganglia caudal to L4 were spontaneously active. The remaining preganglionic neurons were silent. The tonically active preganglionic neurons could be classified into three subsets. Twenty-six of 80 neurons tested were classified as type I. As with postganglionic vasoconstrictor neurons to muscle, they were excited by systemic hypoxia or noxious stimulation of the skin, and displayed cardiac rhythmicity. Forty-eight of the 80 neurons were classified as type II. As with most postganglionic vasoconstrictor neurons to hairy or hairless skin, they were inhibited by systemic hypoxia or noxious stimulation of the skin, and 40% of these neurons displayed cardiac rhythmicity. The remaining six of the 80 neurons were classified as type III. These had no cardiac rhythmicity and, as is characteristic of postganglionic sudomotor neurons, they were activated by vibration, specifically tapping the hind foot. The silent preganglionic neurons were generally resistant to activation by the natural stimuli used, suggesting that they could only be recruited by specific challenges to homeostasis, such as extreme hypoxia or cold (Jänig 1985a,b). In response to stimulation of somatic receptors, reflex patterns have been analyzed in spontaneously active postganglionic neurons, which supply skeletal muscle and hairy skin of the rat hindlimb and are most likely vasoconstrictor in function (Häbler et al. 1994). In principle, these rat reflex patterns show the same differentiation as in the cat.

4.1.3.4
Somatosensory Modulation of Blood Flow in the Genital Organs

It is well known that mechanical stimulation of the penis plays an important role in erection and ejaculation (Dahlöf and Larsson 1976; Larsson and Södersten 1973; Lodder and Zeilmaker 1976; Sachs and Garinello 1980).

In the rat, the dorsal penile nerve is formed exclusively by sensory nerve fibers (Nunez et al. 1986). In the spinal rat, electrical stimulation of the afferent nerve fibers of the dorsal penile nerve elicits tonic reflex erections of the penile body and reflex bulbospongiosus skeletal muscle activity, flips and ejaculations. The tonic erections of the penile body are independent of contractions of the bulbospongiosus skeletal muscle. Tonic erections in the rat are, therefore, more likely to be regulated by the autonomic nerves innervating penile blood vessels, and not by somatic processes (Pescatori et al. 1993). In fact, stimulation of pelvic parasympathetic efferent nerve fibers innervating the penile blood vessels increases intracavernous pressure (Steers et al. 1988).

It is evident that erection occurs because of blood pooling within the cavernosal tissue of the penis, resulting from increased arterial flow, expansion of the spaces of the cavernosa, and decreased venous drainage. Tactile stimulation of the penis can release vasoactive intestinal polypeptide (VIP) from the pelvic parasympathetic efferent nerve terminals in the mammalian penis (Dixson et al. 1984). Retrograde dye staining combined with immunohistochemistry was used to characterize penile neurons in the major pelvic ganglion of the rat. Of the total penile pelvic neurons, 92% were immunoreactive for VIP, while 95% stained intensely for acetylcholinesterase. None of the neurons were immunoreactive for tyrosine hydroxylase, which is an enzyme related to synthesis of catecholamines. Penile neurons in the pelvic plexus receive preganglionic input from the pelvic and the hypogastric nerve, yet the shared histochemical features of the postganglionic neurons suggest that the two pathways have a similar role in penile erectile tissue (Dail et al. 1986).

The clitoris is innervated by the pelvic, pudendal and hypogastric nerves (Langworthy 1965; Purinton et al. 1976). The clitoris is an important sensory organ involved in sexual behavior. Its afferent information may initiate autonomic and somatic reflexes associated with sexual behavior. Clitoral intracavernous pressure and internal pudendal artery blood flow increase in response to stimulation of nerves in the clitoris. This results in tumescence and extrusion of the glans (Diederichs et al. 1991). It is likely that clitoral afferent information produces a reflex vasodilation via activation of the pelvic efferent nerve in a manner similar to that seen in penile erection.

4.1.3.5 Somatosensory Modulation of Regional Cerebral Blood Flow

Global cerebral blood flow (CBF) has been considered to be constant because of the fixed volume of the skull. Since Ingvar (1976), using the Xenon gas method, found an increase in regional cerebral blood flow (rCBF) in the cortex in conscious humans performing assorted tasks, there have been many supporting studies of increases in rCBF. It has long been accepted that an increase in cortical rCBF is caused by the accumulation of metabolites due to regional cortical neuronal activities which are elicited during the various tasks corresponding to these regions.

However, the physiological functions of nerves innervating small cerebral blood vessels have remained obscure. Recently, we have come to appreciate the importance of the neural regulation of rCBF, particularly cholinergic vasodilative regulation of the cortex and hippocampus by the forebrain basal nuclei such as the nucleus basalis of Meynert (NBM) or substantia innominata and the medial septal nucleus (Biesold et al. 1989a; Cao et al. 1989; Hallström et al. 1990; Adachi et al. 1992a, see review by Sato and Sato 1992). Sato and his colleagues have also discovered other neural influences on the regulation of rCBF. These include central noradrenergic fibers originating in the locus coeruleus (Adachi et al. 1991), serotonergic fibers originating in the dorsal raphe nucleus (Cao et al. 1992b), dopaminergic fibers originating in the substantia nigra (Wada et al. 1992) and peripheral sympathetic nerve fibers (Hervonen et al. 1990; Saeki et al. 1990). There is great interest in resolving whether somatic afferent stimulation can modulate rCBF and, if so, whether any neuronal mechanisms, in addition to metabolic vasodilation, are involved in the regulation of rCBF.

4.1.3.5.1
Cutaneous Stimulation

In anesthetized rats, noxious mechanical stimulation of cutaneous areas of the face, forepaw and hindpaw produces significant increases in systemic blood pressure and cortical rCBF (Adachi et al. 1990; Hallström et al. 1990), whereas innocuous mechanical afferent stimulation of the skin has no effect on either parameter (Adachi et al. 1990) (Fig. 47A). The nociceptively induced responses are large, particularly following stimulation of the forepaw and hindpaw (Fig. 48A). These increases in systemic blood pressure and cortical rCBF can occur in the order of a second, so that autoregulation of CBF may be insufficient to compensate for the rapid changes in systemic blood pressure. As such, it is possible that a component of the increase in CBF is due to the elevation of systemic blood pressure. In the above study, in order to prevent increases in CBF secondary to the increase of blood pressure, the spinal cord was transected at the first thoracic level (T1). Following spinal transection at the T1 level, sensory information from the forelimb ascends via the normal circuitry to the brain through the cervical spinal cord. The sensory information arrives in the cardiovascular center in the brain stem and is integrated.

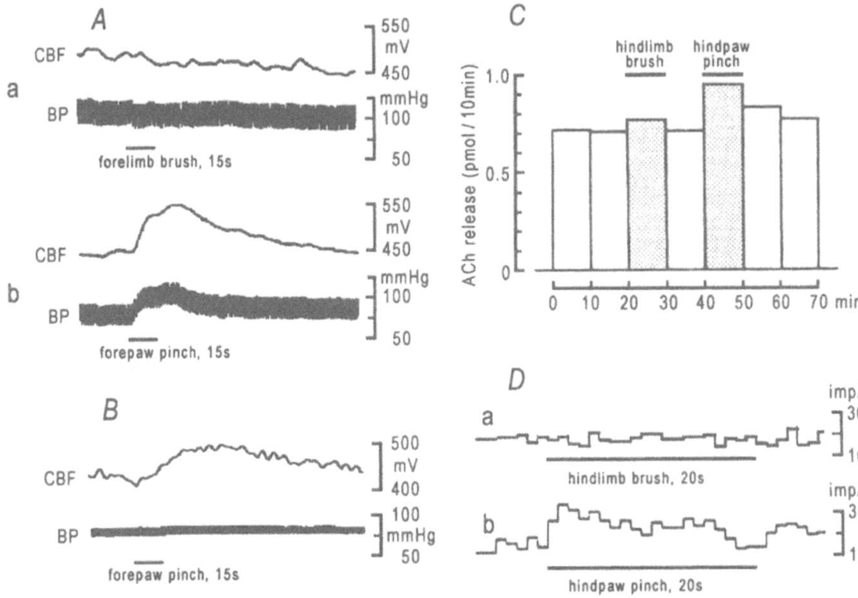

Fig. 47A–D. Effects of somatosensory stimulations on **A,B** cerebral blood flow (*CBF*) and **C** acetylcholine (ACh) release in the parietal cortex in halothane-anesthetized rats, and **D** neuronal activity in the nucleus basalis of Meynert (NBM) projecting their axonal fibers to the ipsilateral parietal cortex in urethane-anesthetized rats. **A** CBF in the parietal cortex (measured by a laser Doppler flowmeter) and blood pressure (*BP*) increased with pinching of the forepaw (**b**) but were not changed by brushing of forelimb (**a**) in CNS-intact rats. **B** In rats spinalized at the T1 level, cortical CBF increased without changes in BP following pinching of a forepaw. **C** Extracellular ACh release in the parietal cortex was increased slightly by brushing of the hindlimb and increased more intensely with pinching of the hindpaw. D The activity of NBM neuron did not change during brushing a hindlimb (a) and increased during pinching a hindpaw (b). (Modified from Adachi et al. 1990; Kurosawa et al. 1992; Akaishi et al. 1990)

However, the integrated information in the brain stem cannot descend to the preganglionic sympathetic neuronal pool in the thoracic and lumbar spinal cord because of the transection of the descending pathways at the upper thoracic spinal cord. Thus systemic blood pressure no longer increases significantly in response to noxious stimulation of the forelimb. There remains only a marginal increase in systemic blood pressure, which is thought to be due to the increase in plasma vasopressin somatically released from the posterior pituitary gland. In these experiments, however, cortical rCBF still increases remarkably following noxious stimulation of

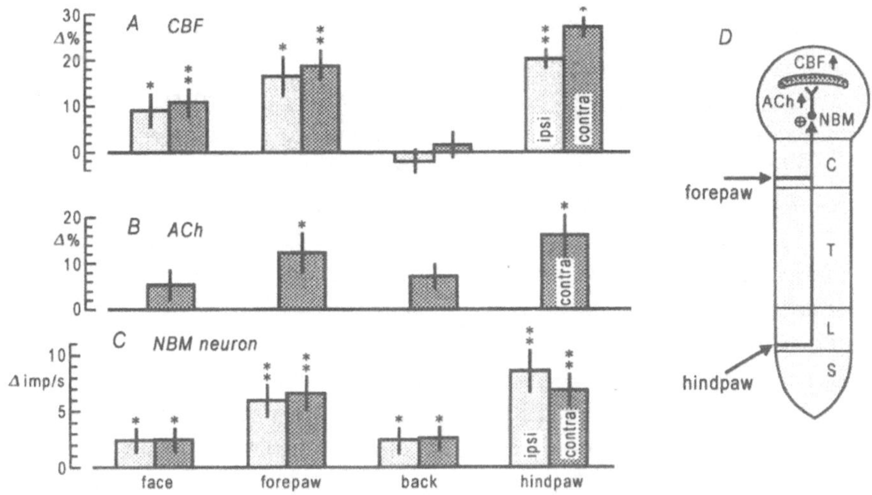

Fig. 48A–C. Summary of the responses of **A** cortical cerebral blood flow (*CBF*) and **B** acetylcholine (*ACh*) release in halothane-anesthetized and **C** nucleus basalis of Meynert (*NBM*) neuronal activity in urethane-anesthetized rats to cutaneous pinching of various segmental areas. *Light-shaded columns* represent responses to ipsilateral stimulation; *dark-shaded columns* represent responses to contralateral stimulation. Both the cortical CBF and ACh release and the NBM neuronal activity were particularly increased following stimulation of the forelimb and hindlimb. **D** Concluding summary suggests that somatosensory stimulation may enhance cortical CBF via activation of cholinergic neurons of the NBM. *$P<0.05$, **$P<0.01$; significantly different from prestimulus control values. (Modified from Adachi et al. 1990; Kurosawa et al. 1992; Akaishi et al. 1990)

the forelimb (Adachi et al. 1990) (Fig. 47B). These data suggest that at least one component of the increase in cortical rCBF following cutaneous noxious stimulation is independent of changes in systemic blood pressure and of any concomitant passive vasodilatation.

Pearce et al. (1981) showed that in anesthetized rabbits, an increase in CBF due to noxious somatic stimulation could be partially blocked by scopolamine, a muscarinic cholinergic receptor antagonist, and suggested that cholinergic mechanisms participated in cerebral vasodilatation resulting from such stimulation.

In anesthetized animals, both electrical (Mitchell 1963; Phillis 1968) and mechanical (Kurosawa et al. 1992) somatosensory stimulation cause an increase in acetylcholine (ACh) release in the cerebral cortex. Characteristically, in anesthetized rats, the responses of ACh release after mechanical somatic stimulation depend upon the types of stimulation and the spinal segments involved. Noxious mechanical stimulation of a fore- or hindpaw induces a significant response, whereas such stimulation to the face or back skin affects no significant change (Kurosawa et al. 1992) (Fig. 47C, 48B). Innocuous stimulation of a hindlimb can produce a significant increase in cortical ACh release, although stimulation of the face, forelimb and back had no effect. The differences in the magnitudes of cortical ACh responses appear to depend either on the different densities of afferent innervation of the various segmental skin areas or on different connections to the central cholinergic system.

In anesthetized rats, most of the neurons in the NBM projecting their axonal fibers to the ipsilateral parietal cortex are not significantly influenced by innocuous mechanical cutaneous stimulation, while they are excited by noxious mechanical cutaneous stimulation (Fig. 47D). The NBM neurons are excited more intensely and frequently by noxious mechanical stimulation of a fore- or hindpaw than by that of the back or face (Akaishi et al. 1990) (Fig. 48C).

Combined with evidence that the cortical rCBF, cortical ACh release and neural activity of the NBM increase in response to noxious somatic stimulation, it seems reasonable to assume that in anesthetized animals noxious somatic stimulation can enhance cortical rCBF via an increase of cortical ACh released from cholinergic nerve fibers originating in the NBM. On the other hand, ACh in the cortex has various other functions such as cortical arousal (Celesia and Jasper 1966; Szerb 1967), excitation of cortical neurons (Lamour et al. 1982), and modulation of the somatically induced cortical neuronal response (Donoghue and Carroll 1987).

Noxious mechanical stimulation of the skin of the face, forepaw, chest or hindpaw also increases hippocampal rCBF in anesthetized rats. After the spinal cord is transected at the T1 level, forepaw pinching causes no change in blood pressure but still increases hippocampal rCBF, as with the responses of cortical rCBF. The increase in hippocampal rCBF due to forepaw pinching in the T1-transected rats is partially reduced by i.v. administration of mecamylamine, while atropine is ineffective, indicating that nicotinic, but not muscarinic, cholinergic receptors are involved in this response (Cao et al. 1992a). Reflecting upon the fact that the activity of a majority of septo-hippocampal neurons increases in response to noxious somatic stimulation (Dutar et al. 1985), we propose that noxious somatic stimulation can influence hippocampal rCBF via the activation of septo-hippocampal cholinergic neurons, resulting in an increase of ACh release and stimulation of nicotinic cholinergic receptors.

4.1.3.5.2
Effect of Slow Walking

As mentioned before, in conscious humans, rCBF changes during various behavioral tasks such as calculation, thinking, speaking, movement of the fingers, etc. (Ingvar 1976). Kurosawa et al. (1993) demonstrated that in conscious rats, extracellular ACh in the cerebral cortex increased during walking at a mild speed (approximately 4 cm/s), and it was proposed that the increase in cortical extracellular ACh was derived from ACh released from cortical interneurons and/or cortical terminals of the cholinergic neurons originating in the NBM. The increase in cortical ACh release was thought to be responsible for an increase in cortical rCBF during walking. A method has recently been developed using laser Doppler flowmetry for the continuous measurement of rCBF in conscious rats kept in a hammock (Sato et al. 1994a), and Kimura et al. (1994b) demonstrated, using this technique, that rCBF in the frontal, parietal, and occipital cortices generally increased during walking at a mild speed (4 cm/s) for a period of 30 s. The responses of rCBF in the three cortices were almost identical, and were accompanied by only a small increase in mean arterial pressure.

The walking-induced increase in cortical rCBF was partially reduced by systemic administration of both blood brain barrier permeable muscarinic and nicotinic cholinergic receptor antagonists (atropine and mecamylamine, respectively), but was unaffected by blood brain barrier impermeable drugs (methylatropine and hexamethonium, respectively). This suggests the involvement of both muscarinic and nicotinic receptors in the rCBF responses in the CNS. Circumstantially, these data suggest that the walking-induced increase in cortical rCBF is, at least in part, due to activation of the cholinergic vasodilative system in the cortex originating in the NBM. However, the incomplete antagonism of the walking-induced response of cortical rCBF after injection of both cholinergic muscarinic and nicotinic receptor antagonists suggests the possible involvement of other transmitters or modulators in the responses. Even though a part of the efferent pathway has been clarified for the walking-induced increase in cortical rCBF, it remains to be determined whether somatic afferent information from skin, skeletal muscles and joints involved in the movement is responsible for the response.

Dutar et al. (1985) have shown, using anesthetized rats, that in most cases the septo-hippocampal neurons are not driven by light cutaneous stimulation such as hair movements or light taps, but are driven only by strong, noxious mechanical or thermal stimuli. The receptive fields of the septo-hippocampal neurons responding to cutaneous stimulation are large, usually involving almost half of the body or even the whole body surface. These results suggest that septo-hippocampal neurons might be involved in cerebral mechanisms related to nociception.

At present, the mechanisms of changes in CBF remain obscure. However, the most likely proposal to date is that blood flow to a specific region is enhanced in response to, and may possibly even precede, an increase in regional neuronal activity. Putatively, an increase in blood flow increases the availability of substrates such as glucose, in addition to facilitating removal of metabolic wastes produced as a consequence of the increased metabolism. Although no direct evidence is available to demonstrate a positive relationship between impaired regulation of cortical rCBF and compromise of memory and learning, a large body of indirect experimental and anatomical evidence suggests that such a relationship may indeed exist.

As described above, it is clear that the cholinergic neurons of the NBM play a vasodilative role in the regulation of cortical rCBF. It is also clear that specific somatic sensory neurons are functional afferents of the NBM (Fig. 48D). This connection presents the possibility that stimulation of somatosensory afferents by massage, acupuncture, or electrical stimulation may be used as a noninvasive, nonpharmacological therapy that will specifically enhance cortical rCBF. If, as suggested previously herein, impairment of the mechanisms associated with regulation of rCBF is associated with or causative of the deterioration of memory in old age or in pathology such as Alzheimer's disease, then further research on somatosensory function could well be considered for prevention or treatment of maladjustment of rCBF regulation.

4.1.3.6
Somatosensory Modulation of Blood Flow in the Peripheral Nerves

Peripheral nerves receive their oxygen and nourishment from blood flowing in the blood vessels in the nerve (vasa nervorum). The blood flow in the vasa nervorum is characteristically dependent on arterial blood pressure (Sundqvist et al. 1985); in other words, autoregulation is not normally evident. The vasa nervorum receives innervation from various fibers containing noradrenaline, acetylcholine, serotonin and several polypeptides (Amenta et al. 1983; Appenzeller et al. 1984; Dhital and Appenzeller 1988; Milner et al. 1992).

Recently, nerve blood flow in the sciatic nerve in rats was found to be regulated by sympathetic vasoconstrictive efferent fibers originating in the T11-L1 segment, by parasympathetic cholinergic efferent vasodilators originating in the L6 segment, and also by peptidergic afferent fibers containing CGRP entering the spinal cord at the L3–S1 segments (Fig. 49A) (Hotta et al. 1991; Sato et al. 1994b). It has also been demonstrated in dogs that electrical stimulation of the cervical sympathetic trunks causes marked decreases in blood flow in the facial nerve (Murakawa et al. 1995).

Budgell and Sato (1994) demonstrated that brief noxious mechanical stimulation of the fore- or hindpaws produced immediate short-term increases in blood pressure and sciatic nerve blood flow (Fig. 49B). Severing the spinal cord immediately above the thoracolumbar junction resulted in a drop in systemic blood pressure but an increase in sciatic nerve blood flow. Furthermore, in spinalized animals, brief pinching of the hindpaws produced no subsequent change in systemic blood pressure but a short-term decrease in sciatic nerve blood flow (Fig. 49C). This apparent reflex decrease in nerve blood flow in response to noxious mechanical stimulation was abolished by severing the lumbar sympathetic trunks and adjacent sympathetic nerves at the level of the L4-L5 intervertebral disk. It was therefore suggested that brief noxious mechanical stimulation of the hindlimbs elicited a somato-sympathetic reflex which tends to reduce sciatic nerve blood flow. When the spinal cord is intact, this reflex might be obscured by a descending tonic influence on the vasa nervorum. Furthermore, with an intact spinal cord, this proposed somato-sympathetic reflex competes with and is usually, but not invariably, subservient to a central pressor response.

Similar to the response elicited by pinching of paws, Budgell et al. (1995) noted that in anesthetized rats, noxious chemical stimulation of interspinous ligaments using capsaicin, a potent chemical algesic, caused a pronounced elevation of mean arterial pressure and a prolonged depression of sciatic nerve blood flow. This decrease in sciatic nerve blood flow was considered to be due to increased activity of sympathetic vasoconstrictor fibers to the sciatic vasa nervorum.

The contribution of parasympathetic efferent vasodilators to the vasa nervorum appears not to have been investigated systematically from the

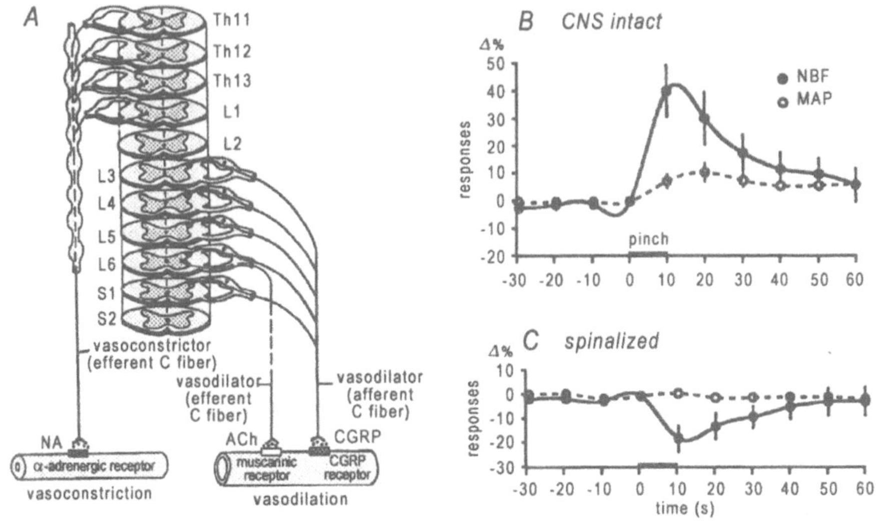

Fig. 49. A Proposed neural mechanisms for regulating sciatic nerve blood flow in the rat. Sympathetic vasoconstrictive efferent fibers originating in the T11–L1 segment, parasympathetic cholinergic efferent vasodilators originating in the L6 segment, and peptidergic afferent fibers containing calcitonin gene-related peptide (*CGRP*) entering the spinal cord at the L3–S1 segments are shown. **B,C** Effect of hindpaw pinching on sciatic nerve blood flow (NBF, *closed circles*) and mean arterial pressure (MAP, *open circles*) in urethane-anesthetized rat **B** with CNS intact and **C** spinalized at thoracolumar junctron (laminectomy at T10). Pinching stimulation produced increases in NBF and MAP in CNS-intact rats. After spinal cord transection, the same stimulation produced decreases in NBF without affecting MAP. (**A** Modified from Sato et al. 1994b; **B,C** Modified from Budgell and Sato 1994)

aspect of somatic afferent stimulation. Concerning afferent peptidergic vasodilators, Hotta et al. (1996) suggested the possibility of vasodilatory mechanisms elicited by electrical stimulation of unmyelinated fibers in the central cut end of the saphenous nerve. An increase in nerve blood flow in the sciatic nerve was independent of changes in mean arterial blood pressure. This increase was abolished by topical application of a CGRP receptor antagonist, hCGRP (8–37). These mechanisms rely on dichotomizing axons innervating both skin and sciatic vasa nervorum, because these responses persist even in rats whose lumbosacral afferent and efferent central connections had been severed close to the spinal cord (Fig. 50) (Hotta et al. 1996). The subject of somatically induced reflex

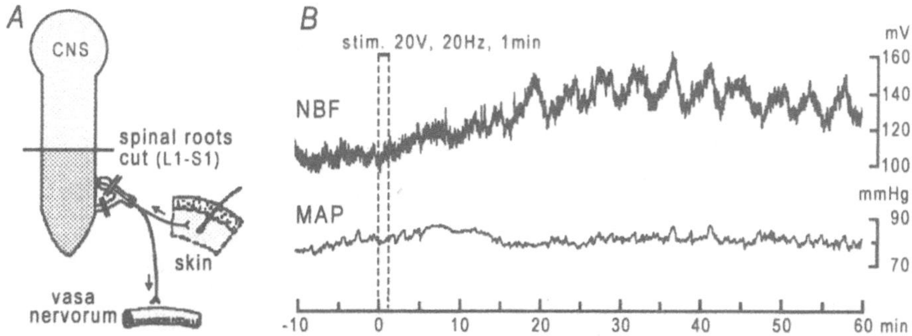

Fig. 50A,B. Stimulation of saphenous afferent nerve produces dilatation of the vasa nervorum via an axon reflex-like mechanism in the sciatic nerve of urethane-anesthetized rats. **A** Experimental method. Spinal roots were cut at L1–S1 level. The responses of nerve blood flow (*NBF*) in the sciatic nerve to electrical stimulation of saphenous nerve afferents were examined using laser Doppler flowmetry. **B** Repetitive electrical stimulation of unmyelinated fibers in the central cut end of the saphenous nerve produced an increase in NBF ipsilateral to the stimulation, independent of changes in mean arterial blood pressure (*MAP*). (Modified from Hotta et al. 1996)

responses of the vasa nervorum is just beginning to be investigated at this time, and further analysis is required.

4.1.4
Conclusion

1. Over the last 20 years, there has been great progress in research on the cardiovascular responses elicited by natural somatic stimulation in anesthetized animals. The effects of different stimulus modalities, as well as stimulus segments on cardiovascular functions, were analyzed by parallel recording of autonomic efferent nerve activity and functions of effector organs. It was shown that sympathetic rather than parasympathetic nerves played a major role in somato-cardiovascular reflexes. It has also been repeatedly confirmed that the brain stem plays a very important role as the reflex center in somato-cardiovascular responses. However, it is noteworthy that there are spinal reflex components elicited by segmental input which, although usually depressed by descending inhibitory pathways from the brain in CNS-intact preparations, are augmented or unmasked in spinalized preparations (Fig. 51).

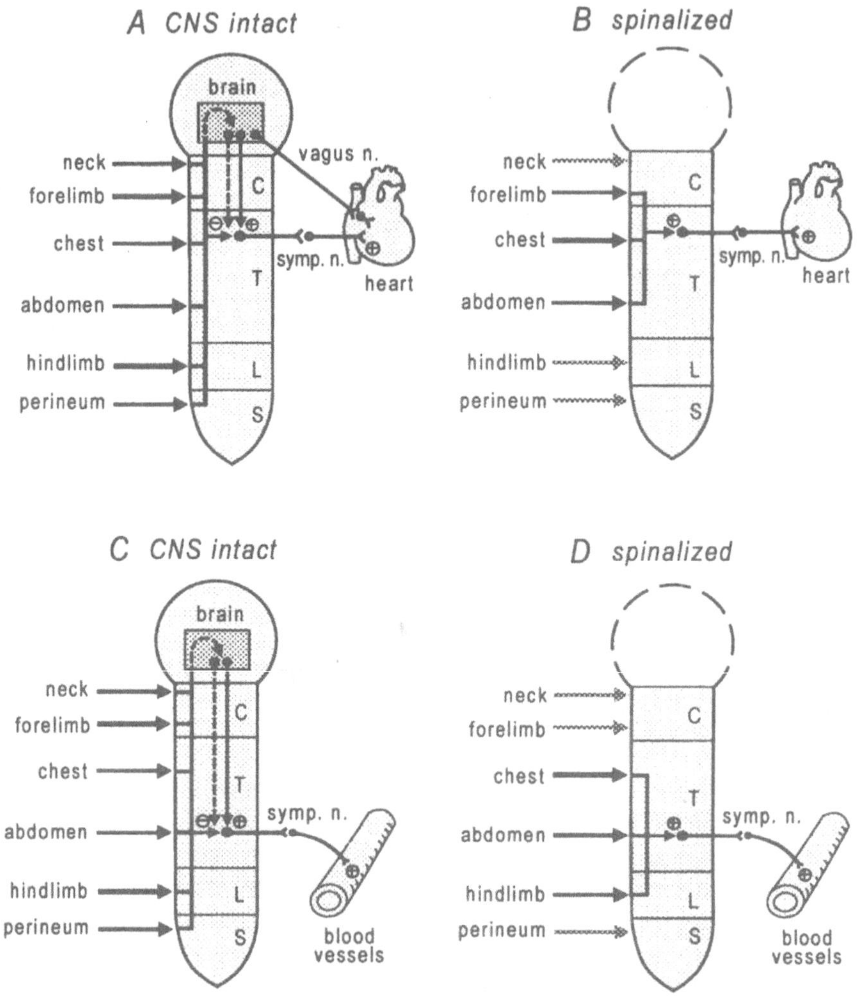

Fig. 51A–D. Reflex pathway **A,B** for the heart and **C,D** blood vessels in **A,C** the CNS-intact and **B,D** spinalized condition. **A,C** Pinching of various segmental skin areas produces an excitatory effect on the cardiovascular system through the sympathetic nerves via a supraspinal reflex pathway. Spinal reflex pathways are tonically inhibited by the brain through an inhibitory descending pathway indicated by *broken line* from the brain. **B,D** After spinal transection, inhibitory descending pathways are eliminated and somatic afferent stimulation at certain areas can produce the segmental spinal reflex. +, excitatory effect; –, inhibitory effect. Somatic stimulation indicated by *shadowed arrows* produces no response. Somatic stimulation indicated by *solid arrows* produces increased responses. *Thickness* of arrow indicates effectiveness of responses to the same amount of stimulation. (Modified from Kimura et al. 1995)

In CNS-intact rats, afferent stimulation of limbs is particularly effective in producing cardiovascular responses. Limb somatic afferents seem to have specific synaptic connections to the central structures in the brain stem related to cardiovascular responses. It remains to be determined whether this is due to the number of somatic afferents from the limb, or to other unknown specific synaptic connections. It is still not known why descending inhibitory influences act strongly on the segmental somato-spinal sympathetic reflex pathways in CNS-intact preparations. New techniques may be necessary to modulate these descending inhibitory functions following somatic sensory stimulation, in order to further elucidate somato-sensory reflex regulation of the cardiovascular system.

2. For years, the neural regulation of rCBF received inadequate attention. Particularly, sympathetic and parasympathetic regulation of rCBF were considered to be only minor factors in comparison to metabolic regulation of rCBF. In the last 8 years, there has been great progress in our understanding of the neural regulation of rCBF by intracerebral nerve fibers innervating the cerebral blood vessels. These pathways include cholinergic nerve fibers originating in the basal forebrain, serotonergic nerve fibers originating in the dorsal raphe, and noradrenergic nerve fibers originating in the locus coeruleus. The cholinergic fibers have been proven to play an important role in regulation of rCBF during somatic afferent stimulation. These intracerebral nerve fibers have not been considered autonomic efferent nerve fibers. However, their actions are similar to those of autonomic fibers, and they thus could be considered to be intracerebral autonomic nerve fibers from the aspect of somato-autonomic reflexes.
3. In the past few years there has been a growing appreciation of the importance of the axon reflex in the control of vascular functions following somatic afferent stimulation independent of involvement of the CNS. The concept of the axon reflex originated as an explanation for vasodilatation of one area of the skin following noxious stimuli delivered to the skin nearby. The concept has recently been applied to explain the vasodilation in other organs, such as the vasa nervorum, following noxious stimulation of the skin. The existence of dichotomizing fibers of single sensory neurons innervating skin, muscle and blood vessels in other organs may explain the clinical phenomena of blood flow modulation in skeletal muscle secondary to physical stimulation of the skin and muscle.

4. Considering all the evidence described, we are closer to being able to use this knowledge of neural mechanisms of somato-cardiovascular reflexes to explain and apply clinical physical therapies. However, before these applications, it is necessary to conduct further studies in pathophysiological experimental animal models displaying autonomic malfunctions.

4.2 Somatosensory Modulation of the Digestive System

4.2.1 Introduction

The digestive system has a broad range of functions starting with mastication of food in the mouth and ending with defecation from the anus. Mastication and defecation are coordinated by somatic motor and autonomic nervous controls. However, a major portion of digestion in the stomach and intestines is regulated by the following: (a) local properties of the gastrointestinal tract itself, (b) gastrointestinal hormones, (c) local enteric nerves, and (d) autonomic nerves.

The effects of somatic stimulation on digestive function were investigated as early as 1913 by Lehmann who used curarized dogs, ventilated and kept under morphine (Lehmann 1913). Gastrointestinal movements were recorded by the balloon method. Sciatic afferent nerve stimulation usually produced inhibition of pendular and local peristaltic movements. Basal tone was also usually reduced, but sometimes was unchanged. Cutting the vagal nerve in the neck region did not change these effects, i.e., they were transmitted via the splanchnic efferent nerves. This, Lehmann claimed, was also true for the excitatory effects sometimes seen. All reflexes disappeared when either the splanchnic nerves or the upper thoracic spinal cord were cut. The same afferent stimuli to the sciatic nerve produced movements of the colon which he interpreted as defecation contractions.

Ruhmann (1927) described his observations, which he made in conscious humans using X-ray techniques. He observed that thermal stimulation to the abdominal wall induced reflex changes in peristaltic movements of the stomach. In the normal stomach, warm abdominal stimuli induced an increase in tone and soft peristaltic movements, and more frequent openings of the pyloric sphincter. This stimulating action accompanying cutaneous vasodilation was changed to a soothing action in cases of pathologically high stomach tone. In cases where the stomach already had a rather high basic tone, cold abdominal stimuli influenced gastric motility, increasing basic tone and producing intense peristaltic movements. In the normal stomach or the stomach with weak tone, the initial reflex response to cold stimulation was a further reduction in tone. Bisgard and Nye (1940) also observed that heat and cold stimulation to the abdominal wall changed gastrointestinal motor activity as monitored by rubber balloons in humans.

In animal models, Kuntz and his colleague (Kuntz 1946; Kuntz and Haselwood 1940) demonstrated the close coupling of localized cutaneous afferent stimulation and circulatory changes in the gastrointestinal tract. They noted that cutaneous vasodilation produced by localized warming of the skin (45°–50°C) or the application of vacuum cups was accompanied by vasodilation in specific segments of the gastrointestinal tract. Similarly, cutaneous vasoconstriction produced by localized cooling of the skin was accompanied by vasoconstriction in corresponding segments of the gastrointestinal tract.

In anesthetized animals, electrical stimulation of limb afferent nerves was used to elicit changes in gastrointestinal motility (Patterson and Rubright 1934; Hodes 1940; Babkin and Kite 1950; Jansson 1969a,b). These authors observed excitation and inhibition of gastrointestinal motility depending upon stimulus parameters (Jansson 1969a,b) and the tonic condition of the effector organs (Patterson and Rubright 1934). Kehl (1975) observed changes in jejunal motility after abdominal skin stimulation in nonanesthetized dogs, and suggested that these changes were due to somato-autonomic reflex responses.

The effects of somatic afferent stimulation of various segmental areas on gastrointestinal motility have been investigated in anesthetized rats (Sato et al. 1975a, 1993; Sato and Terui 1976; Kametani et al. 1978, 1979; Koizumi et al. 1980) and cats (Ito et al. 1979). In anesthetized rats, abdominal stimulation often inhibited gastrointestinal motility, while hindpaw stimulation sometimes facilitated gastrointestinal motility. Furthermore, in anesthetized rats, the neural mechanisms involved in both reflex inhibition and facilitation of gastric motility have been determined by recording gastric motility and autonomic efferent nerve activity (Sato et al. 1975a, 1993; Kametani et al. 1979). These results indicate that the inhibitory gastric response elicited by somatic stimulation of the abdomen is a reflex response. Its afferent neural pathway is composed of abdominal cutaneous and muscle afferent nerves, the efferent neural pathway is the gastric sympathetic nerve, and its reflex center is within the spinal cord. The excitatory gastric response elicited by stimulation of a hindpaw is also a reflex response. Its afferent neural pathway is composed of hindpaw cutaneous and muscle afferent nerves, the efferent neural pathway is the gastric vagal efferent nerve, and its reflex center requires the presence of the brain. Similarly, both the inhibitory somato-duodenal reflex (Sato and Terui 1976) and the inhibitory somato-intestinal reflex (Koizumi et al. 1980) evoked by abdominal cutaneous pinching display a predominantly segmental organization in anesthetized rats.

Somatic afferent stimulation can influence digestive secretion. Neural mechanisms of reflex salivation induced by noxious mechanical somatic stimuli have been determined in anesthetized rabbits (Kawamura and Yamamoto 1977) and rats (Kanosue et al. 1986; Matsuo et al. 1989). Activation of group II, III, and IV afferent fibers in the trigeminal nerve was found to excite salivation-related neurons in the lateral reticular formation of the lower brain stem (Murakami et al. 1983). The efferent pathways of the somato-salivatory reflex include both sympathetic and parasympathetic nerves to the salivary glands (Kawamura and Yamamoto 1977).

4.2.2 Somatosensory Modulation of Gastrointestinal Motility

4.2.2.1 Gastric Motility

4.2.2.1.1 Cutaneous Stimulation

The influence of noxious mechanical cutaneous stimulation of various sites on gastric motility in the pyloric region has been investigated in anesthetized rats (Sato et al. 1975a; Kametani et al. 1978, 1979). When the pressure in an intragastric balloon was maintained at about 100 mmH_2O by expanding the volume of the balloon with water, rhythmic contractions, occurring at a rate of five to six per minute and corresponding to peristaltic movements, could be observed. Pinching of the abdominal skin always produced strong inhibition; pinching the middle to caudal ventral and dorsal thoracic skin produced moderate or weak inhibition; pinching of the paws, nose, forearms and tail in some, but not all cases, produced moderate facilitation; and pinching of the face, ears, neck, legs and sacral

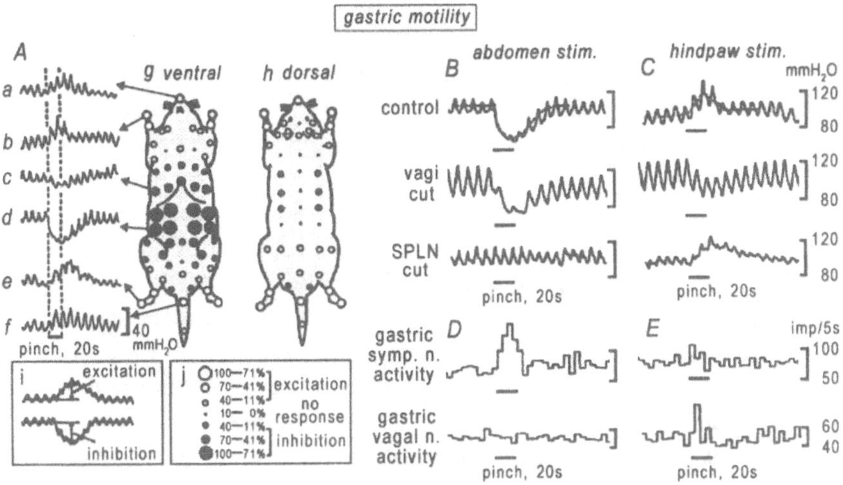

Fig. 52A–E. Effects of noxious mechanical stimulation of various skin areas on gastric motility in chloralose/urethane-anesthetized rats. **A** Specimen records (**a–f**) and schematic diagrams (**g,h**) relating skin areas pinched to reflex changes in gastric motility. The method of estimating the magnitude of the reflex response is illustrated in **i**. **j** *Open circles* indicate excitation, *filled circles* inhibition, and *circle size* magnitude. Pinching of the abdominal skin produced inhibition; pinching of the paws, nose, forearms, and tail produced facilitation. **B,C** Effect of vagal or splanchnic (*SPLN*) denervation on the gastric motility responses to pinching of **B** abdominal skin and **C** hindpaw. The inhibitory reflex response to abdominal skin stimulation disappeared after SPLN denervation, and the facilitatory reflex response to hindpaw stimulation disappeared after vagal denervation. **D,E** Effect on gastric vagal and splanchnic efferent nerve activities of pinching of **D** abdominal skin and **E** hindpaw. The gastric sympathetic efferent nerve activity increased with abdominal skin stimulation, and the gastric vagal efferent nerve activity increased with hindpaw stimulation. (**A** From Kametani et al. 1978; **B–E** Modified from Kametani et al. 1979)

area produced weak facilitation (Fig. 52A). The criteria for reflex facilitation were increased amplitude of gastric contraction and/or increased gastric tone; reflex inhibition was judged by a decrease in those measurements. Both reflex facilitation and inhibition of gastric motility began 1–5 s after the onset of pinching, reached a maximum within 10–30 s and returned to normal within an additional 20–60 s. In these experiments, depth of anesthesia was sufficient to eliminate visible somatic motor reflexes, and remaining effects of skeletal muscle contraction were eliminated, when necessary, using gallamine 10–20 mg/kg i.v. Neither reflex gastric facilitation nor inhibition was significantly influenced by gallamine. Thus the gastric reflex inhibition and facilitation were not due to secondary effects of skeletal muscle contraction or relaxation.

The contributions of vagal and splanchnic nerves to somato-gastric reflexes in rats were examined by transections and recordings of vagus and splanchnic nerves (Kametani et al. 1979). In the absence of vagal innervation, pinching the abdomen still produced significant reflex inhibition of gastric motility, while pinching the hindpaw produced occasional slight inhibition. Bilateral splanchnic transection in vagi intact rats completely abolished reflex inhibition of gastric motility caused by pinching the abdominal skin, but it did not abolish the reflex facilitation produced by pinching the hindpaw (Fig. 52B,C). Abdominal skin stimulation markedly and consistently increased gastric sympathetic efferent nerve activity without significantly affecting the gastric vagal efferent activity (Fig. 52D). Hindpaw stimulation increased gastric vagal efferent nerve activity, whereas gastric sympathetic efferent nerve activity was only slightly increased (Fig. 52E). These experiments, using anesthetized rats, demonstrate that increased reflex activity of gastric vagal efferents is the link between pinching of a hindpaw and facilitation of gastric motility, while the inhibition produced by pinching abdominal skin is through increased reflex activity of gastric sympathetic efferents. The experiments show that noxious stimulation of a hindpaw produces a greater reflex increase in gastric vagal facilitatory efferent fiber activity and a lesser reflex increase in gastric sympathetic inhibitory fiber activity. Consequently, vagal facilitatory fiber activity overcomes sympathetic activity at the gastric level, and finally produces the resultant gastric reflex facilitation.

After spinal transection at the cervical level, the reflex facilitation of gastric motility previously produced by stimulation of the hindpaw was completely abolished, or even reversed to slight reflex inhibition (Kametani et al. 1979). Thus it can be concluded that excitatory somato-gastric reflexes are mediated through supraspinal pathways. On the other hand, reflex inhibition of gastric motility produced by stimulation of the abdomen persisted (Fig. 53C) (Sato et al. 1975a; Kametani et al. 1979). It should be especially emphasized that the inhibitory reflex responses in the CNS-intact and the spinal rats were almost identical. These results suggest that the inhibitory somato-gastric reflex response is primarily a spinal reflex. The different effects of stimulation of abdominal skin or a paw can be explained by the segmental organization of the spinal cord, which is analogous to the segmentation found when sympathetic reflex discharges were

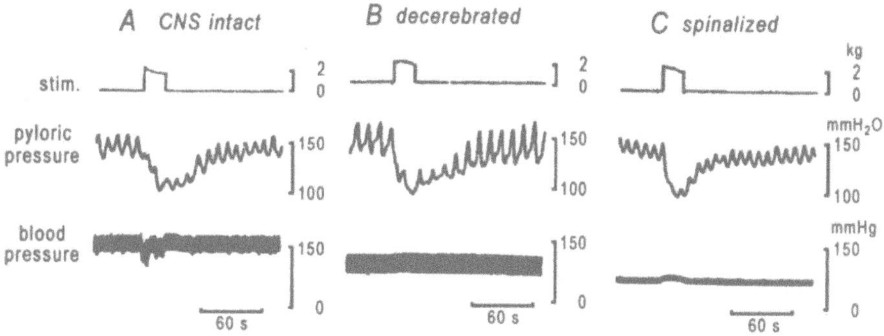

Fig. 53A–C. Effect of pinching stimulation of the abdominal skin on pyloric pressure and blood pressure. The pyloric pressure decreased with abdominal skin stimulation in **A** a chloralose/urethane-anesthetized CNS-intact rat, **B** a decerebrate nonanesthetized rat, and **C** a chloralose/urethane-anesthetized spinal rat (C2 level). (From Sato et al. 1975a)

evoked from the thoracolumbar white rami by electrical stimulation of spinal somatic afferents at various spinal segments (Sato and Schmidt 1971, 1973; see Sect. 3.2.1.1). This inhibitory gastric response was completely abolished after destroying the spinal cord between T5 and T11 by passing a small wire cable back and forth through the vertebral canal (Sato et al. 1975a). Therefore, the inhibitory gastric response was the result of a somato-autonomic nerve reflex, not a dorsal root reflex.

Simultaneous recording of blood pressure showed a depressor response to noxious cutaneous stimulation (Fig. 53A). Similar inhibitory somato-gastric reflex responses were produced in the decerebrate nonanesthetized rat (Fig. 53B) as well as in the spinal rat (Fig. 53C); however, where blood pressure was concerned, in some cases slight pressor responses were seen (Fig. 53B,C). Therefore, a reflex change of blood pressure itself should not be directly responsible for producing the inhibitory somato-gastric reflex responses. The inhibitory somato-gastric reflex response did not change after removal of the adrenal glands, indicating that secretion of adrenal hormones is not essential for this response (Sato et al. 1975a).

Cervero and McRitchie (1981) compared the somato-gastric reflexes of normal rats with those of rats which had been treated neonatally with capsaicin. Noxious mechanical cutaneous stimulation of the abdomen evoked inhibitory gastric reflexes in both normal and capsaicin-treated anesthetized animals. In contrast, noxious thermal stimulation of the abdominal skin, while effective in normal animals, elicited no reflex inhi-

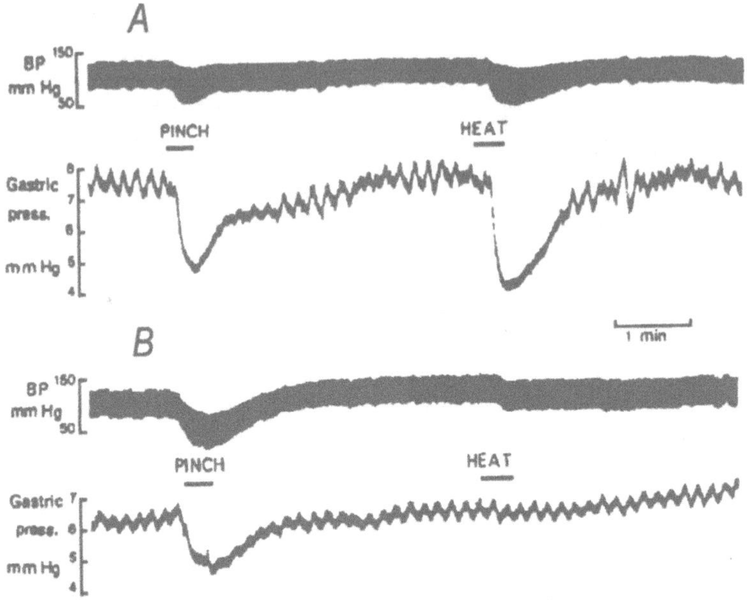

Fig. 54A,B. Effects of noxious stimulation of the abdominal skin on blood pressure (*BP*) and gastric motility of **A** one vehicle -injected rat and **B** one capsaicin-treated rat. The rats were anesthetized with chloralose and urethane. Noxious mechanical stimulation (*PINCH*) produced a drop in blood pressure and a reduction in gastric motility in both animals. Noxious thermal stimulation (*HEAT*) produced the same effects only in the control rat. Note the small change in blood pressure and the absence of any change in gastric motility when noxious thermal stimulation was applied to the skin of the capsaicin-treated rat. (From Cerevero and McRitchie 1981)

bition in the capsaicin-treated rats (Fig. 54). Similarly, Holzer et al. (1992) demonstrated that defunctionalization of capsaicin-sensitive afferent neurons by systemic pretreatment of rats with capsaicin negated the normal gastric motor inhibition elicited by laparotomy. These observations suggest that capsaicin-sensitive neurons serving polymodal nociceptors constitute the afferent limb of the reflex arc which normally mediates inhibitory somato-gastric reflexes evoked by laparotomy or noxious thermal cutaneous stimulation.

In anesthetized cats, the inhibitory response was observed, but it was very difficult to obtain the facilitatory response observed in rats (Ito et al. 1979). Gastric motility was recorded by the balloon method or the strain gauge force transducer method. Pinching either the abdominal skin or a hindpaw evoked a reflex inhibition of gastric motility in two-thirds of the animals. The inhibitory response of gastric motility to abdominal pinching was larger than that evoked by hindpaw pinching, although tachycardia after abdominal pinching was usually less than that after hindpaw pinching. In some instances, the inhibition was followed by increased contractions. In one third of the animals, inhibitory reflex responses in gastric motility were not observed. This was true even more than 12 h after commencing the operation. The reason for this absence of the inhibitory reflex is not clear. No facilitatory reflex response, as reported in rats by Kametani et al. (1978, 1979), could be observed in cats after pinching a hindpaw. When the cat's vagus nerve was severed at the cervical level, electrical stimulation of the peripheral cut branch of the vagus nerve produced a bradycardiac response as well as a transient facilitatory response of gastric motility, which was followed by an inhibitory response. This indicates that the dosage of gallamine triethiodide used did not block vagal neuromuscular transmission to the heart or stomach. Thus a considerable species difference seems to exist between rats and cats with respect to the neural mechanisms for producing reflex activation of vagal gastric efferent discharges after pinching a hindpaw.

4.2.2.1.2
Acupuncture-Like Stimulation

In anesthetized rats, Sato et al. (1993) examined the effects on gastric motility of acupuncture-like stimulation of various segmental areas. An acupuncture needle (diameter 340 μm) was inserted into the skin and underlying muscles to a depth of 4–5 mm, and was twisted manually right and left approximately once every second for 60 s. Stimulation of the abdomen and lower chest region inhibited gastric motility by increasing the activity of gastric sympathetic efferent nerves, and stimulation of the limbs often facilitated gastric motility by increasing the activity of the gastric vagal efferent nerves. Responses of gastric motility were the same regardless of whether stimulation was delivered to the skin and muscles, the skin alone, or the underlying muscles alone (Fig. 55). Abdominal stimulation enhanced the activity of afferent fibers in the lower thoracic spinal nerves, and the inhibitory gastric motility response to abdominal stimulation was abolished by severance of these thoracic spinal nerves. Hindpaw stimulation enhanced the activity of afferent fibers in the femoral and sciatic nerves, and the excitatory gastric motility response to hindpaw stimulation was abolished by severing these hindlimb nerves. Furthermore,

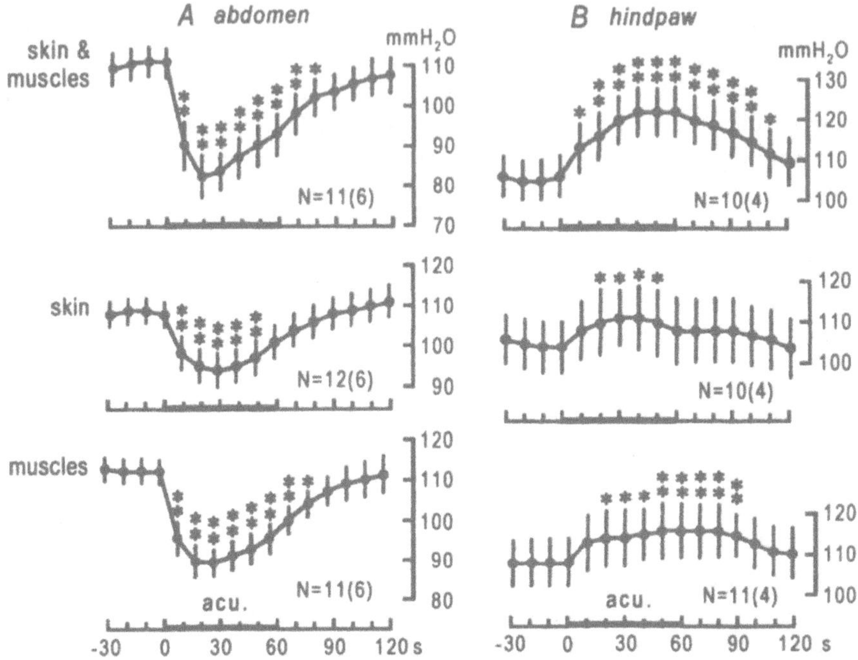

Fig. 55A,B. The effect of acupuncture-like stimulation of **A** the abdomen and **B** the hindpaw on mean gastric pressure in urethane-anesthetized rats. Acupuncture-like stimulation to either the abdominal skin and muscles, abdominal skin alone, or underlying muscles alone produced the inhibitory response, and similar stimulations to hindpaw produced the excitatory response. $^{*}P<0.05$, $^{**}P<0.01$; significantly different from prestimulus control values. (Modified from Sato et al. 1993)

abdomen stimulation-induced inhibition of gastric motility persisted in spinalized rats, while hindpaw stimulation-induced excitation of gastric motility disappeared. It was concluded that the inhibitory and excitatory gastric responses elicited by acupuncture-like stimulation of the abdomen and hindpaw were spinal and supraspinal reflex responses, respectively. The afferent nerve pathways were composed of abdominal and hindpaw somatic afferent nerves, while the main efferent pathways for the inhibitory and excitatory reflexes were sympathetic and parasympathetic, respectively. The excitatory and inhibitory gastric reflex responses induced by acupuncture-like stimulation of the abdomen and hindpaw were not influenced by intravenous administration of naloxone (0.4–4 mg/kg), suggesting that endogenous opioids are not involved in the reflexes.

4.2.2.2
Intestinal Motility

4.2.2.2.1
Cutaneous Stimulation

The influence on intestinal motility of noxious mechanical stimulation of various skin areas has been studied in anesthetized rats (Sato and Terui 1976; Koizumi et al. 1980). In anesthetized rats, using the balloon method, three components of duodenal motility were recorded: (1) the small, fast, rhythmic waves which correspond to the pendular movements of the duodenum, (2) the large, slow, rhythmic waves which correspond to peristaltic movement and (3) the baseline component of duodenal pressure (Sato and Terui 1976). When noxious stimulation was applied to the skin of the neck, chest and abdomen, only abdominal stimulation produced changes in duodenal pressure. These changes were: (a) a decrease in the amplitude of the small, fast, rhythmic waves, (b) a decrease in frequency of the large, slow, rhythmic waves, and (c) a decrease in the baseline of duodenal pressure (Fig. 56, rows II, III). These inhibitory responses persisted after bilateral denervation of the vagi. However, in animals whose splanchnic nerves were cut but whose vagi were kept intact, the responses were abolished. These results suggest that a reflex increase in splanchnic nerve activity caused the responses seen in intact animals. In spinal rats whose splanchnic nerves were kept intact, the responses were also present. This suggests predominantly segmental organization for the cutaneo-duodenal reflex, similar to the organization of the cutaneo-gastric reflexes in rats (Sato et al. 1975a; Kametani et al. 1979).

The effect on intestinal (jejunal) motility of pinching various skin areas was examined in anesthetized rats (Koizumi et al. 1980). Pinching the abdominal skin always inhibited rhythmic contractions and decreased the basal tone of the jejunum (Fig. 56A, row IV). Pinching the skin of the upper chest, neck, forepaws, or hindpaws produced augmentation of the pendular contractions and an increase in the basal tone (Fig. 57A). Thus, as with the cutaneo-gastric reflex, stimulation of distant segmental cutaneous afferents produced a facilitatory response, while stimulation of segmental cutaneous afferents produced inhibition. Cutaneous stimulation of the right and left sides was equally effective. The size of the effective dorsal skin area was similar but the response was much smaller than that obtained by stimulating an equivalent area on the ventral surface. This may be due to differences between the ventral and the dorsal skin areas

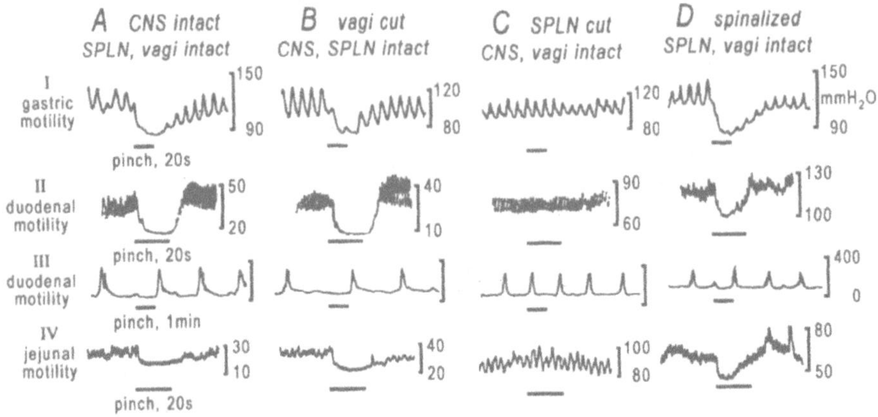

Fig. 56A–D. Inhibitory reflex responses of gastric (*I*), duodenal (*II*, small, fast rhythmic waves; *III*, pendular movements), and jejunal (*IV*) motility to pinching of abdominal skin and effects of denervation of autonomic nerves or spinal transection (C1 level) on these responses in chloralose/urethane-anesthetized rats. Calibration of III in **D** applies to other three specimens of duodenal pendular movements. **A** CNS, splanchnic nerve (SPLN), vagi intact. **B** Vagi cut, CNS and SPLN intact. **C** SPLN cut, CNS and vagi intact. **D** Spinalized SPLN, vagi intact. The inhibitory responses were abolished only after SPLN denervation. (Modified from Kametani et al. 1979; Sato and Terui 1976; Koizumi et al. 1980)

in terms of density of receptors, distribution of cutaneous nerve fibers, or thickness of the skin. Tactile stimulation, such as rubbing the skin, produced no significant changes in jejunal motility.

The role of extrinsic innervation to the jejunum in the cutaneo-intestinal reflex was examined. It was found that both the inhibitory and excitatory cutaneo-intestinal reflexes were elicited via the sympathetic efferent pathway. Pinching the abdomen induced an increase in mesenteric sympathetic efferent nerve output to the small intestine and a simultaneous decrease in jejunal activity (Fig. 57D). Pinching the upper chest, neck or paws produced a decrease in mesenteric sympathetic activity and a simultaneous increase in basal jejunal tone (Fig. 57E). Changes in jejunal motility and mesenteric efferent nerve activity caused by skin stimulation were not modified after bilateral section of the vagus nerves at the cervical level (Fig. 56B, row IV). However, after the splanchnic nerves were severed bilaterally, changes in jejunal motility did not occur in response to pinching the abdomen, chest or hindpaw (Fig. 56C, row IV). Thus the inhibitory

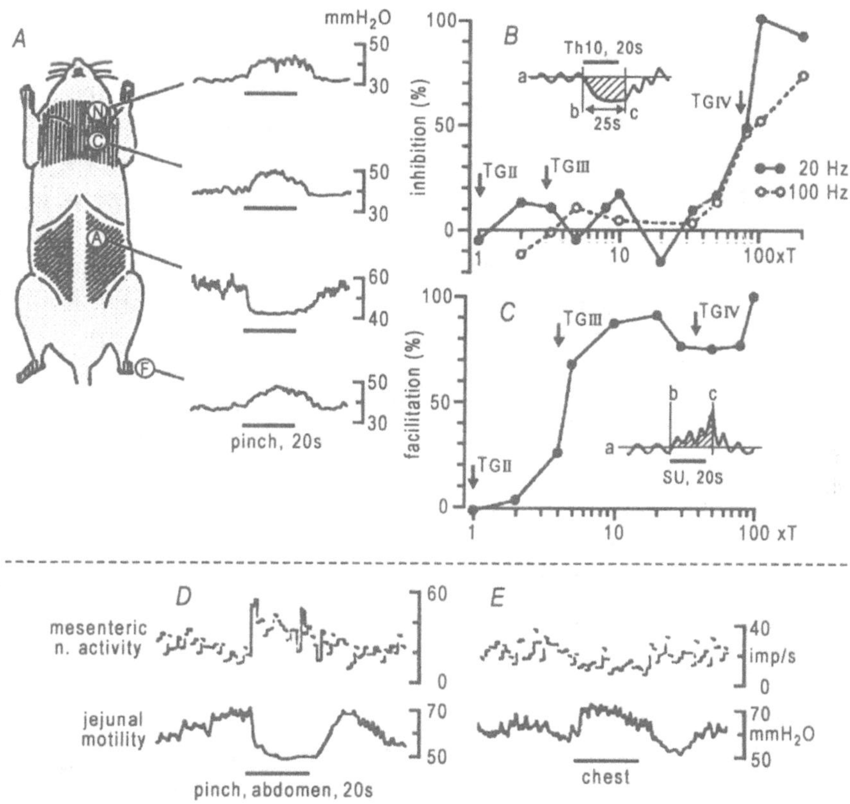

Fig. 57A–E. Somatosensory regulation of jejunal motility in chloralose/urethane-anesthetized rats. **A** Specimen records of jejunal motility. Pinching of the abdominal skin produced inhibition and pinching of neck, chest, and hindpaw produced facilitation. **B,C** Relationships between intensities of electrical stimulation of cutaneous afferent nerves and magnitude of intestinal responses. **B** Stimulation of a cutaneous afferent nerve branch of the tenth thoracic spinal nerve (T10) produced inhibition, and **C** stimulation of left sural nerve (*SU*) produced facilitation. Stimulus intensities are expressed as multiples of threshold strength for group II fibers in each nerve. **D,E** Simultaneous recordings of mesenteric efferent nerve activity and jejunal pressure. **D** Pinching of abdomen increased mesenteric sympathetic nerve activity and decreased jejunal pressure. **E** Pinching the chest skin produced opposite responses. (Modified from Koizumi et al. 1980)

cutaneo-intestinal reflex was largely due to an increase in intestinal sympathetic efferent activity, and the facilitatory cutaneo-intestinal reflex was due to a decrease in intestinal sympathetic efferent nerve activity.

The inhibition of jejunal motility and the increase in mesenteric efferent nerve activity produced by pinching the abdominal skin were not significantly influenced by spinal transection at the second cervical level (Fig. 56D, row IV). However, the excitatory response of jejunal motility and the reflex decrease in mesenteric efferent nerve activity caused by pinching the skin of the upper chest were completely abolished following spinal transection in the same preparation. Compared with results obtained in cats and dogs, spinal transection at the cervical level in rats, when properly done, produced fewer signs of spinal shock. Not only the cutaneo-intestinal reflex but also other spinal reflexes appeared within 30 min following spinal cord transection. The facilitatory reflex, on the other hand, did not appear even after several hours of observation. It was concluded that excitatory cutaneo-intestinal reflexes were mediated through supraspinal pathways, while inhibitory cutaneo-intestinal reflexes were propriospinal.

When a somatic nerve is stimulated, a "dorsal root reflex" response (Barron and Matthews 1938) might be recorded as antidromic potentials in the afferent fibers running in the mesenteric nerves. It has been reported that bicuculline (0.2 mg/kg), a GABA receptor antagonist, abolishes the dorsal root reflex recorded from the mesenteric nerve in cats (Miyamoto 1976). However, intravenous injection of bicuculline (0.45 mg/kg) had no effect on the inhibitory cutaneo-jejunal reflex. Hexamethonium bromide, a nicotinic cholinergic receptor antagonist, which is known to block synaptic transmission at the celiac ganglion but does not eliminate the dorsal root reflex, abolished the inhibitory cutaneo-jejunal reflex completely. These results seem to indicate that the dorsal root reflex does not contribute to cutaneo-jejunal reflexes (Koizumi et al. 1980).

4.2.2.2.2
Electrical Stimulation

By using electrical stimulation of various intensities, Koizumi et al. (1980) systematically analyzed the relationship between the magnitude of the cutaneo-intestinal reflex response and the afferent fiber groups stimulated. Either the sural nerve or a cutaneous branch of the T10 spinal nerve was used for electrical stimulation. Stimulation of group II and III cutaneous fibers of the T10 spinal nerve did not produce any changes in jejunal motility, while stimulation of group IV fibers in addition to the group II and III fibers always produced inhibitory responses (Fig. 57B). Stimulation of the sural afferent nerve of a hindlimb elicited the facilitatory jejunal reflex. Facilitatory responses were obtained when the stimulus intensity reached approximately four times the threshold (4T) (i.e., near the threshold for group III afferents) and the maximal response was obtained when the intensity was about twenty times the threshold (20T, i.e., below the threshold for group IV afferents; Fig. 57C). Thus it was concluded that the inhibitory cutaneo-intestinal reflex was caused by excitation of group IV cutaneous afferent nerve fibers, while the facilitatory cutaneo-intestinal reflex was evoked mainly by excitation of group III cutaneous afferent nerve fibers.

Camilleri et al. (1984) studied the effects of transcutaneous electrical nerve stimulation (TENS) on upper gastrointestinal phasic pressure activity in conscious humans. TENS applied to either the hand (afferents entered the spinal cord at the C8-T1 levels) or the upper abdomen (afferents entered the spinal cord at the T5-T10 levels) produced a significant reduction in the antral motility index, and no qualitative changes in proximal intestinal pressure activity. In as much as essentially the same responses were elicited from sustained somatic afferent stimulation of these different dermatomal regions, it would appear that the dominant reflex center for these responses was supraspinal.

Frost et al. (1993) investigated whether surface dermatomal electrical stimulation could be used to alter colonic motility and bowel emptying in patients with spinal cord injuries. In spinal patients and control subjects, the effects of electrical stimulation of the second sacral (S2) dermatome were monitored with rectal manometry. Additionally, electrical stimulation was used to supplement the bowel-care programs of spinal patients. In fact, the activity of the recto-sigmoid colon did respond to afferent stimulation of the S2 dermatome. In both the spinal patients and the control group, there was a significant rise in the number of rectal pressure spikes in response to electrical stimulation. There was no difference between the two groups in terms of the amplitude of spike waves in the colon. Regarding the clinical application of this work, the methods employed did not achieve any significant change in the time required for spinal patients to initiate or complete bowel movements. It remains to be seen whether a different method of applying electrical stimulation, using a different frequency, stimulus setting, or electrode placement, might be found to exert a more powerful modulating effect on spinal cord reflexes of colonic motility.

4.2.3
Somatosensory Modulation of Digestive Secretion

4.2.3.1
Salivary Secretion

Salivation always occurs during mastication. The mechanism of salivation during mastication had been considered to be one of the following: (a) central commands activating efferent pathways of parasympathetic and sympathetic nerves or (b) excitation of afferent nerves from the oral tissues involved in mastication and subsequent reflex excitation of salivary parasympathetic and sympathetic efferent activity. Both mechanisms appear to have a physiological background and meaning.

Noxious mechanical stimuli to the oral area induced reflex salivation in anesthetized rabbits (Kawamura and Yamamoto 1977) and rats (Kanosue et al. 1986; Matsuo et al. 1989). Kawamura and Yamamoto (1977) employed various kinds of taste stimuli and innocuous and noxious mechanical stimuli to various orofacial regions, and compared the reflex salivary secretion in anesthetized rabbits. Moderate concentrations of various chemicals applied for 1 min produced salivation within a fairly narrow range of 2.5–10.0 µl. However, pinching the tongue induced at least twice as much salivation as the application of 0.5 mol tartaric acid. Thus noxious mechanical stimulation was much more effective than gustatory stimulation in inducing salivation in anesthetized rabbits. Regarding the reflex responses to mechanical stimulation, the volume of salivation was extremely sensitive to the modality of stimulation. Noxious mechanical stimulation was much more effective than innocuous stimulation. Hence, innocuous touch stimulation of the lower lip for 20 s induced 8 µl secretion, whereas pressure stimulation of the same site produced 20 µl secretion. However, pinching of this location produced 130 µl secretion (Fig. 58A). Noxious stimulation thus induced more salivary secretion than other kinds of mechanical stimulation. The magnitude of the reflex response was also influenced by the site of stimulation. The front of the oral cavity was more sensitive than other facial regions, and the face was more sensitive than other regions of the body (Fig. 58B). For example, pricking stimulation of the upper incisor gingiva for 20 s induced 95 µl secretion per minute, whereas comparable noxious stimulation of the hindlimb produced only 5 µl secretion. Cervical sympathectomy produced an 18% decrease in salivation induced by pinching of the oral tissues

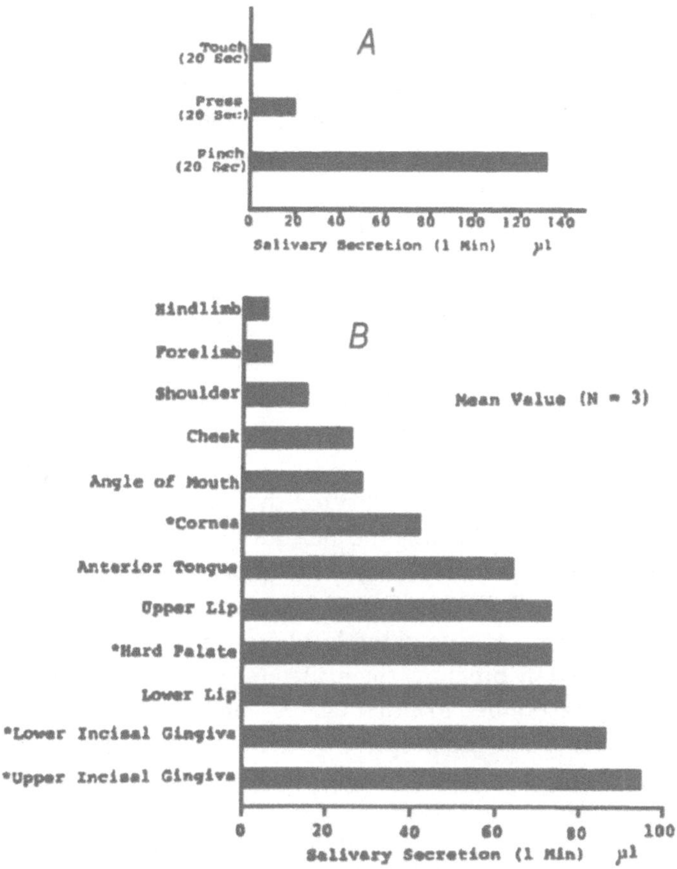

Fig. 58A,B. Salivary secretion induced by various stimuli in chloralose/urethane-anesthetized rabbits. **A** Effects of mechanical stimulation of the ipsilateral lower lip for 20 s. Noxious pinching stimulation was much more effective than innocuous touch or press stimulation for producing salivary secretion. **B** Effects of pinching of various ipsilateral regions of the body for 20 s. Pinching of oral and facial region is more effective than pinching of other regions of the body. (From Kawamura and Yamamoto 1977).

from the ipsilateral submandibular gland, and transection of the ipsilateral chorda tympani that contains the parasympathetic nerves to the salivary glands produced an approximately 80% decrease in the salivary response. These results suggest that the pain-salivary reflex depends heavily upon the parasympathetic innervation of the chorda tympani and somewhat upon sympathetic innervation.

The secretory and vasodilatory responses evoked in the parotid gland of anesthetized cats by electrical stimulation of the central cut end of the lingual nerve after cutting the chorda tympani nerve (hence, somatic stimulation) were also mediated via activation of the autonomic nervous system, particularly by a parasympathetic mechanism (Takahashi et al. 1995).

In urethane-anesthetized rats, Kanosue et al. (1986) investigated the effects of body temperature on salivary secretion, reflexly induced by noxious stimulation of the tongue or perioral region. They found that the magnitude of the reflex response to pinching stimulation increased with increasing rectal temperature. Hence, at 38.7°C, pinching scarcely evoked any salivary secretion. However, at a rectal temperature of 39.5°C, there was an obvious salivary response to pinching, which was further increased when the rectal temperature was raised to 40.2°C. In anesthetized, decerebrate rats, Matsuo et al. (1989) investigated the neurological mechanisms of salivation induced by pinching the anterior oral cavity. Lesioning of the caudal or interpolar trigeminal sensory nuclei substantially reduced the reflex salivation induced by noxious mechanical stimulation. Furthermore, of the preganglionic parasympathetic fibers innervating salivary glands, 63% were activated by pinching the ipsilateral anterior oral region, but none of these fibers responded to either taste or innocuous mechanical stimulation. This suggests that nociceptive somatosensory input to the superior salivary nucleus is relayed through the caudal or interpolar trigeminal sensory nuclei.

In anesthetized cats, Murakami et al. (1983) used antidromic electrical stimulation of the chorda tympani to identify salivary neurons in the lateral reticular formation of the lower brain stem. Of the 71 salivary neurons, 72% responded with spike potentials to single-shock stimulation (0.1 ms) of afferent branches of the trigeminal nerve, indicating, as one would expect, the importance of afferent input from the oral region in reflex salivary secretion. Input from group II and group III afferent fibers of the trigeminal nerve have all been found to activate salivary efferent neurons.

4.2.3.2
Gastrointestinal Secretion

In conscious dogs, via esophagostomy and gastric cannulation, Zhou and Chey (1984) studied the effects of acupuncture and electroacupuncture on gastric secretion during the interdigestive period. Bilateral stimulation of three points, Tsu-San-Li (peroneal nerve region in hindlimb), Pishu (paravertebral region of T11), and Neiguan (median and radial nerve region in forelimb), led to increased secretion of bicarbonate and sodium, and decreased secretion of acid. Electroacupuncture was more effective than manual acupuncture without electrical stimulation in inducing gastric secretion. The reflex response of bicarbonate and sodium secretion was completely abolished by local anesthetic applied to the stimulation areas or by intravenous administration of anticholinergic drugs, establishing the neurological basis for these somato-visceral reflexes.

Noguchi and Hayashi (1996) demonstrated that electroacupuncture stimulation of the hindlimb of anesthetized rats increased gastric acid secretion with somatic nerves serving as the afferent pathway and branches of the vagus nerve to the stomach as the efferent pathway. Based on pH values of the perfusate, there was marked enhancement of gastric acid secretion following electroacupuncture stimulation of the hindlimb. The electroacupuncture-specific response was not observed in rats after sciatic nerve denervation or after vagotomy.

4.2.4
Somatosensory Modulation of Gastrointestinal Blood Flow

Kuntz and Haselwood (1940) observed reflex changes in the diameters of gastrointestinal blood vessels after thermal stimulation of the abdominal skin of decerebrate cats. The changes in caliber of the gastrointestinal vessels corresponded to those produced in the matching vessels in the cutaneous areas stimulated. Moderate, localized warming of the skin (45°–50°C) and the application of vacuum cups produced vasodilation in the cutaneous area stimulated and in the corresponding segments of the gastrointestinal tract. Localized warming of the skin to 52°C or more caused an initial vasoconstriction in the gastrointestinal tract. Moderate, localized cooling of the skin resulted in vasoconstriction in the cutaneous area stimulated and in the corresponding segments of the gastrointestinal tract. The caliber changes were more marked in smaller gastrointestinal vessels than in larger ones.

In unanesthetized rats with the spinal cord transected in the lower cervical region and in anesthetized rats with the cerebrospinal axis intact, localized warming of the skin in the caudal half of the thoracic region to approximately 45°C consistently elicited reflex vasodilation in the small intestine. Localized cooling of the skin in the same area consistently elicited reflex vasoconstriction in the small intestine (Kuntz 1946).

Intestinal blood flow responses to stimulation of limb nerves have been studied by measuring the venous outflows in anesthetized cats (Johansson 1962; Winsö et al. 1985) and the perfusion pressure in the superior mesenteric artery in anesthetized dogs (Nutter and Wickliffe 1981). In anesthetized cats, stimulation of mixed somatic nerves resulted in diminished venous outflow from muscle, skin, kidney, and splanchnic beds, with a concomitant increase in systemic arterial pressure, suggesting a generalized vasoconstrictor response (Johansson 1962).

4.2.5
Central Processing of Somato-gastrointestinal Reflexes

4.2.5.1
Somato-splanchnic Sympathetic Reflex

Splanchnic efferent reflex discharges caused by electrical stimulation of limb afferent nerves or intercostal afferent nerves were studied in anesthetized rats (Nosaka et al. 1980). Stimulation of the limb afferent nerve produced the supraspinal reflex discharges via group II and III afferent excitation with a latency of 48± 6 ms (Fig. 59C). These reflex discharges were abolished by spinal transection at the C1 level (Fig. 59D). Stimulation of the intercostal afferent nerves, with a sufficient intensity, evoked an early, ipsilateral splanchnic nerve reflex discharge with a relatively short latency of 9 ± 2 ms (Fig. 59A). The early splanchnic nerve reflex discharge was not abolished, but rather was often facilitated by spinal transection at the C1 level (Fig. 59B). This early sympathetic reflex discharge with a short latency was considered to be identical to the early spinal sympathetic reflex component reported in cats by Sato and Schmidt (1973). When the stimulus intensity was increased enough to excite the group IV or C afferent fibers of an intercostal nerve, the stimulation elicited another reflex discharge component with a long latency of 51 ± 6 ms, just behind the short latency reflex discharge. Since the late reflex discharge remained after spinal transection at the C1 level, it was evident that this late reflex discharge was also a spinal reflex.

The stimulus strength–reflex response curve obtained after spinal transection at the C1 level indicates that the excitation of the group II and group II + III afferent fibers of the intercostal nerve was involved in

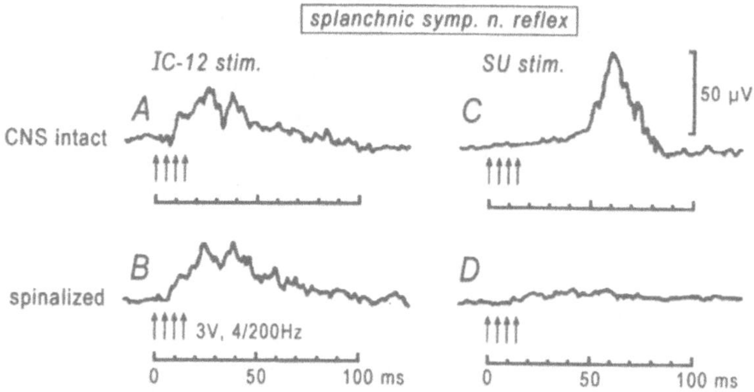

Fig. 59A–D. Reflex discharges evoked in the splanchnic sympathetic nerve by stimulation of group II and III afferent fibers (3 V, four pulses at 200 Hz) of **A,B** the 12th intercostal nerve and **C,D** the sural nerve in urethane/chloralose-anesthetized rats with **A,C** CNS intact and **B,D** spinalized at the C1 level. Stimulation of the intercostal nerve produced an early spinal reflex with a relatively short latency, and stimulation of the sural nerve produced a late supraspinal reflex with a long latency. (Modified from Nosaka et al. 1980)

eliciting the early reflex response, i.e., the A reflex (see Sect. 3.2.1), while the late response was elicited by group IV afferent fiber excitation, i.e., the C reflex (see Sect. 3.2.2). The magnitude and pattern of the early spinal splanchnic reflex discharges were identical before and after administration of bicuculline, a GABA receptor antagonist, indicating that the early spinal reflex discharge in rat splanchnic nerves was composed of splanchnic efferent reflex discharges only and was not related to the dorsal root reflex reported by Miyamoto (1976). In the rat, both of the intercostally elicited splanchnic sympathetic A and C reflexes are of spinal origin, and seem to contribute to the spinal inhibitory reflex responses of gastrointestinal motility via a reflex increase in gastrointestinal sympathetic nerve activity after pinching stimulation of the abdominal skin.

4.2.5.2
Somato-vagal Reflex

Kimura et al. (1996a) were the first to record two gastric vagal efferent reflex discharges, i.e., A and C reflexes, using single-shock stimulation of the tibial afferent nerve in the hindlimb in urethane-anesthetized rats. Single shock of group II and III afferents produced the A-reflex discharge with a latency of about 120 ms, while single shock of group IV afferents also produced a C-reflex discharge with a latency of about 360 ms (Fig. 60). A single shock to a first lumber spinal afferent nerve produced only a weak reflex component with a latency of about 120 ms in gastric vagal efferent nerves. Limb afferents appear to have stronger central pathways functionally connecting to gastric vagal efferent preganglionic neurons in the brain stem, than do abdominal afferents.

In anesthetized rats, noxious stimulation of the hindlimb skin by pinching elicits a facilitatory response of gastric motility (Kametani et al. 1979). As noxious stimulation is well known to excite thin myelinated and unmyelinated C afferent fibers, it appears that the reflexly induced increase in gastric vagal efferent activity following pinching of a limb skin was the result of accumulation of the single shock-elicited gastric vagal A- and C-reflex discharge responses using single electrical shock of a limb afferent nerve.

4.2.6 Conclusion

1. In anesthetized animals, motility of the gastrointestinal tract has been shown to be inhibited or facilitated by natural somatic stimulation, especially by noxious mechanical stimulation, and both inhibitory or facilitatory responses have been shown to depend on the segmental areas stimulated. The inhibitory response was due to stimulation to the abdominal area, whereas stimulation to the limb or paw produced the facilitatory response. The excitation of the thin myelinated group III and unmyelinated group IV afferent fibers in somatic nerves was proven to be necessary for producing gastrointestinal reflex responses in anesthetized preparations. The central reflex pathways, both spinal and supraspinal, were identified by analyzing the latency of the reflex discharges from sympathetic and parasympathetic efferent nerves following single electrical shock or train stimuli to somatic afferent nerves, and also by comparing autonomic nerve responses to natural somatic afferent stimuli in CNS-intact and spinalized preparations. The inhibitory reflex response was a consequence of a somatically induced increase in sympathetic efferent activity, which was characterized by dominant spinal segmental organization. Its reflex center was in the spinal cord. The facilitatory response of the gastrointestinal tract was a consequence of either a somatically induced increase in vagal activity

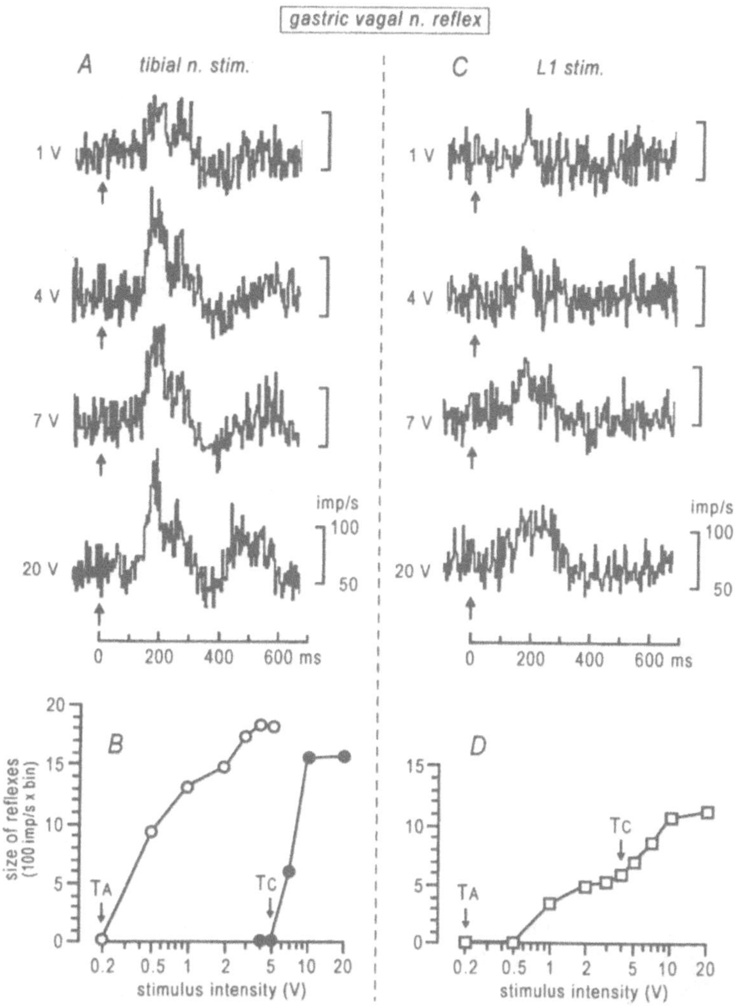

Fig. 60A–D. Single electrical shock stimulation of somatic afferent nerves can evoke a reflex response in vagal efferent nerves innervating the stomach in urethane-anesthetized rats. **A,B** A single shock to a tibial nerve produced A-reflex discharges with a latency of about 120 ms and C reflex discharges with a latency of about 360 ms. **C,D** A single shock to a L1 spinal nerve produced only a weak reflex component with a latency of about 120 ms. *TA*, threshold intensity of A fiber; *TC*, threshold intensity of C fiber. (From Kimura et al. 1996a)

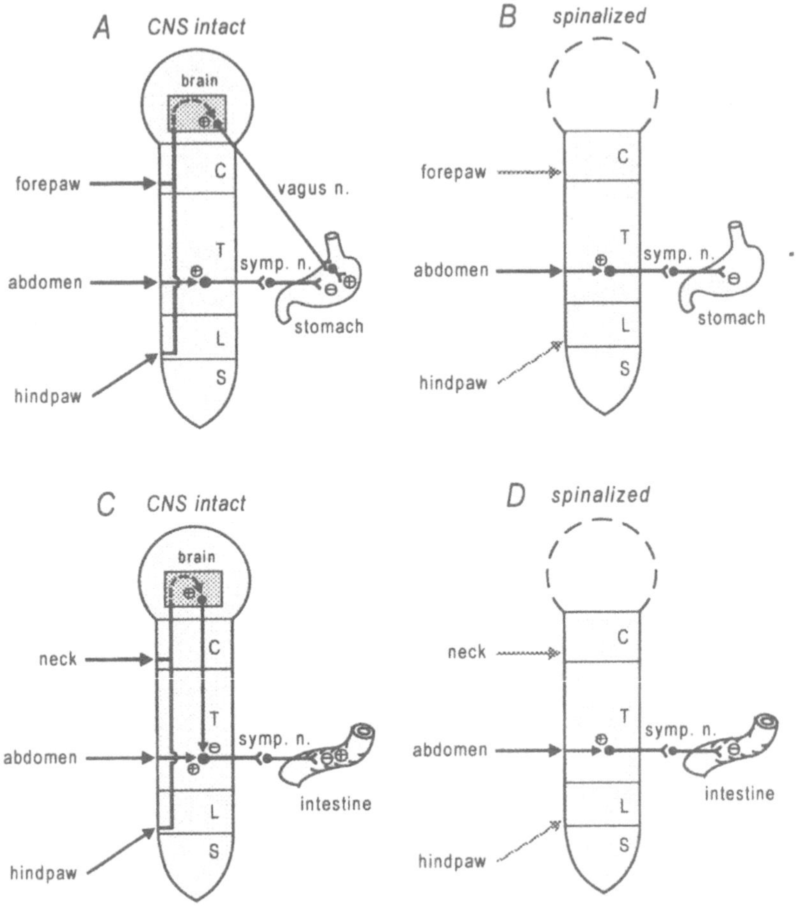

Fig. 61A–D. Reflex pathway **A,B** for the stomach and **C,D** for the intestine in **A,C** the CNS-intact and **B,D** spinalized condition. **A,C** Pinching of abdominal skin produces an inhibitory effect on the gastrointestinal system through the sympathetic nerves via a spinal reflex pathway. On the other hand, pinching of neck, forepaw or hindpaw produces an excitatory effect on the gastrointestinal system through excitation of parasympathetic nerve activity (in the case of stomach) or inhibition of sympathetic nerve activity (in the case of intestine) via a supraspinal reflex pathway. **B,D** After spinal transection only the inhibitory spinal reflex is produced. Other details are the same as in Fig. 51

or a decrease in sympathetic activity, which was characterized by supraspinal and nonsegmental organization (Fig. 61).

The well-known clinical evidence that gastrointestinal motility and secretion in humans change following somatic sensory stimulation appears to involve the same sort of somato-autonomic reflexes which have been demonstrated in anesthetized preparations.

2. In conscious humans and animals, somatic sensory stimulation may induce not only the reflexes mentioned above, but may also induce or modulate gastrointestinal or digestive functions via emotion, in which case the efferent paths are still autonomic efferent nerves. Furthermore, a conditioning reflex may be involved in somatic regulation of digestive functions in conscious preparations. Studies in conscious animals are the next necessary step.

4.3 Somatosensory Modulation of the Urinary System

4.3.1 Introduction

Urinary bladder function consists mainly of storage (or continence) and voiding (or micturition) of urine. The bladder muscle relaxes for storage and contracts for micturition. The urethral sphincter contracts during storage and relaxes during micturition. Therefore, both functions of the bladder are achieved by close cooperation with the urethral sphincter. This cooperative relationship between the bladder and urethral sphincter is called a synergic relationship. The bladder is innervated by both sympathetic (hypogastric) and parasympathetic (pelvic) nerves. Each nerve contains both efferent and afferent fibers. Stimulation of the efferent components of the pelvic nerve evokes vesical contractions, while stimulation of the efferents of the hypogastric nerve may have some inhibitory effect on bladder contraction. Afferents of both pelvic and hypogastric nerves provide signal transmission paths for vesical stretch information to the CNS. The urethral sphincter is innervated by a somatic (pudendal) nerve, which contains both efferent and afferent fibers. Pelvic and pudendal nerves come out of the sacral spinal cord, while the hypogastric nerves emerge from the lumbar spinal cord. Normal continence and micturition are characteristically regulated by the interaction of both autonomic and somatic nervous systems. In the early stages of development, micturition is mainly regulated by reflex mechanisms, but in humans gradually comes under voluntary control, while reflex components of storage and micturition remain present.

Barrington (1914, 1925) was the first to systematically challenge the study of the neural mechanisms of continence and micturition reflexes arising in the bladder and urethra. Barrington (1928) described reflex regulation of micturition in anesthetized cats as follows: "If the bladder volume was increased by small amounts at regular intervals of time, a certain volume was found, which was more or less constant for each individual but varied widely in different individuals, at which the bladder pressure suddenly rose 40 to over 90 cm of water. The pressure which excited this contraction was always less than 10 cm of water. The reflex persisted after division of the hypogastric or of the pudic nerves, but was abolished by division of the pelvic nerves, transection of the spinal cord, or by cocainizing the interior of the bladder. The reflex must therefore arise in the brain, and have both its afferent and efferent paths in the pelvic nerves."

In patients whose spinal cords are injured, neural regulation of micturition is clinically important. Denny-Brown and Robertson (1933) observed patients with injured spinal cords, and found that automatic micturition presented some striking characteristics both in reaction to distension and to external somatic sensory stimuli to some skin areas, such as the sacral cutaneous segments. Further, using spinalized cats, Langworthy and Hesser (1937) noted a spinal micturition reflex whose efferent path involved parasympathetic fibers. After transection of the spinal cord in the lower lumbar region, the activity of the vesical musculature can be influenced by somatic sensory stimuli applied to the tail or perineal region. Stimulation of the tail or the skin and deep tissues of the perineum in almost all cases caused a contraction of the bladder. Thus micturition following transection of the cord indicates control by means of reflex arcs mediated through the sacral portion of spinal cord. The efferent pathway for these reflexes is by way of parasympathetic fibers.

McPherson (1966) first noted the inhibitory effects of somatic sensory stimulation on micturition contractions of the bladder using anesthetized cats. He observed that in the CNS-intact anesthetized cat, cold stimulation of the skin decreased the pressure of bladder contractions and shortened the interval between them. There was no evidence that this effect was reflex in nature. The fore- and hindpaws of cats were placed in water at 45°C for 15 min or the skin of the hindlimbs was pinched vigorously for 3 min. Neither stimulus had any effect on the peak intravesical pressures or on the intervals between contractions. Ice applied to the hindpaws decreased the peak pressures and also the intervals between the contractions. These effects did not occur until ice had been applied for 2–5 min.

Electrical stimulation of the sural afferent nerve sometimes produced contraction of the bladder or increased the pressure developed by spontaneous contractions. Stimulation of any hindlimb afferent nerves with conduction velocities of approximately 50 m/s serving muscles inhibited bladder contractions.

DeGroat and Ryall (1969) further studied the effects of somatic afferent nerve stimulation on parasympathetic neurons in anesthetized cats. It was found that stimulation of sacral somatic afferent nerve fibers caused inhibition of the firing of pelvic parasympathetic neurons, similar to the inhibition produced by stimulation of the pelvic afferent nerve. This inhibition of pelvic parasympathetic neurons was followed by excitation with a variable latency sometimes similar to, and at other times much longer (up to 100 ms) than that following stimulation of the pelvic afferent nerve. An excitation with a long latency also followed stimulation of various leg afferent nerves, including the biceps-semitendinosus, sural, gastrocnemius, peroneal and the whole sciatic nerve. Activation of somatic afferent fibers by light tactile stimulation of the perineal region usually inhibited bladder contractions.

Considering the historical background of reflex regulation of vesical function by somatic afferent stimulation, Sato and his colleagues have systematically studied the neural mechanisms of such regulation of vesical function using anesthetized animals (Sato et al. 1975b, 1977, 1979a, 1980, 1992). Their main interests were the modality of somatic stimulation, afferent fiber characteristics, central nervous divisions of spinal and supraspinal structures involved in the reflexes, efferent paths including sympathetic and parasympathetic nerves to the bladder and the somatic motor nerve to the urethral sphincter. They also focused on reflex modulation under the two opposing conditions of the nonexpanded and the expanded bladder. It has been emphasized that the response pattern of somato-vesical reflexes depends strongly on the condition of the bladder: the same afferent input which evokes reflex contractions of a nearly empty and quiescent bladder may induce an inhibition of the large, rhythmic micturition contractions of a distended bladder. Using neurophysiological techniques for recording efferent nerve activity in the sympathetic and parasympathetic nerves, in addition to monitoring bladder behavior, the two responses of facilitation and inhibition of the vesical micturition contractions following somatic afferent stimulation have been proven to be reflexes whose main efferent path is in the parasympathetic pelvic nerve.

These studies resolved two reflex pathways: spinal and supraspinal (Sato et al. 1975b, 1977, A. Sato et al. 1983; Morrison et al. 1996a, b).

Similar to the response of the bladder, activity of the urethral sphincter is regulated by somatic afferent stimulation. Activity in the external urethral sphincter or in pudendal sphincteric motoneurons has been reported to increase (McMahon et al. 1982; Jolesz et al. 1982; Rampal and Mignard 1975a,b; Sasaki et al. 1994) or decrease (Morrison et al. 1995a) after somatic afferent stimulation.

Recent studies of the neural mechanisms of somatic sensory reflex regulation of vesical and urethral sphincter functions will be reviewed.

4.3.2 Cutaneous Stimulation

4.3.2.1 Mechanical Stimulation

Excitatory Reflex Responses (At Low Vesical Pressure). Sato et al. (1975b) studied the effect of stimulation of various skin areas on the function of the bladder in anesthetized rats with the CNS-intact and in decerebrated nonanesthetized rats. The tone and contraction of the bladder were measured by the intravesical balloon method. When the volume of the intravesical balloon was slightly expanded so that the resting vesical pressure was increased to approximately 40 mmH_2O, the bladder revealed only small spontaneous contractions but not large and rhythmic micturition contractions. Under these conditions, tactile or noxious mechanical stimulation of the perineal skin, but not stimulation of chest or abdomen, caused an abrupt increase in intravesical pressure (Fig. 62A). This excitatory cutaneo-vesical response was obtained in anesthetized as well as in decerebrated nonanesthetized rats. This excitatory response remained even after transection of hypogastric nerve branches, but it was completely abolished after either surgically destroying the sacral segment of the spinal cord or bilaterally severing the pelvic nerve branches to the bladder. The influence of the efferent nerves on this excitatory cutaneo-vesical reflex was further examined by directly recording the efferent nerve discharge activity from the vesical nerve branches of the pelvic and hypogastric nerves. Tactile or noxious stimulation of the perineal skin always markedly increased the efferent discharge activity of the pelvic nerve branches. Therefore, it was concluded that the excitatory cutaneo-vesical reflex was produced by

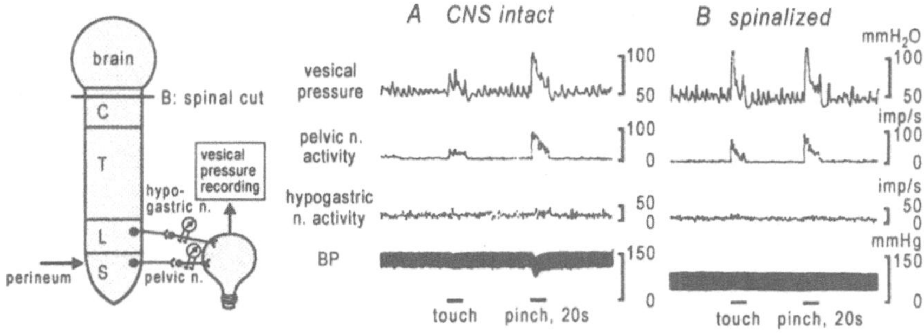

Fig. 62A,B. Effect of perineal cutaneous stimulation by touching and pinching on vesical pressure, pelvic and hypogastric efferent nerve activity of quiescent bladder, and blood pressure (*BP*) in chloralose/urethane-anesthetized rats **A** before and **B** after spinal transection (C2 level). Perineal stimulation produced increases in vesical pressure and pelvic nerve activities in CNS-intact and spinalized rats. (Modified from Sato et al. 1975b)

the reflexly increased efferent activity of the pelvic nerves. However, the same stimulation, especially noxious, sometimes produced a slight increase in hypogastric efferent activity.

Transection of the spinal cord at the upper cervical level or the middle thoracic level in anesthetized rats did not cause any significant change in the excitatory cutaneo-vesical reflex (Fig. 62B). It is noteworthy that an almost identical excitatory cutaneo-vesical reflex was observed only 5 min after spinal transection. In spinalized animals, the skin area which produced this excitatory cutaneo-vesical reflex was also localized exclusively to the perineal region. This evidence strongly suggests that the excitatory cutaneo-vesical reflex is a segmentally organized propriospinal reflex whose efferent pathway is contained in the pelvic nerves. The reflex contraction of the bladder produced specifically by perineal stimulation was present both before and after acute cervical spinal transection in anesthetized cats (Fig. 63A–D) (Sato et al. 1977; Sasaki et al. 1994).

Maggi et al. (1986a) also observed in urethane-anesthetized rats that pinching of the perineum induced bladder contractions of spinal cord origin. It was noted that the amplitude of perineal pinching-induced bladder contractions was decreased by urethane in a dose-dependent manner. Furthermore, they showed that the pinching-induced bladder contractions were suppressed by topical tetrodotoxin, a Na^+ channel blocker, and by intravenous hexamethonium, a nicotinic cholinergic receptor antagonist.

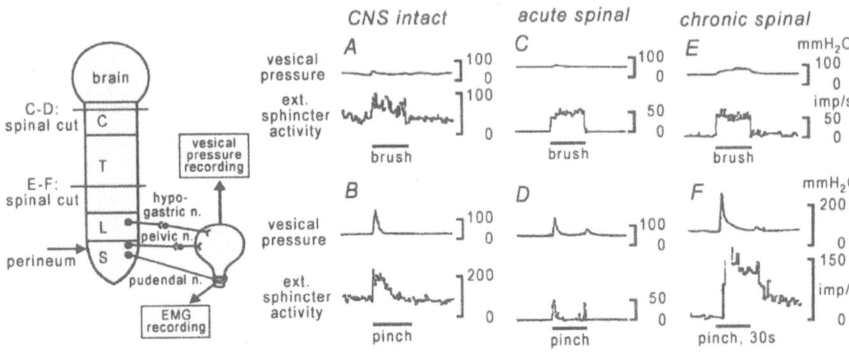

Fig. 63A–F. Effects of **A,C,E** brushing and **B,D,F** pinching the perineal skin on the intravesical pressure and the external urethral sphincter activity when the bladder showed a quiescent steady tonus in **A,B** CNS-intact, **C,D** acute spinal (at the cervical level), and **E,F** chronic spinal cats (at the lower thoracic level) anesthetized with chloralose and urethane. Both the pinching and the brushing stimulation caused increases in vesical pressure and external sphincter activity in CNS-intact, acute, and chronic spinal cats. (Modified from Sasaki et al. 1994)

This indicated that the response was neurogenic. This reflex contraction appears not to be sufficient to produce urination.

In some patients who have difficulty with micturition, for example in paraplegic or cerebral stroke patients, the excitatory cutaneo-vesical reflex contraction of spinal origin appears to be clinically useful for producing micturition by physical manipulation of the perineum or associated areas. This excitatory cutaneo-vesical reflex appears to have a physiological meaning for micturition during development, for example in neonatal kittens and rats (De Groat et al. 1975; Maggi et al. 1986b; Thor et al. 1989; Kruse and De Groat 1990, 1993). Maggi et al. (1986b) studied the amplitude of the pinching-induced phasic bladder contraction in relation to age in anesthetized rats. In rats less than 10 days of age, micturition was not elicited in response to bladder distension, but was triggered by perineal stimulation.

Perineal stimulation can also influence activity of the external urethral sphincter (EUS) or periurethral sphincter in anesthetized animals. Sasaki et al. (1994) clarified the effects on the reflex activity of the bladder and EUS of noxious and innocuous mechanical cutaneous stimuli to different sites of the body surface in anesthetized CNS-intact cats, and in acute, and chronic spinal cats. Brushing or pinching perineal skin produced tran-

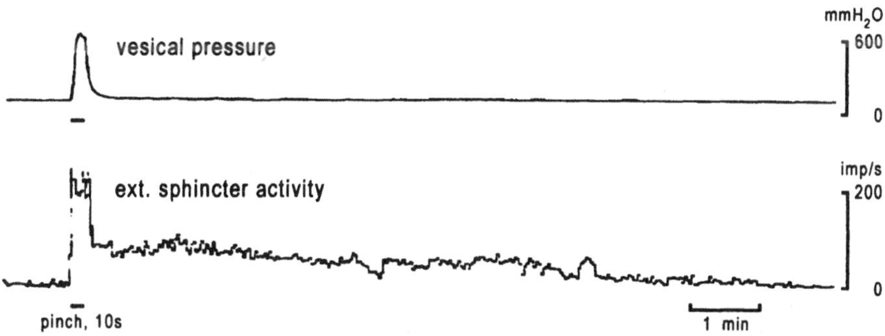

Fig. 64. Example of the long-lasting response of external urethral sphincter activity with a short-lasting increase in intravesical pressure in response to pinching stimulation of perineal skin in chronic spinal cats anesthetized with chloralose and urethane. (From Sasaki et al. 1994)

sient reflex-induced increases in bladder pressure, and elevation of EUS activity in all groups of animals (Fig. 63). The effects in the chronic spinal preparations were larger in amplitude and longer in duration than in animals with the CNS intact. It is noteworthy that the changes were more marked in the EUS than in the vesical contraction. For example, the elevated EUS activity induced by a pinch of 10 s took almost 10 min to return to the control level (Fig. 64). Of the stimulation of other skin areas (brushing and pinching stimulation to the forepaw, neck, lower chest, and hindpaw) in CNS-intact, acute spinal, and chronic spinal animals, only pinching the hindpaw in chronic spinal animals was effective in producing a marginal reflex increase in EUS activity. In anesthetized CNS-intact rats, an increase in the periurethral electromyographic activity (EMG) was also elicited by pinching of the perineal skin (Morrison et al. 1995b).

Inhibitory Reflex Responses (At High Vesical Pressure). When the intravesical pressure was increased and set at 100–300 mmH$_2$O by injecting water into the intravesical balloon, the bladder revealed fairly large rhythmic micturition contractions with an amplitude of 400–1000 mmH$_2$O and a frequency of one to three per minute. The rhythmic micturition contractions did not alter significantly after transection of the hypogastric nerve branches to the bladder, but these contractions were abolished completely after bilateral transection of the pelvic nerve branches to the bladder and after acute spinal transection.

Simultaneous recordings of intravesical pressure and efferent activity in the pelvic and hypogastric nerve branches to the bladder in CNS-intact anesthetized animals always showed rhythmic burst discharges of the pelvic nerves synchronized with the micturition contractions. This synchronization of the pelvic nerve branches with the micturition contractions of the bladder was abolished completely after spinal transection. This indicates that rhythmic contractions of the bladder are activated mainly via the pelvic efferent nerves due to supraspinal organization (De Groat and Ryall 1969; Sato et al. 1975b, 1977).

Tactile stimulation had virtually no effect, but pinching of the perineal skin always caused an inhibition of micturition contractions in both CNS-intact anesthetized and decerebrated nonanesthetized rats (Sato et al. 1975b) (Fig. 65A,B). The inhibition of the micturition contractions by perineal stimulation was not influenced by total destruction of the hypogastric nerve branches. Pinching of the perineal skin inhibited the burst discharges of the pelvic nerve branches. These burst discharges of the pelvic nerves were not inhibited by tactile stimulation of the perineum or by pinching of other segmental skin areas, such as the chest or abdomen. Pinching of perineal skin sometimes facilitated the discharge activity of the hypogastric nerve branches. These results suggest that the efferent path for the reflex inhibition of the large micturition contractions is in the pelvic nerves.

In anesthetized cats, similar results were observed as in rats (Fig. 65C,D). In cats, the rhythmic micturition contractions were inhibited by pinching the skin of the perineum, abdomen and chest, and the degree of inhibitory efficacy of stimulation was in the same order (Sato et al. 1977). Rubbing the perineal skin for a period of 30 s also inhibited the micturition contractions, although the degree of the inhibitory response was much less than that after pinching the skin. The inhibitory effects on the micturition contractions of rubbing the skin of the neck, chest or abdomen were unconvincing.

The inhibitory effects of somatic afferent stimulation on the distended bladder suggest that vesical afferent activity and somatic afferent volleys interact in the CNS. The activation of pelvic nerve afferents at high pressures was able to reverse reflex behavior, switching the pelvic parasympathetic system on or off in a somewhat graded manner within the physiological range of bladder pressures and volumes (McMahon and Morrison 1982c). Possible sites for the integration of afferent input might be the micturition center of the brain stem, or the spinal sacral parasympathetic

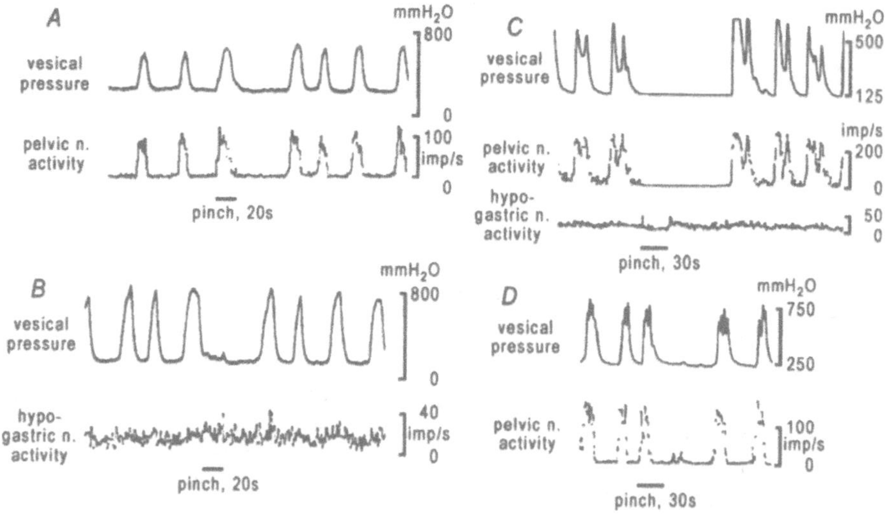

Fig. 65A–D. Effect of pinching of perineal skin on large micturition contractions of the bladder and vesical pelvic and hypogastric efferent nerve activity in urethane/chloralose-anesthetized **A,B** rats and **C,D** cats. Pinching stimulation inhibited the rhythmic vesical contraction and pelvic efferent nerve activities, but sometimes increased (in rats) or did not change (in cats) the hypogastric efferent nerve activities. (Modified from Sato et al. 1975b, 1977)

neuronal circuits. Within the sacral spinal cord, the mixing of somatic and visceral signals inevitably plays a part in these mechanisms, and somato-vesical convergence in the dorsal horn and intermediolateral column of the spinal cord has also been recognized (Cervero 1985; Fields et al. 1970a,b; Milne et al. 1981; McMahon et al. 1982; McMahon and Morrison 1982a–c).

In CNS-intact cats, during rhythmic micturition contractions, the EMG of the EUS and efferent nerve activity in the urethral branches of the pudendal nerve revealed a characteristic rhythm which was reciprocal to the rhythmic bladder activity (Fig. 66A) (Sasaki et al. 1994). On the other hand, the periurethral EMG showed a marked oscillatory increase during bladder contraction in CNS-intact rats (Morrison et al. 1995b). In both animals, noxious stimulation of the perineal skin slowed the frequency of the large rhythmic micturition contractions of the bladder and increased the EUS activity (Sasaki et al. 1994; Morrison et al. 1995b) (Fig. 66A).

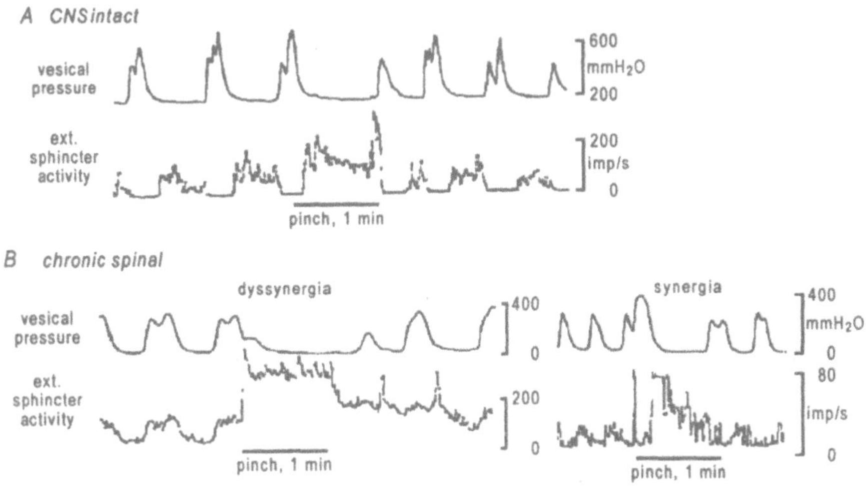

Fig. 66A,B. Effects of pinching of perineal skin on the intravesical pressure and the external urethral sphincter activity when the bladder showed large, rhythmic micturition contractions in **A** CNS-intact and **B** chronic spinal cats anesthetized with chloralose and urethane. In the chronic spinal cat, the *left panel* shows dyssynergia and the *right panel* shows synergia. Pinching stimulation inhibited the vesical contraction and increased the external sphincter activity in CNS-intact and chronic spinal cats. (Modified from Sasaki et al. 1994)

After spinal transection, the micturition contractions of the expanded bladder disappeared (Barrington 1914, 1925, 1928; Kuru 1965; Ruch 1960). Therefore, the spinal components of the inhibitory somato-vesical reflex could not be explored in acute spinal animals. However, it is known that vesical functions slowly recover after spinal transection. In chronic spinal cats, Langworthy and Hesser (1937) recorded rhythmic contractions of the bladder 3 days after the transection. In chronic spinal cats, brushing the perineal skin increased the amplitude and/or frequency of micturition contractions, while pinching the perineal skin gave rise to an immediate contraction of the bladder followed by a reduction in the frequency of rhythmic vesical micturition contractions (Fig. 66B) (Sasaki et al. 1994).

In some chronic spinal animals, the EUS rhythmic activity was seen to be synchronous with the micturition contractions of the bladder. The simultaneous contractile activity of the bladder and EUS excitation have been called "dyssynergia." In the remaining chronic spinal animals, the EUS activity revealed rhythmic activity which was characteristically reciprocal to the vesical micturition contractions, as observed in CNS-intact animals. Such a reciprocal relationship between the rhythmic vesical contractions and EUS activity has been described as "synergia." In all of the animals showing synergia or dyssynergia, brushing or pinching stimulation of the perineal skin increased the EUS activity (Sasaki et al. 1994) (Fig. 66B).

It has been proposed that transcutaneous electrical stimulation applied to the perianal skin (Nakamura et al. 1986) or anus (Eriksen and Mjölneröd 1987) can ameliorate detrusor instability in patients with incontinence. The excitatory cutaneo-EUS reflex appears to be essential for this treatment of incontinence.

4.3.2.2
Thermal Stimulation

In anesthetized cats, the responses of the bladder to thermal stimulation of the skin were examined at different temperatures for periods of 60 s (Sato et al. 1977). When the bladder was in its quiescent state, an excitatory reflex could only be obtained by thermal stimulation of the perineal skin. For cold stimuli, the threshold temperature for producing a change in vesical pressure was around 13°C, and for warm stimuli the threshold was around 43°C. Thermoprobe temperature between 10° and 44°C did not produce noxious thermal stimulation to cats (Kaufman et al. 1977). Therefore, the results suggest that innocuous thermal stimulation of the perineal skin could elicit the excitatory responses. When the temperature of the thermoprobe was adjusted to noxious levels, such as to 4°C, 46°C, or 49°C, the responses became much larger, suggesting that activation of cutaneous thermal nociceptors at these temperature ranges could also evoke an additional response (Fig. 67A–E).

During rhythmic micturition contractions, thermal stimulation of the skin evoked an inhibitory vesical response. An inhibitory reflex response was obtained by thermal stimulation of, in order of descending efficacy, the perineum, abdomen and chest. Thermal stimulation of the skin of the neck was ineffective. The perineum was the most effective skin area for producing reflex responses. When the temperature of the thermoprobe on the perineum was changed from the control temperature of 34°C to 5°C, 11°C, 43°C, or 47°C, micturition contractions were inhibited. Temperatures of 5°C or 47°C produced a longer-lasting inhibition than temperatures of 11°C or 43°C, respectively (Fig. 67F–I).

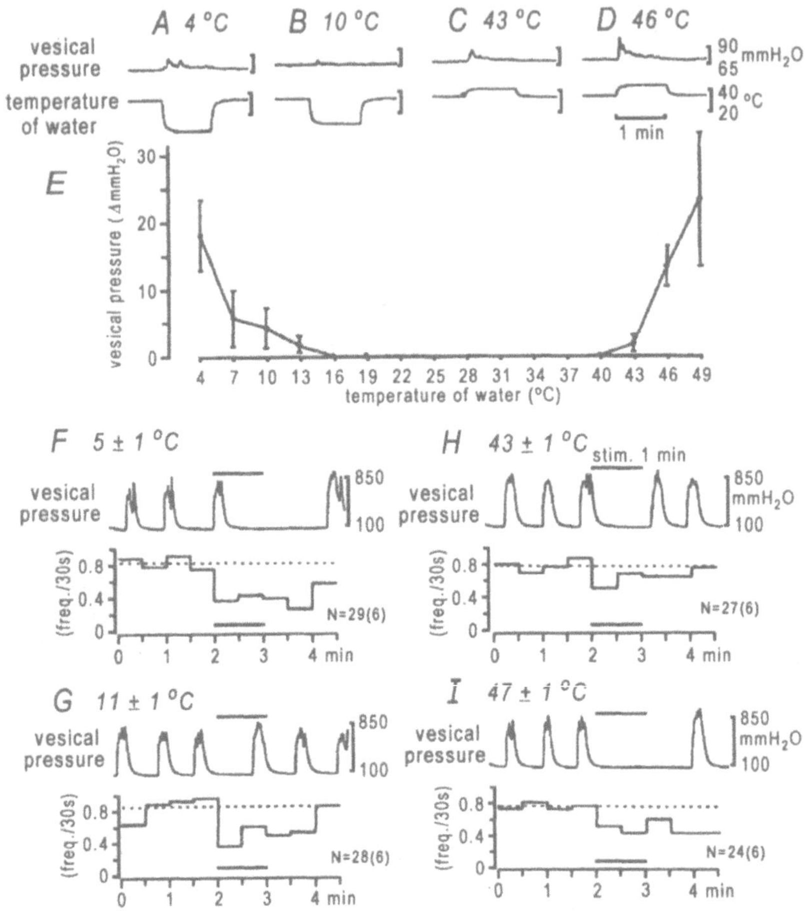

Fig. 67A–I. The effect of thermal stimulation of the perineal skin on bladder function in chloralose/urethane-anesthetized cats. The water temperature in the thermoprobe was changed from the control temperature of 34°C to 4–49°C. Thermal stimulation of the perineal skin **A–E** increased vesical pressure of the quiescent bladder and **F–I** inhibited the large rhythmic micturition contractions of the expanded bladder. (Modified from Sato et al. 1977)

4.3.3 Muscle Stimulation

In anesthetized cats, selective activation of fine muscle afferents (group III and IV fibers) of the hindlimb by potassium chloride (KCl) or bradykinin induced reflex changes in the urinary bladder (Sato et al. 1979a). The nature of the reflex responses depended on the functional state of the bladder: they were excitatory when the bladder was quiescent and inhibitory during rhythmic micturition contractions (Fig. 68). All reflexes observed disappeared after cutting the sciatic or the pelvic nerves. This indicates that the afferent and efferent limbs of the reflexes are in sciatic and pelvic nerves, respectively. Serotonin (5-HT) did not produce any change in vesical functions, but augmented the responses elicited by KCl or bradykinin. Selective activation of large muscle primary afferents (group I fibers) by succinylcholine was ineffective.

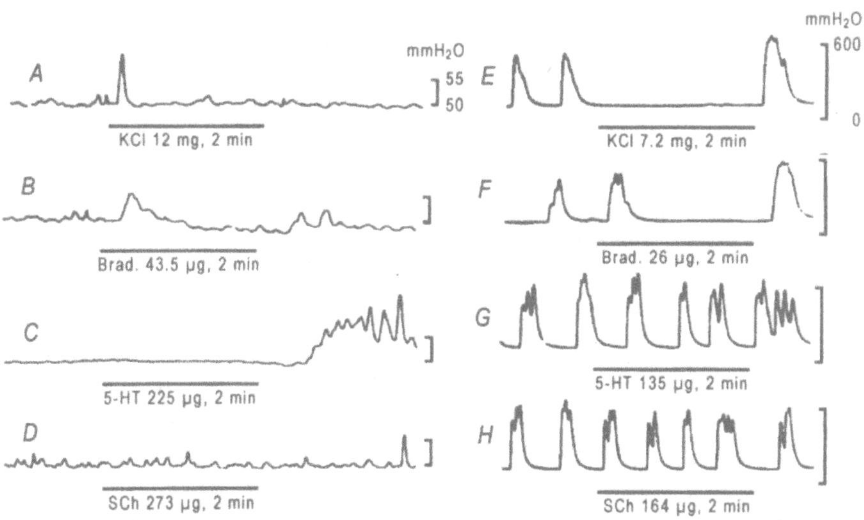

Fig. 68A–H. Effects of chemical stimulation of muscle afferents on **A–D** vesical pressure of the quiescent bladder and **E–H** the micturition contractions of the expanded bladder in chloralose/urethane-anesthetized cats. Muscle afferents were stimulated by injecting various chemicals into the muscle artery. KCl and bradykinin (*Brad.*) injections produced reflex responses of the urinary bladder, but serotonin (*5-HT*) and succinylcholine (*SCh*) injections had no effect. **C** The vesical pressure was increased after removing the occulsion of the femoral vessels. This response was due to the 5-HT released into the general circulation. (Modified from Sato et al. 1979a)

4.3.4 Acupuncture-Like Stimulation

The effects of acupuncture-like stimulation of various segmental areas on the rhythmic micturition contractions of the urinary bladder were systematically studied in anesthetized rats (Sato et al. 1992). An acupuncture needle having a diameter of either 160 or 340 μm was inserted into the skin and underlying muscles, and then manually twisted left and right at a frequency of approximately 1 Hz. Acupuncture-like stimulation applied to the perineal area inhibited both the rhythmic micturition contractions and the discharges of vesical pelvic efferent nerves, without any significant changes in hypogastric efferent nerve activity (Fig. 69A). Stimulation anywhere within an area of about 5 mm^2 of the perineum induced identical responses. In contrast, stimulation applied to the face, neck, forelimb, chest, abdomen, back, and hindlimb areas was ineffective. After surgically separating the perineal skin from the underlying muscles while preserving the main cutaneous nerve branches, stimulation of either the perineal skin or the perineal muscles inhibited the rhythmic micturition contractions. Stimulation of the perineal muscles produced a stronger inhibition of the rhythmic micturition contractions than did stimulation of the perineal skin (Fig. 69B). Stimulation of the perineal area increased afferent nerve activity recorded from the pudendal nerve branches innervating the perineal skin or underlying muscles, or recorded from the pelvic nerve branches innervating the perineal muscles. The stimulation-induced inhibition of rhythmic micturition contractions was abolished after surgically severing both the pudendal and pelvic nerve branches that innervated the perineal skin and underlying muscles. These findings indicate that the inhibition of rhythmic micturition contractions following acupuncture-like stimulation of the perineal area is a reflex response characterized by segmental organization. The afferent limbs of the reflex involve both pelvic and pudendal nerve branches innervating the perineal skin and underlying muscles, while the efferent limbs of the arc are in pelvic nerve branches innervating the urinary bladder.

The effects of acupuncture-like stimulation on the periurethral EMG were examined by Morrison et al. (1995a) in anesthetized rats whose bladders were kept partially filled. When acupuncture-like stimuli were applied to the skin and underlying structures, either together or separately, in the rostral half of the body, the hindpaw or the urethra, these stimuli usually induced excitation of the periurethral EMG. Depression of the

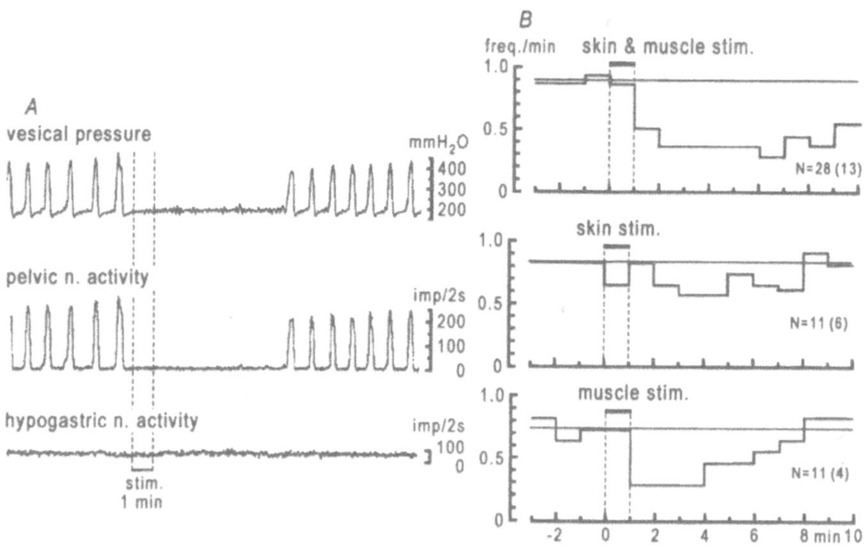

Fig. 69A,B. Effects of acupuncture-like stimulation of the perineal area on rhythmic micturition contractions in urethane-anesthetized rats. **A** Rhythmic micturition contractions and vesical pelvic efferent nerve activity were inhibited and hypogastric efferent nerve activity was not affected following acupuncture-like stimulation of the perineal area. **B** Stimulation of the perineal muscles alone produced a stronger inhibition of the rhythmic micturition contractions than that of the perineal skin alone. (Modified from Sato et al. 1992)

EMG was seen predominantly during stimulation of muscle close to the urethra, or stimulation of the bulbocavernosus, pubococcygeus, or the dorsal or ventral sacrococcygeal muscle. In spinalized animals, the inhibitory pattern of the EMG induced by acupuncture-like stimulation of these muscles was similar to that seen in spinal cord-intact animals.

4.3.5 Electrical Stimulation

In CNS-intact anesthetized cats, electrical stimulation of the group III and IV afferent fibers of cutaneous (sural, superficial peroneal) and muscle (gastrocnemius and soleus, peroneal and deep peroneal) nerves of the hindlimb elicits reflex responses of the bladder (Sato et al. 1980). When the bladder was quiescent, stimulus effects of somatic afferents were excitatory (Fig. 70A), by reflexly increasing efferent discharges of the pelvic

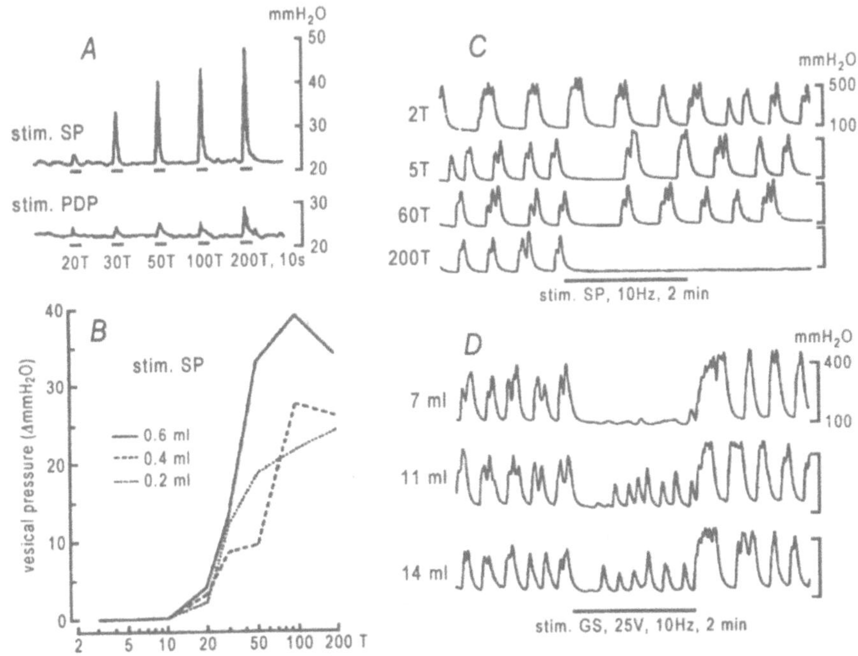

Fig. 70A–D. Electrical stimulation of hindlimb cutaneous (superficial peroneal nerve, *SP*) or muscle nerves (muscle branches of peroneal and deep peroneal nerves, *PDP*; nerves to gastrocnemius and soleus muscles, *GS*) **A,B** increased vesical pressure of the quiescent bladder and **C,D** inhibited the micturition contraction of the expanded bladder in chloralose/urethane-anesthetized cats. **A** Effects on the quiescent bladder of stimulating the SP (*upper trace*) and PDP (*lower trace*) afferent nerve at various intensities. The strength of stimulation of nerves is given relative to the threshold of group I (muscle) or group II (cutaneous) afferents, which is expressed as 1.0T. **B** Relations between intensity of electrical stimulation of the SP nerve (*abscissa*) and amplitude of the vesical reflex responses (*ordinate*) in a quiescent bladder filled with various volumes of water. **C** Effects on the rhythmic micturition contractions of stimulating the SP nerve at various intensities. **D** Effects on the rhythmic micturition contractions of stimulating the GS nerve, with various volumes in an intravesical balloon. (Modified from Sato et al. 1979a)

nerves. The stimulus thresholds for the reflex contractions were in the range of those for group III fibers regardless of the intravesical balloon volume. However, the reflex increases in vesical tonus evoked by the additional group IV afferent stimulation were stronger when the balloon volume was greater (Fig. 70B). The optimal volume for the quiescent vesical condition varied from animal to animal as well as at different times in the same animal. Superficial peroneal afferent stimulation was more effective than sural and peroneal and deep peroneal afferent stimulation in producing reflex increases in vesical tonus. Gastrocnemius and soleus afferent stimulation was always ineffective in this respect.

During rhythmic micturition contractions, stimulus effects were inhibitory (Fig. 70C), reflexly attenuating the rhythmic efferent burst discharges of the pelvic nerves. Sometimes there were transient increases in the magnitude of the micturition contraction at the beginning of afferent nerve stimulation. It was often noted that the contraction immediately after stimulus cessation was larger than the prestimulus control contractions. Stimulation of group IV afferents in addition to group III afferents was always more effective in producing vesical reflexes than stimulation of group III afferents alone. No effective contribution of the group I and II somatic afferents to these vesical functions was observed. The inhibitory effects of superficial peroneal, gastrocnemius and soleus, and peroneal and deep peroneal afferent stimulation were about the same, while the inhibitory effects of sural afferent stimulation were less pronounced. Volume-dependence of inhibition of vesical contractions by hindlimb afferent stimulation was also observed. When the volume was small, the inhibitory effect was very pronounced. However, when the volume was increased stepwise, the rhythm of the micturition contractions became faster and inhibitory effects progressively decreased (Fig. 70D). However, optimal volumes for producing the micturition contractions were variable in different animals as well as time-dependent in the same animal.

In anesthetized 2- to 19-month-old chronic spinal cats, the excitatory and inhibitory reflex responses of the bladder were essentially the same as those obtained in CNS-intact cats (Sato et al. 1979b, A. Sato et al. 1983). Volleys in group III and IV fibers produced transient increases both in intravesical pressure and in the efferent discharges of pelvic nerves of the quiescent bladder (Fig. 71A). The same volleys inhibited the rhythmic micturition contractions and associated pelvic nerve discharges of the expanded bladder (Fig. 71B). The magnitude of the excitatory somatovesical reflexes appeared to be larger in spinal than in intact cats. For example, in the CNS-intact animals, gastrocnemius and soleus nerve stimulation was almost ineffective (Sato et al. 1980), whereas in the chronic spinal animals the same stimulation produced a definite excitatory response. In addition, in spinal animals the excitatory effect of superficial peroneal nerve and sural nerve stimulation was stronger than that in CNS-intact animals. These results suggest that in normal cats, the exci-

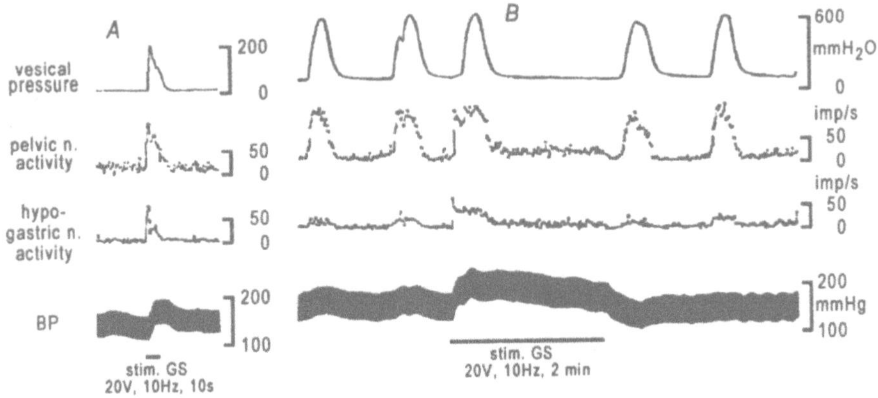

Fig. 71A,B. Effects of electrical stimulation of hindlimb muscle nerve (nerve to gastrocnemius and soleus muscles, *GS*) on bladder function in chronic spinal cats anesthetized with chloralose-urethane. Repetitive electrical stimulation of the GS nerve at group IV stimulus strength **A** produced an excitatory reflex response in the quiescent bladder and **B** an inhibitory reflex effect on the micturition contractions of the expanded bladder. Vesical pressure, pelvic and hypogastric efferent nerve activity, and blood pressure (*BP*) were recorded simultaneously. (From Sato et al. 1979b)

tatory spinal reflex from the hindlimb somatic afferents is depressed by descending activity from supraspinal structures. Alternatively, spinal transection increases the reflex excitability of the spinal pathways of the somato-vesical reflexes.

The inhibition of rhythmic micturition contractions by hindlimb afferent volleys was observed clearly and consistently in the 9- to 19-month-old chronic spinal cats, while the responses were inconsistent and very weak in the 2- to 5-month-old chronic spinal cats. This indicates that the inhibitory effect of hindlimb somatic afferent stimulation on the pelvic rhythmic burst discharges related to micturition contractions can be organized within the spinal cord after it has been separated from the higher CNS for a long time.

4.3.6 Central Processing of Somato-vesical Reflexes

The effects of afferent volleys in hindlimb cutaneous (sural, superficial peroneal) and muscle (gastrocnemius and soleus) nerves on the discharges in pelvic nerves to the bladder were measured in anesthetized CNS-intact

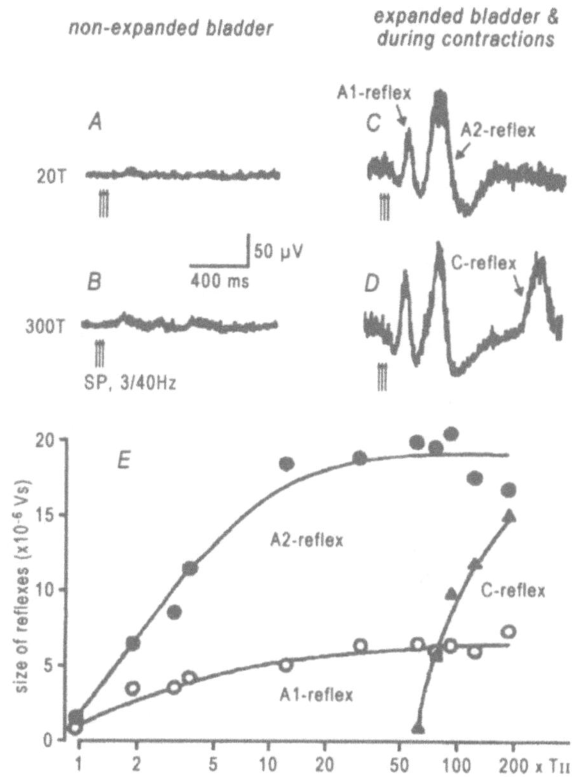

Fig. 72A–E. Reflex discharges evoked in vesical branches of the pelvic nerve by electrical stimulation of superficial peroneal nerve (*SP*) in a CNS-intact cat anesthetized with chloralose-urethane. **A,B** Only marginal reflex discharges were observed when the bladder was quiescent and empty. **C,D** Three distinct reflex discharges (A1, A2, and C reflexes) were observed when the bladder was expanded and showed rhythmic micturition contractions. **E** Strength–response curves. *Abscissa*, stimulus intensity of SP; *ordinate*, magnitude of reflex discharge. A1 and A2 reflexes (*open* and *closed circles*) were evoked by group II and III afferent volleys, and C reflex (*closed triangles*) was evoked by group IV afferent volleys. (From A. Sato et al. 1983)

and 2- to 19-month-old chronic spinal cats (A. Sato et al. 1983). In CNS-intact cats, single or short tetanic volleys induced a reflex discharge in pelvic vesical nerve branches with three distinct components. These reflexes could be observed during micturition contractions; not markedly between the contractions or when the bladder was empty and quiet (Fig. 72). The latencies of the three components were 90 ms (A1 reflex),

320 ms (A2 reflex), and 770 ms (C reflex). The two early components (A1 and A2 reflexes) were evoked by volleys in both group II and III hindlimb afferents. Separation of A1 and A2 reflexes was not due to the different afferent conduction velocities of the group II and III fibers, but appeared to be due to separation in the central reflex pathways. The late component (C reflex) was induced by group IV volleys. The inhibition of spontaneous efferent discharges appeared first about 50 ms after the onset of stimulation, and lasted until the A1 reflex was elicited. This inhibition was independent of the subsequent A1 reflex excitation, and can be called preexcitatory inhibition. The inhibition appeared again between the A1 and A2 reflexes, and immediately after the A2 reflex, and this inhibition appeared to represent postexcitatory depression. In chronic spinal cats, stimulation of group II and III hindlimb afferents induced an A reflex (latency, 90 ms), and stimulation of group IV hindlimb afferents induced a C reflex (latency, 340 ms).

The A1 reflex, with a latency of about 90 ms in CNS-intact anesthetized cats, corresponds to the reflex discharges recorded in the pelvic vesical branches after stimulation of myelinated afferents in the pelvic or pudendal nerves (De Groat and Ryall 1969; McMahon and Morrison 1982c). De Groat and Ryall (1969) reported that the pelvic nerve reflexes recorded about 100 ms after the stimulation of pelvic afferents in anesthetized cats were of supraspinal origin, because they were absent in acute or chronic spinal animals. In chronic spinal animals, an A reflex with a latency of about 90 ms was regularly induced by hindlimb afferent volleys (A. Sato et al. 1983). This finding suggests that the A1 reflex in the pelvic vesical branches of CNS-intact cats may include a spinal component. A2 reflexes in pelvic vesical branches with a latency of about 300 ms in CNS-intact cats appeared to be of supraspinal origin, since no reflex component with a similar latency was observed in chronic spinal cats. The magnitude of the A2 reflex usually did not increase when group IV afferent activity was included in the centripetal volley. Therefore, any contribution of the hindlimb group IV afferents to the A2 reflex seems negligible. However, the possibility could not be excluded that the spinal C reflex with a latency of about 340 ms overlaps somewhat with the A2 reflex. McMahon and Morrison (1982c) recorded both early and late waves from the pelvic nerve after stimulation of the ventrolateral part of the cervical cord. The latency of the early wave was 39 ms, while that of the late wave was 250 ms. The A2 reflex may use a slow descending tract which is related to the late wave, and the A1 reflex may use a fast descending pathway related to the early wave of McMahon and Morrison (1982c).

Bradley and Teague (1968) reported that stimulation of the pudendal nerve produced a supraspinal reflex potential with a latency of 150–200 ms. This latency is longer than that of the A1 reflex and shorter than that of the A2 reflex. De Groat et al. (1981) recorded a late reflex with a latency of 180–200 ms in the pelvic nerve after stimulation of unmyelinated pelvic afferents in CNS-intact cats. In chronic spinal cats, they recorded similar reflexes of a slightly shorter latency (150–180 ms). They suggested that the late reflex in CNS-intact cats was of spinal origin, mainly because its latency was nearly as short as that of the C reflex in spinal cats. In the experiments by A. Sato et al. (1983), the C reflex of CNS-intact cats had a latency of about 770 ms, whereas that in chronic spinal cats appeared after 340 ms. This latency difference of more than 400 ms makes it likely that the late C reflex induced in normal cats by hindlimb group IV volleys takes a supraspinal course. In the CNS-intact cats, the spinal C reflex was not seen. Either it was absent or it overlapped with the A2 reflex.

The effects of various somatic afferent volleys on the discharges in pelvic nerves to the bladder were measured in anesthetized CNS-intact and acute spinal rats (Morrison et al. 1996b). In CNS-intact rats, stimulation of segmental inputs to the sixth lumbar (L6) and first sacral (S1) segments from the perineo-femoral branch of a pudendal nerve had excitatory A- and C-reflex discharges if the bladder pressure was low i.e., when there was almost no background resting activity in the parasympathetic efferents (Fig. 73A,B, labeled as Ae and Ce). When the bladder was distended, i.e., when there were rhythmic burst discharges in efferent pelvic nerves corresponding to the vesical micturition contractions, the same input evoked additional inhibitory effects of efferent discharges (Fig. 73E,F, labeled as Ai and Ci). Acute spinal transection resulted in the abolition of micturition contractions. Electrical excitation of A or A and C afferents in the perineo-femoral branch of a pudendal nerve elicited excitatory A- and C-reflex discharges in the pelvic nerve which were of the same latency as those observed before the spinal transection, but were larger and of longer duration (Fig. 73C,D,G,H).

In CNS-intact rats, stimulation of the tibial nerve caused no effect on pelvic postganglionic fibers if the bladder pressure was low (Fig. 74A,B). When the bladder was expanded, electrical stimulation of the same afferent nerve produced early reflex inhibition (Ai) and excitation (Ae) after A-fiber activation, and late reflex inhibition (Ci) and excitation (Ce) of efferent

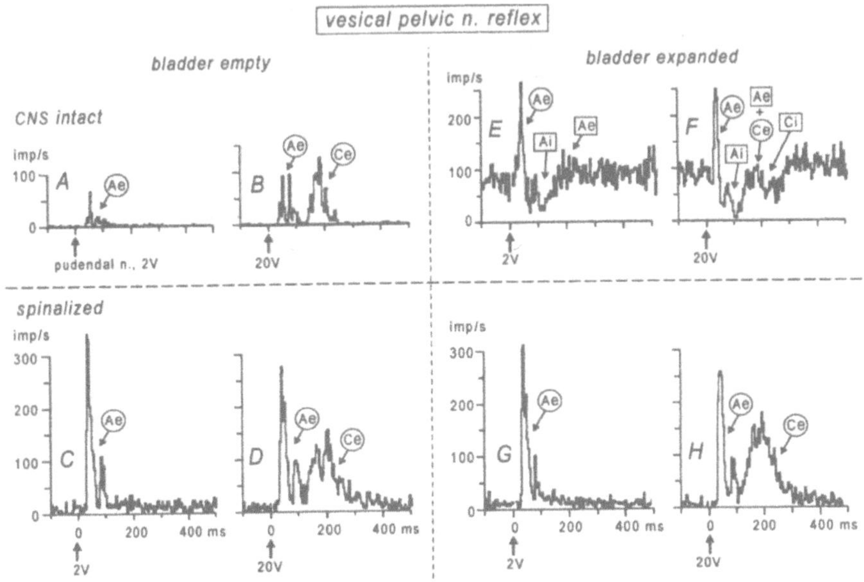

Fig. 73A–H. Poststimulus-time histograms (PSTH) of pelvic efferent nerve activity produced by a single electrical stimulation of a perineo-femoral branch of a pudendal nerve in urethane-anesthetized rats **A,B,E,F** before and **C,D,G,H** after spinal transection (C1 level). Stimulus intensities are indicated: 2 V was subthreshold for the group IV fibers, while 20 V was supramaximal for the group IV fibers. **A–D** Quiescent, empty bladder. **E–H** Expanded bladder. Single-shock stimulation produced excitatory A- and C-reflex discharge components of spinal origin (labeled as *Ae* and *Ce*, enclosed by *circles*) in the pelvic efferent nerve when the bladder was not expanded. When the bladder was expanded, there were additional inhibitory components of both A and C reflexes (labeled as *Ai* and *Ci*, enclosed by *squares*). These inhibitory components were of supraspinal origin. (From Morrison et al. 1996b)

pelvic discharges (Fig. 74E,F) after additional C-fiber activation. The Ai and Ae responses represent the A reflexes, while the Ci and Ce responses represent the C reflexes. All of these reflex responses were easily demonstrable in the vesical branches of the pelvic nerve when pressure in the bladder was elevated nearly to the point at which vesical micturition contractions appeared, and the response magnitudes were dependent on bladder pressure over the range of 100–500 mmH_2O. After acute spinal transection, only marginal reflexes were observed in same rats (Fig. 74C,D,G,H).

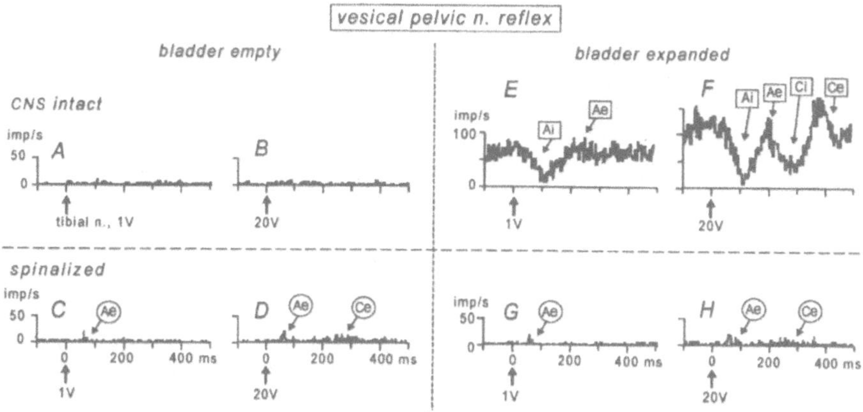

Fig. 74A–H. Poststimulus-time histograms (PSTH) of pelvic efferent nerve activity produced by a single electrical stimulation of a tibial nerve in urethane-anesthetized rats **A,B,E,F** before and **C,D,G,H** after spinal transection (C1 level). Stimulus intensities are indicated: 1 V was subthreshold for the group IV fibers, while 20 V was supramaximal for the group IV fibers. **A–D** Quiescent, empty bladder. **E–H** Expanded bladder. Single-shock stimulation caused no effect when the bladder pressure was not expanded, but when the bladder was distended, it produced early A-reflex inhibition (preexcitatory) and excitation (labeled *Ai* and *Ae*, enclosed by *squares*), and late C-reflex inhibition (preexcitatory) and excitation of efferent pelvic discharges (labeled *Ci* and *Ce*, enclosed by *squares*). All of these reflexes were abolished following spinalization. (From Morrison et al. 1996b)

4.3.7
Conclusion

1. In anesthetized animals, it was demonstrated that natural somatic afferent stimulation applied particularly on the perineal area changes bladder function as a consequence of reflex changes in parasympathetic efferent activity. It is characteristic that the response is strongly organized at the spinal segmental level, and the response can be changed from excitatory to inhibitory by filling the bladder. This switch of the bladder response can be explained by the interaction of somatic afferent and pelvic afferent excitation at the sacral cord. When the bladder is not filled, somatic afferent fibers in the sacral area excite parasympathetic pelvic efferent nerves at the sacral cord, irrespective of the connection to the brain. When the bladder is filled, information of bladder expansion is carried by pelvic afferents to the sacral spinal cord, and

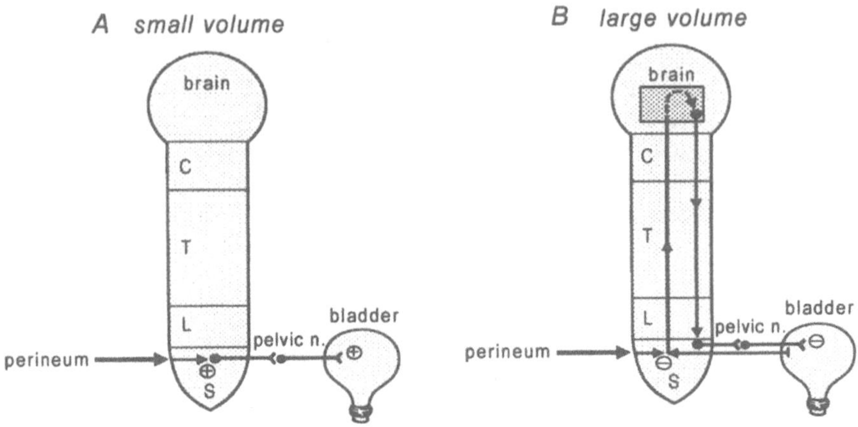

Fig. 75A,B. Reflex pathway for the urinary bladder **A** in the quiescent condition (small volume) and **B** showing rhythmic micturition contractions (large volume). **A** Pinching of perineal skin areas produces an excitatory effect on the quiescent bladder through excitation of the pelvic parasympathetic nerve activities via a spinal reflex pathway. **B** Pinching of perineal skin areas produces an inhibitory effect on micturition contractions of the expanded bladder through inhibition of the pelvic parasympathetic nerve activities. Sacral somatic sensory afferent stimulation seems to inhibit the pelvic afferent transmission at the spinal level. +, excitatory effect; –, inhibitory effect

ascends to the micturition center in the brain stem; the signal for micturition then descends the spinal cord to the sacral level and excites parasympathetic pelvic efferent neurons innervating the bladder. Sacral somatic sensory afferent stimulation seems to inhibit pelvic afferent transmission at the spinal level, because even in chronic spinal preparations, function of the expanded bladder is inhibited by sacral somatic afferent stimulation (Fig. 75).

2. When the bladder is not expanded, a single electrical shock to the sacral somatic afferent nerve produces only excitatory reflex discharges, which are divided into A- and C-reflex discharge components. In the expanded bladder, pelvic efferent discharges show rhythmic activity, and a single shock to the sacral somatic afferents produces inhibitory and excitatory reflex discharge components. Accumulation of the inhibitory components during natural sacral somatic stimulation may explain the inhibitory response of the filled bladder.
3. These excitatory and inhibitory bladder responses were also associated with sphincter muscle activity. All of the somatically induced bladder

and sphincter muscle reflex responses appear to be applicable to clinical use for modulation or therapy of micturition, by stimulating the skin and muscle or joints whose afferents enter the spinal cord at the sacral segments. Effective techniques to excite the relevant afferent fibers and elicit these responses have yet to be validated.

4.4 Somatosensory Modulation of the Sudomotor System

4.4.1 Introduction

Thermal sweating originates with excitation of warm sensory receptors in the skin and their afferent fibers. This afferent information is sent in part to the thermoregulatory center in the hypothalamus, where the warm sensation is integrated and processed, engaging the autonomic and somatic nervous systems, and the endocrine system to prevent an increase in body temperature. The autonomic component of this regulation includes inactivation of cutaneous sympathetic vasoconstrictor fibers and activation of cutaneous sympathetic sudomotor fibers. This regulation is a typical cutaneo-sympathetic reflex. A directly opposite response occurs when cold receptors in the skin are stimulated.

Kuno (1956) classified human perspiration into "thermal" and "mental" sweating. Thermal sweating occurs over the body surface when body temperature exceeds certain values, and is also seen during moderate and severe exercise, but does not affect sweat glands of the palms of the hands and soles of the feet. In conscious humans and some animals, sweat glands in the palms and soles sweat even in the absence of thermal stimulation, and mental excitement increases the rate of sweating with a short latency in the order of seconds. This latter type of sweating is called mental sweating.

Under conditions of moderate thermal sweating, gentle pressure upon certain localized skin areas evokes sweating in some regions and depresses that in others (Takagi and Sakurai 1950). For example, application of pressure on the right side of the chest causes a marked increase in sweating on the left half of the body, and equally inhibits sweating on the right side of the body. This evidence indicates that pressure sensory information delivered unilaterally to a restricted skin area depresses activity of sympathetic sudomotor nerves on the ipsilateral side, and facilitates activity of sudomotor nerves on the contralateral side. Which parts of the CNS, i.e., spinal cord or supraspinal structures, are responsible for these excitatory and inhibitory responses has not yet been clarified.

In conscious humans, Hagbarth and Vallbo (1968) recorded bursts of postganglionic sympathetic mass activity using microneurography. Single-unit activity in sympathetic nerve fibers was later recorded from intact cutaneous nerves (median or peroneal nerve) in alert humans (Hallin and Trebjörk 1974). Touching the skin could induce reflex discharges. Painful or intense stimuli were not necessary to induce reflex activity. Some of the units probably had a sudomotor function, since the unit activity correlated well with changes in galvanic skin resistance recorded within the innervation zone of the explored nerve fascicle.

Many neurophysiological studies have been carried out on cats. The cat has sweat glands only in the footpads. The sweat gland activity resembles human mental sweating, but not human thermal sweating. Activation of sweat glands produces changes in both the voltage and impedance of the skin (Wang 1957, 1958). Reflex changes in skin potential or impedance produced by various stimuli, including emotional ones, have been called the "galvanic skin reflex" (GSR) (Gildemeister 1928; Wang 1957, 1958) or "electrodermal reflex" (EDR) (Wang 1964). In the cat, it is known that the EDR consists of simple monophasic waves which can be quantitatively analyzed, and many studies of the central mechanisms of the sympathetic nervous system have been made using the EDR as an indicator. Wang and his colleagues did extensive work on the EDR evoked by somatic afferent stimulation. Their conclusions about the neural control mechanisms of the EDR can be summarized as follows: (a) only excitatory EDRs are produced at the spinal level and (b) the spinal excitatory center for the EDR is controlled by excitatory and inhibitory centers in the brain (see Wang 1964). Ito et al. (1978) found that not only excitatory but also inhibitory EDRs were produced in acute spinal cats, and that occurrence of the excitatory and inhibitory EDRs depended on both the segmental position and the laterality of the stimulated area.

4.4.2
Somatosensory Modulation of the Sweat Gland

In cats anesthetized with chloralose, electrodermal activity was recorded from the central footpads of the forelimb and hindlimb, and electrodermal reflexes were evoked by stimulating the cutaneous (sural nerve, superficial peroneal nerve) and muscle (peroneal and deep peroneal nerve, gastrocnemius and soleus nerve) afferent nerves of the hindlimb (Karl et al. 1975). Single afferent volleys in cutaneous nerves evoked by stimulus strengths of less than two times threshold (2T) were sufficient to elicit electrodermal reflexes, and almost maximal reflex responses could be obtained at 2–5T, i.e., stimulus strengths at which almost only group II (or Aβ) afferent fibers were excited. At stimulus strengths extending into the range of group III (or Aδ) and IV (or C) fiber stimulation, no regular increase of the electrodermal reflex has been found. Low-threshold muscle afferents (group I) did not produce electrodermal reflexes. The stimulus strength had to be extended well into the group II fiber range before electrodermal reflexes were recorded. The addition of group III fiber stimulation definitely increased the reflex further, but only small increases of the electrodermal reflex were observed when group IV fibers were stimulated.

Bernthal and Koss (1984) found that transection of the cervical spinal cord in unanesthetized decerebrated cats greatly increased the amplitude of the sympathetic-cholinergic electrodermal reflex, and that systemic administration of the α_2-adrenoceptor antagonist, yohimbine hydrochloride, also greatly facilitated the EDR evoked from intact as well as decerebrate preparations. Depletion of monoamines in the CNS by pretreatment with reserpine and α-methyl-*p*-tyrosine reduced the concentrations of noradrenaline, dopamine and serotonin in the thoracic spinal cord. In monoamine-depleted preparations, yohimbine no longer facilitated the reflex amplitude, whereas the effect of spinal transection was not altered. These results suggest that there are two distinct sympathoinhibitory systems in the lower brain stem that converge on spinal sympathetic neurons; one system is monoaminergic and one is not (Bernthal and Koss 1984).

When attempts were made in nonanesthetized acute spinal cats to elicit EDRs by mechanically stimulating sensory receptors of the skin and also by electrically stimulating cutaneous afferent nerves in the trunk and limbs, both inhibitory and excitatory EDRs were observed. It was found that in the spinal cord, the EDR was either inhibitory or excitatory according to the level of the spinal segment stimulated and the laterality of the stimulation with respect to the recording site (Ito et al. 1978). Excitatory EDRs in the forepaw were produced more frequently from the caudal thoracic, lumbar and sacral skin than from the cervical or rostral and midthoracic skin. Inhibitory EDRs were produced much less frequently than excitatory ones. Excitatory EDRs in the hindpaw were produced most frequently from the rostral and midthoracic skin. Pinching of the caudal thoracic and lumbar skin elicited inhibitory EDRs. Excitatory EDRs were produced more often by stimulating the skin area contralateral to the paw being observed, whereas there was a tendency for inhibitory EDRs to result from ipsilateral stimulation. Excitatory and inhibitory EDRs similar to those produced by pinching the skin were also evoked by electrical stimulation of the lateral cutaneous nerves of the sixth thoracic to fourth lumbar (T6-L4) spinal segments (Fig. 76). Generally speaking, there is a tendency for the EDR to be inhibited when the afferent input enters

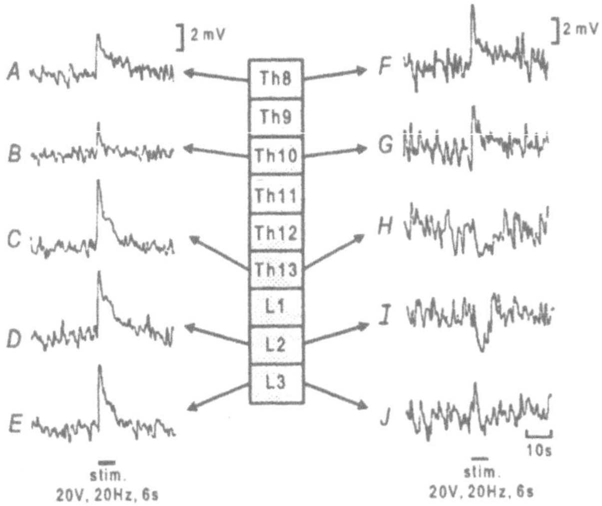

Fig. 76A–J. Excitatory and inhibitory electrodermal reflexes (EDRs) produced by electrical stimulation of cutaneous afferent nerves at indicated spinal levels in a nonanesthetized acute spinal cat (C1–2 level). EDRs recorded form **A–E** left forepaw and **F–J** right hindpaw. Right lateral cutaneous branches of the T8, T10, T13, L2, or L3 spinal nerves were stimulated with 20 V (supramaximal for all nerve fibers). The EDR recorded from the hindpaw was either excitatory or inhibitory, depending on the segmental level stimulated. Though the EDRs recorded from the forepaw were all excitatory in this case, their magnitude was also a function of the segmental level stimulated. (From Ito et al. 1978)

spinal segments near the outflow of the sudomotor sympathetic nerve fibers to the corresponding paw, whereas excitation occurs when the input is to distant segments. Interestingly, an inhibitory EDR is produced more frequently by ipsilateral than by contralateral cutaneous stimulation. It seems reasonable to assume that this phenomenon results from a mechanism similar to the one which produces the hemihidrosis discovered by Takagi and Sakurai (1950) in human beings. Hemihidrosis in humans is a condition in which there is excessive perspiration contralateral to the side pressed, and reduced perspiration ipsilateral to the side pressed. As both the inhibitory and excitatory EDRs in spinalized cats disappeared after intravenous (i.v.) administration of atropine or after denervation of the sudomotor nerves, these responses would seem to be produced by changes in sudomotor activity at the spinal level. It is especially interesting to find both excitatory and inhibitory autonomic reflex mechanisms at the spinal level.

In nonanesthetized acute spinal cats, maximal inhibitory EDRs were elicited by stimulation of the group II afferent fibers of the cutaneous nerves only, whereas maximal excitatory EDRs were elicited after stimulation of the group II, III and IV afferent fibers of cutaneous origin (Ito et al. 1978) (Fig. 77). As mentioned above, in CNS-intact cats anesthetized with chloralose, electrical stimulation of the group II cutaneous afferents of hindlimb nerves was noted to be sufficient to produce the maximal excitatory EDR with no inhibitory EDRs (Karl et al. 1975). One reason why inhibition was not seen in the experiments of Karl et al. (1975) is apparently because only limb afferents, and not trunk afferents, were stimulated.

The EDRs recorded from the central pads on the hind- and forepaws were also elicited by innocuous and noxious stimulation of skin in ketamine-anesthetized CNS-intact cats (Jänig and Räth 1977). Stimuli which excite the Pacinian corpuscles in the paws (air jet stimuli applied to the paw, vibrational stimuli produced by tapping on the experimental frame) and the cutaneous nociceptors (mechanical and thermal noxious stimuli) can elicit EDRs. Stimuli exciting hair follicle receptors on the trunk or legs or slowly adapting receptors in the feet were noted to be ineffective. Neurophysiological investigations of postganglionic neurons in the medial planter nerve revealed two populations of postganglionic neurons (Jänig and Spilok 1978). Type 1 neurons were excited by noxious cutaneous stimuli and by vibrational stimuli and were classified as sudomotor neurons. Type 2 neurons were inhibited or excited by noxious

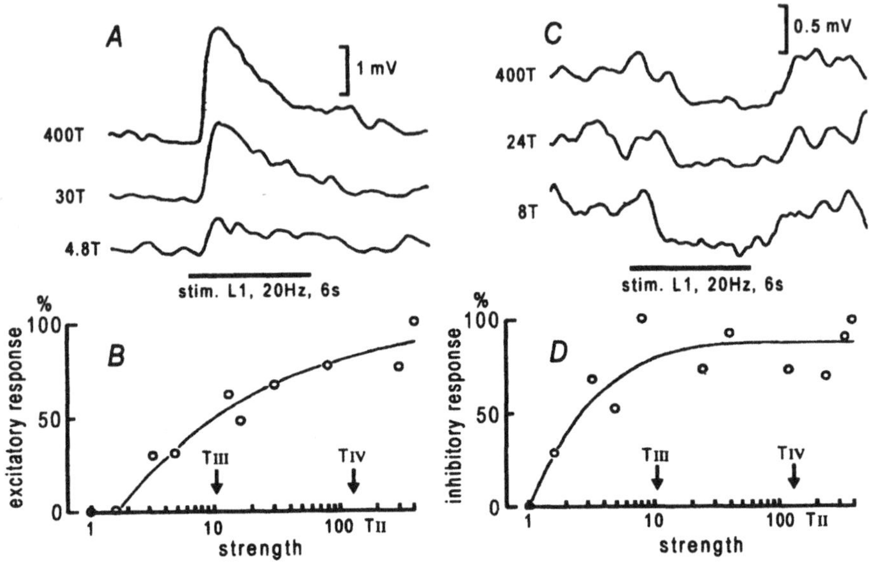

Fig. 77A–D. Relation between electrodermal reflex (EDR) magnitude and intensity of electrical stimulation of cutaneous nerve in a spinal cat. Excitatory and inhibitory EDRs recorded from **A,B** right forepaw and **C,D** left hindpaw. Right lateral cutaneous nerve branch of the L1 spinal nerve was stimulated. The stimulus intensity was expressed as multiples of threshold intensity for group II fibers. Both excitatory and inhibitory EDRs were produced by stimulation of group II fibers. The excitatory EDR continued to increase with increases in stimulus intensity beyond the threshold intensity for the group IV fibers, while the inhibitory EDR reached its maximum before the stimulus intensity reached the threshold for the group III fibers (From Ito et al. 1978)

stimuli of the hindfoot, but not affected by vibrational stimuli, and were classified as being most likely vasoconstrictor neurons.

Saper and DeMarchena (1986) reported the case of a cervical spinal cord-injured man who developed unilateral pupillary dilatation and sweating from the head to the midthoracic region as a response of rib fractures. This regional sympathetic reflex response was similar in many ways to that seen in spinally transected animals. Regional sympathetic reflex responses of the sudomotor system may provide clinically useful signs of an otherwise painful and perhaps dangerous condition which is located below the analgesic level in spinal cord-injured patients.

4.4.3
Conclusion

The existence of the somato-sudomotor reflex is well established. The spinal component of the somato-sudomotor reflex has been clearly demonstrated, and both excitatory and inhibitory spinal reflex components manifest from within the spinal cord depending upon the segmental level of the skin which is stimulated.

4.5
Somatosensory Modulation of Hormonal Secretion

4.5.1
Introduction

Visceral functions, generally speaking, are influenced not only by the autonomic nerves but also by various hormones. Although many studies have examined the effect of somatic afferent stimulation on autonomic nerve activity, the possibility of somato-hormonal secretion reflexes remains a relatively untested hypothesis.

The mechanisms controlling the secretion of hormones are diverse, but can be classified into the following three main types:

1. Neural regulation of hormonal secretion by autonomic nerves, e.g., secretion of catecholamines from the adrenal medulla, and secretion of insulin and glucagon from the pancreas.
2. The hypothalamo-anterior pituitary hormone system. This system can be further divided into the following two types:
   a) Hormones secreted from the anterior pituitary lobe which act on the peripheral endocrine glands and regulate their secretion, e.g., adrenal cortical hormones, sex hormones, thyroid hormones, etc.
   b) Anterior pituitary hormones such as growth hormone and prolactin, which can act directly on effector organs.
3. The posterior pituitary hormone system; neurosecretory cells originating in the hypothalamus and projecting their axons to the posterior pituitary lobe to secrete oxytocin and vasopressin. These neurons are sometimes compared to the parasympathetic neurons (see Fulton 1949).

Although the importance of negative feedback for control of hormonal secretion has been emphasized, it must be kept in mind that all of these hormonal secretions are integrated in the brain. As some somatic afferent information also arrives in the area of the brain which integrates hormonal secretion, it must be recognized that hormonal secretion is influenced by somatic afferent stimulation. In this text, the effects of somatic sensory regulation on various hormonal secretory functions will be divided into the three main themes: (1) the adrenal medullary and pancreatic hormones, (2) the hypothalamo-anterior pituitary hormone system, and (3) the posterior pituitary hormone system.

4.5.2 Adrenal Medullary and Pancreatic Hormones

The first extensive physiological study of the effects of somatic afferent stimulation on hormonal function was initiated early in this century by W.B. Cannon, who produced evidence that adrenal medullary secretion was stimulated by emotional excitement, by pain, and by asphyxia (Cannon and Hoskins 1912; Cannon 1919). He observed that stimulation of sensory fibers in one of the larger nerve trunks resulted in discharges along sympathetic paths. Such stimulation thus caused dilatation of the pupils, inhibition of gastric peristalsis and gastric secretion, and contraction of arterioles. Since the adrenal glands are activated by sympathetic impulses, it seemed possible that they would be affected like some of the other organs innervated by the sympathetic system, and secrete in greater abundance when sensory nerves were irritated (Cannon and Hoskins 1912). Increased adrenal medullary secretion was considered to be part of the sympathetic discharge provoked in emergency situations. Cannon called this the "emergency function of the sympatho-adrenal system." In addition to the "emergency function" of the adrenal medulla, it was shown that peripheral sensory stimulation could increase adrenaline secretion in the absence of emotional excitement. Hartman and Hartman (1923) showed that cooling the skin of cats with ice could increase adrenaline output independent of excitement and struggling. Later, with the use of radioenzymatic assays, it became possible to measure separately the concentrations of adrenaline and noradrenaline in small samples of plasma. It has been shown that even handling of rats produces moderate increases in plasma catecholamine levels, while immobilization and decapitation produce much greater increases (Popper et al. 1977; Kvetnansky et al. 1977).

The precise contribution of the sympathetic nervous system to the somatosensory regulation of adrenal medullary hormonal secretion has been clarified by parallel recording of adrenal efferent nerve activity and secretion of catecholamines from adrenal glands in anesthetized animals (Araki et al. 1981, 1984; Kurosawa et al. 1985; Sato et al. 1986a, see review Sato 1987). They used spectrofluorometry at beginning and then electrochemical detector for measuring a small amounts of catecholamines after separating them using high performance liquid chromatography. Their approach using anesthetized animals eliminated emotional factors, and has made it possible to correlate responses of nerve activity and catecholamine secretion after somatic afferent stimulation. It has been shown that adrenal sympathetic efferent nerve activity and catecholamine secretion are changed by innocuous and noxious cutaneous stimulation in the anesthetized rat, and that these changes in adrenal catecholamine secretion result from changes in adrenal sympathetic nerve activity. It is noteworthy that innocuous mechanical stimulation of the skin produced a decrease, while noxious mechanical stimulation of the skin produced an increase in sympatho-adrenal medullary function in anesthetized rats. The study was also extended to examine the effect of afferent stimuli to muscles and joints in anesthetized cats (Sato et al. 1986a).

The possibility of somatic modulation of pancreatic hormones via autonomic nerves was also tested in anesthetized animals (Uvnäs-Moberg et al. 1986; Kurosawa et al. 1994). Excitation of cutaneous nociceptive afferents from various spinal segments could regulate glucagon secretion from the pancreas as a reflex response whose efferent limb was composed of both sympathetic and parasympathetic nerves (Kurosawa et al. 1994).

4.5.2.1 Adrenal Medullary Hormones

The adrenal medulla, derived from neuroectodermal cells from the neural crest, synthesizes and secretes catecholamines. The cells of the adrenal medulla are homologous to postganglionic neurons. They are directly innervated by preganglionic sympathetic nerve fibers. The cells of the adrenal gland, when activated, release the catecholamines of adrenaline and noradrenaline into the circulatory system (adrenaline-synthesizing cells represent 80% of the cells in the adrenal gland in humans and rats).

4.5.2.1.1 Cutaneous Stimulation

Innocuous Mechanical Stimulation. Araki et al. (1984) attempted to correlate adrenal nerve activity with adrenaline and noradrenaline secretion rates from adrenal glands. Mass discharge activity of efferent nerves was recorded from the severed central segment of the adrenal nerve (Fig. 78A). To measure the secretion rates of both catecholamines from the adrenal gland, adrenal venous blood samples were collected directly from the adrenal vein (Fig. 78B). The catecholamines were separated by high performance liquid chromatography, and measured either by an electrochemical detector or by spectrofluorometry. In CNS-intact rats, brushing of the lower chest skin produced a reflex inhibition of both nerve activity and catecholamine secretion limited to the stimulation period (Fig. 78C,D). Some slight increase in nerve activity and secretion rates followed cessation of the stimulation as a rebound response. The decreases in nerve activity, adrenaline and noradrenaline secretion rates reached 74%, 73%, and 78% of control levels, respectively, during the stimulation period. Brushing of the hindlimb also caused similar reflex decreases in nerve activity, as well as in adrenaline and noradrenaline secretion rates.

These results demonstrate that cutaneous stimulation causes reflex responses in adrenal nerve activity and catecholamine secretion which are proportional in magnitude and duration. After severing the sympathetic connection which innervates the adrenal gland, spontaneous secretion rates of both catecholamines decreased significantly. Following such denervation, lower chest brushing did not produce any significant change in secretion. The evidence from nerve recording and severing suggests that the cutaneously evoked reflex response of adrenal nerve activity is responsible for the cutaneous stimulation-induced catecholamine secretion response.After acute spinal transection at the first to second cervical (C1-2) level, similar reductions in adrenal sympathetic nerve activity and catecholamine secretion were noted, suggesting that some tonic excitatory supraspinal control of adrenal medullary function exists in CNS-intact animals. Contrary to the inhibition observed in CNS-intact animals, in spinalized animals brushing of the lower chest caused reflex increases in nerve activity and adrenaline and noradrenaline secretion rates during the stimulation period (Fig. 78E,F).

When the effects of brushing various segmental skin areas (forehead, cheek, neck, lower chest, abdomen and thigh) on adrenal sympathetic

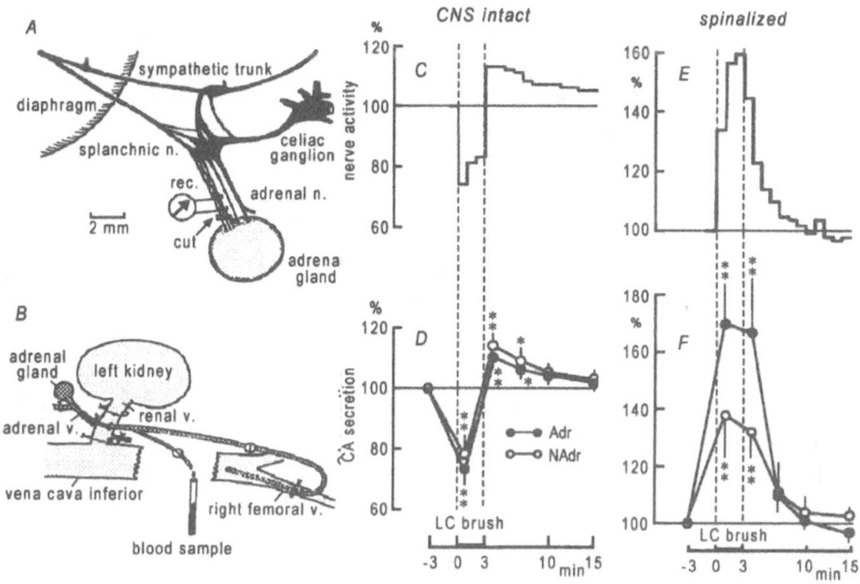

Fig. 78A–F. Effect of cutaneous brushing on adrenal efferent nerve activity and catecholamine secretion rate. **A** Retroperitoneal approach of the anatomical arrangement of adrenal nerves on the left side of a rat. Efferent nerve mass discharge activity was recorded from the central cut segment of the nerve with bipolar platinum-iridium wire electrodes. **B** Adrenal venous blood sampling procedure as viewed from a ventral surgical approach. One end of a polyethylene catheter fitted with a Y connector is inserted into the left adrenal vein. Blood samples are removed from one end of the connector. The remaining end is connected to a silicon tube inserted into the right femoral vein. During collection of blood samples, the silicon tube is closed. **C,D** CNS-intact and **E,F** spinalized (C1-2 level) rats anesthetized with chloralose and urethane. **C–F** Brushing the lower chest (*LC*) skin decreased the adrenal nerve activity, and secretion rates of adrenaline (*filled circles*) and noradrenaline (*open circles*) in CNS-intact rats, but increased these variables in spinalized rats. $^{*}P<0.05$, $^{**}P<0.01$; significantly different from prestimulus control values. (Modified from Araki et al. 1984)

efferent nerve activity were examined in anesthetized CNS-intact rats, brushing of all areas produced a similar reflex decrease in the efferent nerve activity (Kurosawa et al. 1982; Tsuchiya et al. 1991a). In spinalized rats, brushing the lower chest or abdominal skin area produced an increased response, while brushing the neck or thigh skin area produced no significant changes in the efferent nerve activity (Kurosawa et al. 1982). These results imply that some segmental organization of the excitatory reflex may exist in spinalized animals, and that the segmental organization of these excitatory reflexes within the spinal cord is modified into a non-segmental or generalized inhibitory response by supraspinal structures.

Brushing did not produce any significant responses in anesthetized cats (Sato et al. 1986a).

Noxious Mechanical Stimulation. Pinching of the skin produced a reflex increase in adrenal sympathetic nerve activity in anesthetized rats (Araki et al. 1980, 1981, 1984; Togashi et al. 1987; Tsuchiya et al. 1991a) and cats (Sato et al. 1986a). The duration and the magnitude of the response varied depending upon the site stimulated (Araki et al. 1980). The response to pinching the lower chest, abdomen, forepaw, or hindpaw was relatively large, and that to pinching the neck or thigh was small (Fig. 79A–F). Pinching the lower chest or abdomen for 20 s produced responses lasting for several minutes, while pinching the forepaw, hindpaw or neck produced shorter responses.

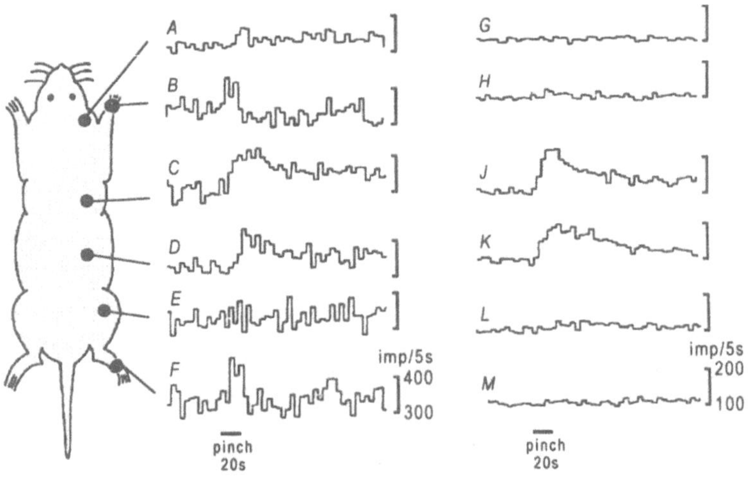

Fig. 79A–M. Effects of pinching various segmental skin areas on adrenal efferent nerve activity in a urethane/chloralose-anesthetized rat with **A–F** the CNS intact and **G–M** spinalized at the C1 level. **A,G** Neck, **B,H** forepaw, **C,J** lower chest, **D,K** abdomen, **E,L** thigh, and **F,M** hindpaw were pinched. In CNS-intact rats, stimulation of various skin areas caused increases in nerve activity. After spinal transection, only lower chest and abdominal stimulation caused the increased response. (Modified from Araki et al. 1980)

Proportional reflex increases in both nerve activity and secretion rates of adrenaline and noradrenaline from the adrenal medulla were observed following noxious cutaneous stimulation in anesthetized rats (Araki et al. 1984). Pinching of the lower chest for 3 min caused a marked reflex increase in nerve activity, with a maximum response reaching 136% of the control activity. The secretion rates of both catecholamines showed similar increases, reaching about 125% of control levels during the stimulation period (Fig. 80A,B).

It has already been reported that stimulation of higher centers, such as the hypothalamus and the cerebral cortex, elicits selective secretion of adrenal adrenaline and noradrenaline. There has been some discussion as to whether adrenaline and noradrenaline are secreted selectively or

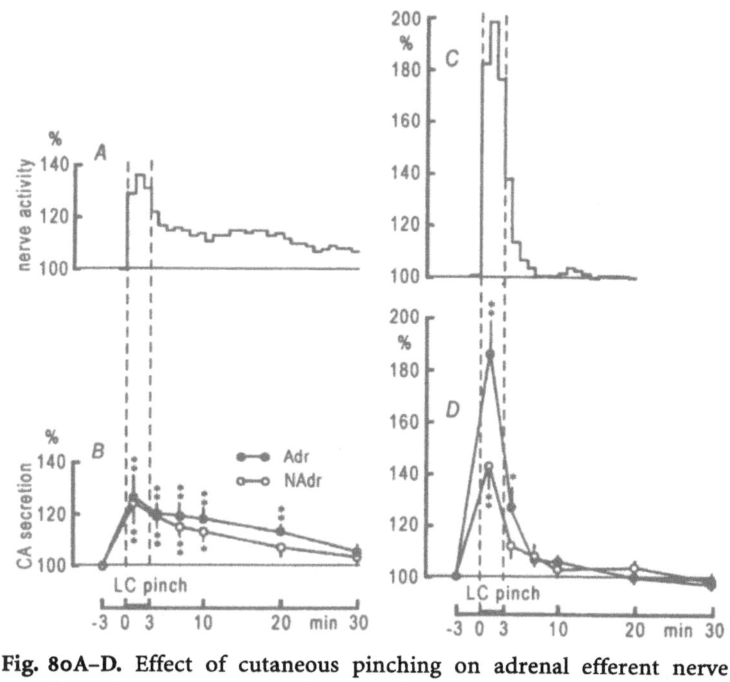

Fig. 80A–D. Effect of cutaneous pinching on adrenal efferent nerve activity and catecholamine secretion rate in **A,B** CNS-intact and **C,D** spinalized (C1-2 level) rats anesthetized with chloralose and urethane. Pinching the lower chest (*LC*) skin increased the adrenal nerve activity, and secretion rates of adrenaline (*filled circles*) and noradrenaline (*open circles*) in both CNS-intact and spinalized rats. Duration of responses in spinalized rats was shorter than in CNS-intact rats. *$P<0.05$, **$P<0.01$; significantly different from prestimulus control values. (Modified from Araki et al. 1984)

nonselectively from the adrenal gland in response to different types of stimulation (e.g., carotid occlusion, asphyxia, hypoglycemia, central nervous stimulation) in CNS-intact anesthetized animals. In the case of somatic stimulation, von Euler and Folkow (1953) reported that, in CNS-intact anesthetized cats, a greater secretion of adrenaline than noradrenaline occurred after electrical stimulation of either the sciatic nerve or brachial plexus. In anesthetized cats, approximately half of the animals demonstrated greater increases in adrenaline than noradrenaline, whereas in the remaining animals, there was no preferential secretion following any kind of somatic noxious stimulation (Sato et al. 1986a). In contrast, there was nonselective adrenal secretion of adrenaline and noradrenaline in response to cutaneous stimulation in CNS-intact anesthetized rats (Araki et al. 1984). Selectivity or nonselectivity of adrenal adrenaline and noradrenaline secretion may be a species difference or may be influenced by general experimental conditions (including anesthesia).

In acutely spinalized rats (transected at the C1-2 level), lower chest or abdominal skin stimulation elicited a marked reflex response, while pinching the neck, forepaw, thigh, or hindpaw was often ineffective or only slightly effective in increasing efferent nerve activity (Fig. 79G–M). The adrenal sympathetic efferent nerves are known to emerge from the spinal cord mainly at the T7-T10 levels (Schramm et al. 1975). Cutaneous sensory information from the lower chest and abdomen enters the spinal cord close to the segments from which the adrenal sympathetic outflow emerges. The input–output relationship suggests a strong spinal segmental organization in spinalized animals. The spinal segmental organization can then be modified to a less segmental, generalized organization by supraspinal structures. Pinching of the lower chest caused reflex increases in nerve activity as well as in adrenaline and noradrenaline secretion rates during the 3-min stimulation period (Fig. 80C,D). Although the percentage increases in activity and catecholamine secretion were larger in spinalized animals than in CNS-intact animals, the absolute value of the reflex response was less in the spinalized animals. This apparent discrepancy is due to the much lower prestimulus control level in spinalized rats (Araki et al. 1984).

Convergence of Nociceptive and Innocuous Cutaneous Afferents and Baroreceptor Afferents onto Single Adrenal Sympathetic Neurons. In order to ascertain the response properties of units contributing to the overall reflex, a study was initiated in anesthetized rats to determine the effects of cutaneous pinching and brushing, and arterial baroreceptor stimulation on the activity of individual adrenal sympathetic neurons (Ito et al. 1984).

The frequency of the ongoing spontaneous discharges of single adrenal nerve fibers in anesthetized and resting rats averaged 1.6 Hz. The majority of single fibers responded to all three stimuli in a typical fashion, i.e., with increases in response to cutaneous pinching, and with decreases in response to either cutaneous brushing or arterial baroreceptor stimulation produced by intravenous phenylephrine administration (Fig. 81A,B).

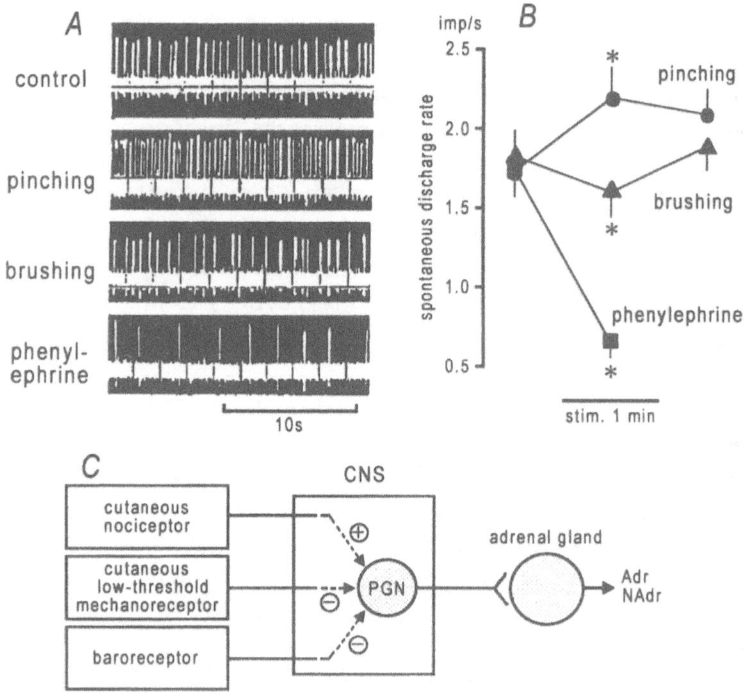

Fig. 81A–C. Convergence of nociceptive and innocuous cutaneous afferents and baroreceptor afferents onto single adrenal sympathetic neurons. **A** Sample recordings of single adrenal nerve filament activity without stimulation (control), during lower chest pinching and brushing, and phenylephrine administration (50 μg/kg, i.v.) in a chloralose/urethane-anesthetized rat. **B** Mean activity of all nerve filaments was increased by pinching and decreased by brushing of lower chest skin and by phenylephrine administration. **C** Convergence of somatic cutaneous and visceral baroreceptor inputs onto a single adrenal sympathetic preganglionic neuron (*PGN*). + and –, indicate excitatory and inhibitory influences to PGN, respectively. $^{*}P<0.01$; significantly different from prestimulus control values. (From Ito et al. 1984; Sato 1987)

However, although the nature of the responses was quite consistent from fiber to fiber, there was little correlation between the magnitudes of individual fiber responses to these three different afferent stimuli. These findings indicate that adrenal sympathetic responses to stimuli do not result from the activation of separate sets of fibers in the adrenal nerve, and that every fiber receives an input from each of these reflex pathways (Fig. 81C), although the input differs substantially from fiber to fiber with regard to its potency. The system appears to involve relatively independent reflex pathways acting directly or indirectly on the adrenal sympathetic preganglionic neurons, leading to a response of variable magnitude depending upon the particular reflex relationship with the preganglionic neuron.

As can be seen in Fig. 81B, pinching of the lower chest produced increases in the mean activity of single adrenal nerve fibers to 127 ± 11% of control levels, which was quite close to the 136% level reported for mass adrenal activity. The decreases in mean activity of single fibers during lower chest brushing and arterial baroreceptor stimulation by intravenous phenylephrine administration were to 88 ± 8% and 38 ± 6% of control, respectively, as compared with decreases in mass activity to 74% and 54% of control levels.

Thermal Stimulation. Responses of nerve activity and catecholamine secretion rates to a 1-min stimulation of the abdominal skin with various temperatures ranging from 4° to 49°C were investigated in anesthetized rats (Kurosawa et al. 1985). Although neither the nerve activity nor the catecholamine secretion rates responded to thermal stimulation between 13° and 40°C, they did show increases to stimulation at temperatures above 43°C and temperatures below 10°C (Fig. 82). Following low-temperature stimulation, nerve activity and secretion rates of adrenaline and noradrenaline showed parallel and proportional increases to, for example, approximately 120% of control levels for stimulation at 4°C. High-temperature stimulation (more than 43°C, especially above 46°C) caused larger increases in nerve activity and adrenaline secretion than did low temperature stimulation. Of note, however, is the fact that adrenaline increased much more than noradrenaline in the case of high temperature.

Both cold stimulation below 10°C and heat stimulation above 43°C produced the same qualitative responses, i.e., increased responses. These thermal stimulations are considered to be among the noxious or near-noxious stimulations capable of activating nociceptors.

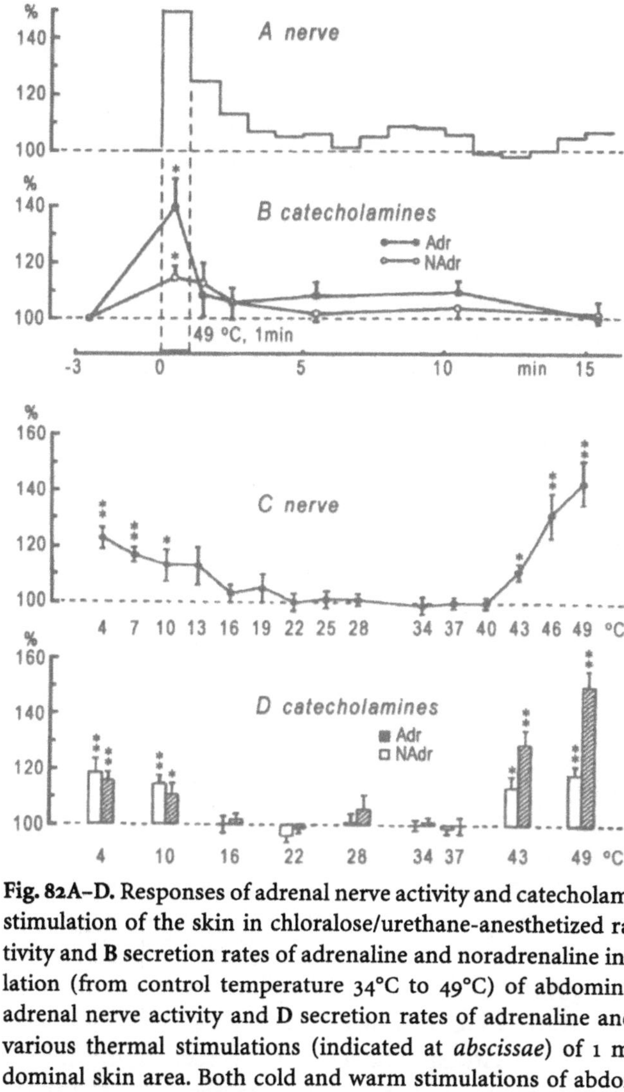

Fig. 82A–D. Responses of adrenal nerve activity and catecholamine secretion to thermal stimulation of the skin in chloralose/urethane-anesthetized rats. **A** Adrenal nerve activity and **B** secretion rates of adrenaline and noradrenaline increased by warm stimulation (from control temperature 34°C to 49°C) of abdominal skin. **C** Responses of adrenal nerve activity and **D** secretion rates of adrenaline and noradrenaline against various thermal stimulations (indicated at *abscissae*) of 1 min duration to the abdominal skin area. Both cold and warm stimulations of abdominal skin either below 10°C or above 43°C produced reflex increases in both nerve activity and catecholamine secretion rates. *$P<0.05$, **$P<0.01$; significantly different from prestimulus control values. (Modified from Kurosawa et al. 1985)

Chemical Stimulation. In anesthetized rabbits, stimulation of the nasal mucous membrane with ether vapor or water at 20°C caused an increase in adrenal medullary catecholamine secretion. Increased catecholamine secretion was abolished by local anesthesia of the nasal passages and by surgical division of the splanchnic nerve (Allison and Powis 1971).

4.5.2.1.2
Muscle Stimulation

To test the hypothesis that sympatho-adrenal function during exercise is stimulated by a reflex arising in skeletal muscle, Vissing et al. (1991) induced skeletal muscular contraction of the hindlimb while simultaneously recording adrenal sympathetic nerve activity in anesthetized rats. A static muscular contraction during 1 min stimulation of the tibial nerve at two times motor threshold caused reflex increases in sympathetic discharges to the adrenal gland, with a latency of <1 s from the onset of contraction. During the static contraction, adrenal sympathetic nerve activity rapidly increased to a maximum of 89 ± 12% above basal level and then declined, reaching basal levels after 30 s of muscle contraction. Tibial nerve stimulation during a neuromuscular blockade had no effect on adrenal sympathetic nerve activity.

4.5.2.1.3
Acupuncture-Like Stimulation

The effects of acupuncture-like stimulation of abdomen and hindlimb on adrenal sympathetic efferent nerve activity and secretion rates of the adrenal medullary hormones (adrenaline and noradrenaline) were studied in urethane-anesthetized rats (A. Sato et al. 1996). Acupuncture needles (diameter of 340 μm) were inserted into the skin and underlying muscles of an abdomen and hindlimb to a depth of 1 cm and were twisted at about 1 Hz for 90 s. The stimuli induced three types of responses in both adrenal sympathetic nerve activity and catecholamine secretion, i.e., decrease, increase, no change. These different responses corresponded with the three similar types of response of mean arterial pressure. After spinal transection at the T2 level, stimuli only increased in both nerve activity and catecholamine secretion. In spinalized animals abdominal stimulation elicited a larger response than hindlimb stimulation did. The response was abolished after surgically severing afferent nerves that innervated the abdomen and hindlimb. These findings indicate that the secretion of adrenal medullary hormones can be controlled reflexly by acupuncture-like stimulation via somatic afferent nerves and adrenal sympathetic efferent nerves.

4.5.2.1.4
Joint Stimulation

Knee Joint Stimulation. The effects of various passive knee joint movements on adrenal catecholamine secretion and adrenal sympathetic efferent nerve activity have been studied using halothane-anesthetized cats (Sato et al. 1986a).

Rhythmic flexions and extensions, as well as rhythmic inward and outward rotations of the knee joint within its physiological ranges of motion did not change nerve activity or adrenal catecholamine secretion. Static outward rotation in the normal working range also had no effect. However, as soon as this static rotation was extended into the noxious range, significant increases in both of these variables were elicited (Fig. 83A–D). Catecholamine concentrations in systemic blood were also increased by noxious movements of the normal knee joint. The responses were virtually abolished by the severance of all nerves innervating the knee joint and its surrounding tissues.

Movements of the inflamed knee joint generally produced much larger responses of both adrenal nerve activity and adrenal catecholamine secretion than movements of normal joints (Fig. 83E–H). For example, rhythmic flexion and extension, which did not change adrenal variables when delivered to the normal joint, produced reflex increases both in nerve activity and secretion of catecholamines during the 1 min stimulation of the inflamed joint.

Vertebral Joint Stimulation. In anesthetized rats, lateral stress stimulation of the spine at the lower thoracic (T 10–13) or lumbar (L2-5) level with 0.5–3.0 kg force produced initial decreases to about 90% of control levels, followed by gradual increases in adrenal sympathetic nerve activity to about 120% of control levels (Fig. 84B,C). These stimuli produced clear and consistent decreases in blood pressure and renal nerve activity (Sato and Swenson 1984). After baroreceptor denervations, only initial decreases in adrenal nerve activity were observed following mechanical stimulation of the spine (Fig. 84D). Cutting the dorsal roots below L3 had no effect on the response to lower lumbar stimulation; however, severing roots T10 to L2 bilaterally totally abolished all responses.

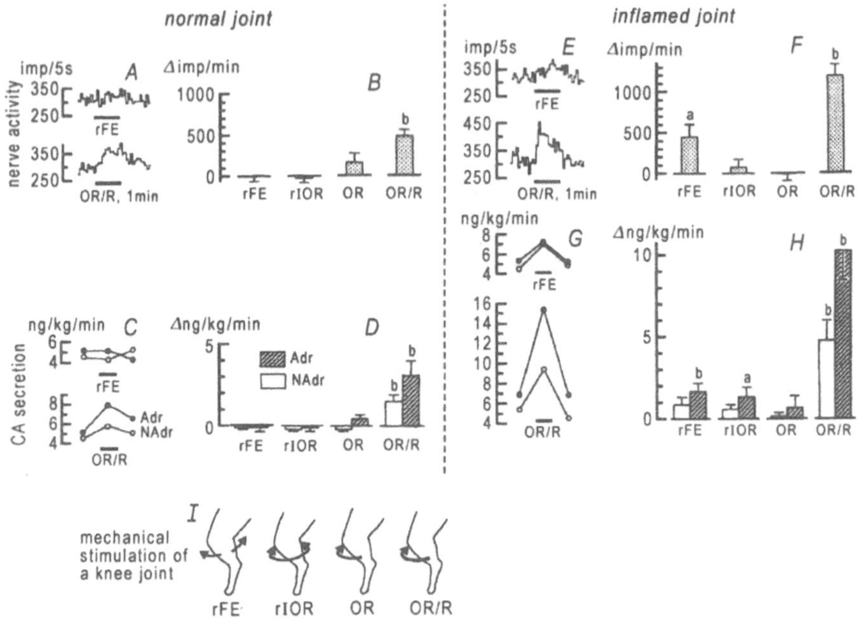

Fig. 83A–I. Changes in **A,B,E,F** adrenal nerve activity and **C,D,G,H** catecholamine secretion rate induced by movements of **A–D** the normal knee joint and **E–H** the inflamed knee joint in halothane-anesthetized cats. **I** Method of knee joint movement. *rFE*, rhythmic flexion and extension movement; *rIOR*, rhythmic inward and outward rotations; *OR*, static outward rotation; *OR/R*, static outward rotation against resistance. Noxious movements (OR/R) of normal knee joints produce increases in both adrenal sympathetic nerve activity and adrenal catecholamine secretion. Experimental arthritis augments articularly induced sympathoadrenal medullary responses and, further, certain motions (rFE) which are ineffective when performed in the normal joint increase adrenal nerve activity and catecholamine secretion if applied to the inflamed joint. $^{a}P<0.05$, $^{b}P<0.01$; significantly different from prestimulus control values. (Modified from Sato et al. 1986a)

4.5.2.1.5
Electrical Stimulation

Repetitive electrical stimulation of somatic afferent (T13 spinal or sural) nerves was used to demonstrate the categories of afferent fibers involved in adrenal sympathetic efferent nerve responses in anesthetized rats (Isa et al. 1985). In general, low intensity repetitive electrical stimulation, activating myelinated (Aβ and Aδ) afferent fibers alone, produced inhibitory responses, similar in character and duration to the responses evoked by

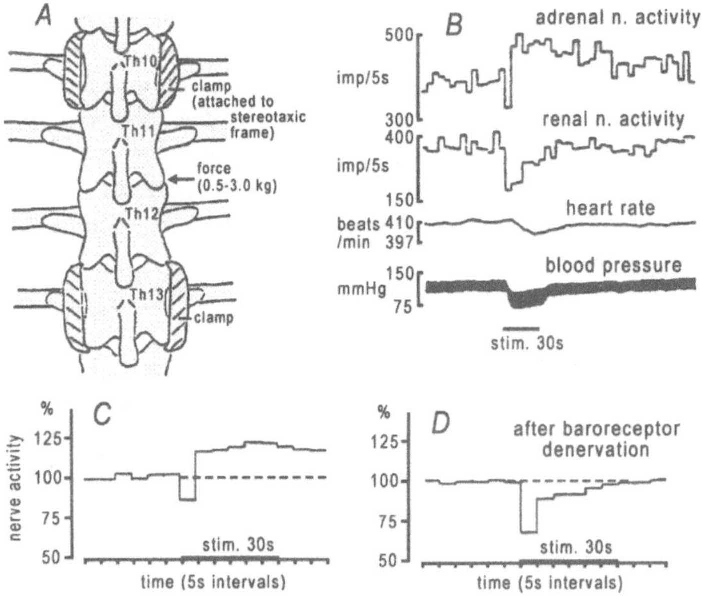

Fig. 84A–D. Changes in adrenal nerve activity induced by thoracic spine stimulation in chloralose/urethane-anesthetized rats. **A** Stimulation procedure (thoracic spine shown). Segments are isolated from skin and muscle, upper and lower segments fixed in a spinal stereotaxic device, force exerted on mobile segments. **B** Simultaneous recordings of adrenal and renal nerve activity, heart rate, and blood pressure in a CNS-intact animal with thoracic spine stimulation. **C,D** The mean response in adrenal nerve to stimulation of the thoracic spinal column (3.0 kg) in **C** baroreceptor-intact and **D** baroreceptor-denervated animals. Lateral stress stimulation of the spine produces consistent decreases in blood pressure and renal nerve activity. In the case of adrenal sympathetic nerve activity, initial decreases are followed by gradual increases. The pattern of adrenal nerve response dramatically changes after arterial baroreceptor denervation, with only a decrease in activity observed. (Modified from Sato and Swenson 1984)

innocuous mechanical stimulation of the skin in rats. In contrast, high intensity repetitive electrical stimulation, activating both myelinated and unmyelinated C afferent fibers, produced increases in adrenal nerve activity nearly identical to those induced by noxious mechanical or thermal stimulation of the skin innervated by the nerves stimulated electrically.

In spinalized rats, repetitive stimulation of T13 spinal nerve always increased adrenal nerve activity, regardless of whether A fibers alone or A plus C fibers were stimulated, just as brushing and pinching of the lower chest skin always increased nerve activity.

The effects of sciatic nerve stimulation on adrenal venous catecholamines were examined in anesthetized cats (Gaumann et al. 1990). Sciatic nerve stimulation for 10 min at 50 times minimal motor threshold, which excites Aδ and C fibers, evoked a significant increase in adrenal venous catecholamine levels: fivefold for adrenaline, fourfold for noradrenaline, and twofold for dopamine.

4.5.2.1.6
Central Processing of Somato-adrenal Medullary Reflexes

Changes in adrenal sympathetic nerve activity resulting from single-shock electrical stimulation of various spinal afferent nerves, especially the T13 spinal nerve and the sural nerve, were examined in anesthetized rats (Araki et al. 1981; Isa et al. 1985). Single-shock stimulation of spinal afferent nerves evoked various reflex components in the adrenal nerve: (a) an initial depression of spontaneous activity (the early depression); (b) a reflex discharge due to activation of A afferent fibers (the A reflex); (c) a subsequent reflex discharge due to activation of C afferent fibers (the C reflex); and (d) postexcitatory depressions (Fig. 85A,B,E,F). These reflexes seem to be mediated mainly via supraspinal pathways, since they were abolished by spinal transection at the C1-2 level. Although the supraspinal A and C reflexes could be elicited from stimulation of a broad band of spinal segmental afferent levels, the early depression was more prominent when afferents at spinal segments closer to the level of adrenal nerve outflow were excited (Fig. 86). It is suggested that the decreased responses of the adrenal nerve during repetitive electrical stimulation of A afferent nerve fibers are attributable to summation of both the early depression and postexcitatory depression evoked by single-shock stimulation, while the increased responses during repetitive stimulation of A plus C afferent fibers are attributable to summation of the C reflex after single-shock stimulation. After spinal transection, two other spinal A- and C-reflex components with much shorter latencies can be observed following stimulation of myelinated A and unmyelinated C somatic afferent fibers in the T13 spinal nerve or the sural nerve (Fig. 85C,D,G,H). The spinal A- and C-reflex discharge components seem to be suppressed by supraspinal influences, including early depression.

Studies on anesthetized animals have proven that sympatho-adrenal medullary functions are influenced by various somatic afferent stimuli,

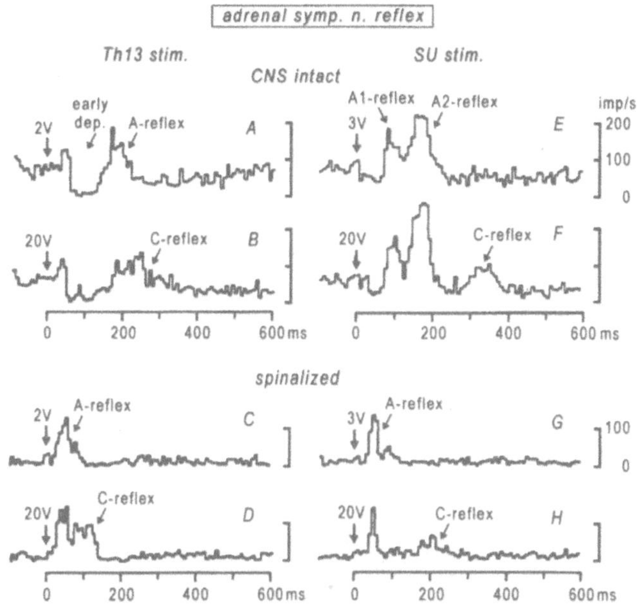

Fig. 85A–H. Poststimulus-time histograms of adrenal nerve activity following a single shock (0.5 ms) to **A–D** the 13th thoracic spinal nerve (T13) and **E–H** the sural nerve (SU) in chloralose/urethane-anesthetized rats **A,B,E,F** with CNS intact and **C,D,G,H** spinalized at the C2 level. In CNS-intact rats, single-shock stimulation of both somatic afferent nerves evokes a reflex response in the adrenal sympathetic nerve consisting of various reflex components: (a) early depression, (b) A reflex (in some cases the A reflex is separated into A1- and A2-reflex components), (c) C reflex, and (d) postexcitatory depression. These reflex components seem to be mediated mainly via supraspinal pathways, since they are abolished by spinal transection at the C1–2 level. After spinal transection, two other spinal A- and C-reflex components with much shorter latencies can be observed. (From Isa et al. 1985)

either natural or electrical, and that the somatically induced adrenal sympathetic reflex responses result in changes in catecholamine secretion from the adrenal gland. Somato-adrenal sympathetic reflexes contain both spinal and supraspinal reflex components, the spinal component having a segmental reflex organization, whereas the supraspinal component is generalized and nonsegmental (Fig. 87). In CNS-intact condition, the reflexes are excitatory or inhibitory, depending upon the stimuli and species of animals, e.g., excitatory by pinching and inhibitory by brushing in rats. The excitatory responses are mainly mediated by somatic unmyelinated

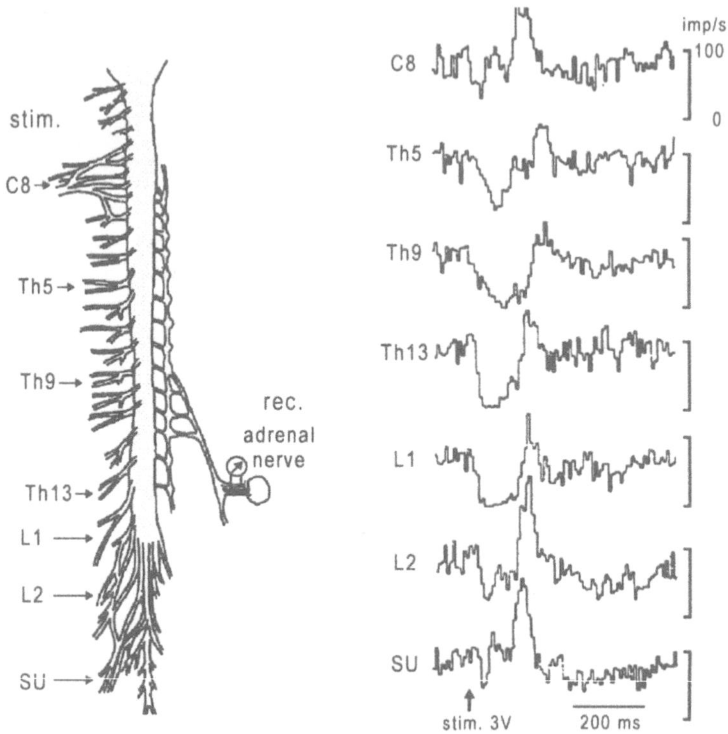

Fig. 86. Poststimulus-time histograms taken from one chloralose/urethane-anesthetized rat following single-shock, supramaximal stimulations of myelinated afferent A fibers of the various spinal nerves indicated. Note the large early depression from stimulation of lower thoracic and upper lumbar nerves. Stimulated spinal afferent nerves are schematically drawn at the left side of sample recordings. *SU*, sural nerve. (From Isa et al. 1985)

C afferent fiber activation, whereas the inhibitory responses are due to somatic myelinated A afferent activation. In spinalized animals, the reflexes are only excitatory irrespective of the stimulus modality, noxious or innocuous.

4.5.2.2 Pancreatic Hormones

The pancreas acts as both an exocrine and endocrine organ. Pancreatic hormones are produced in the islet cells of Langerhans, insulin from the

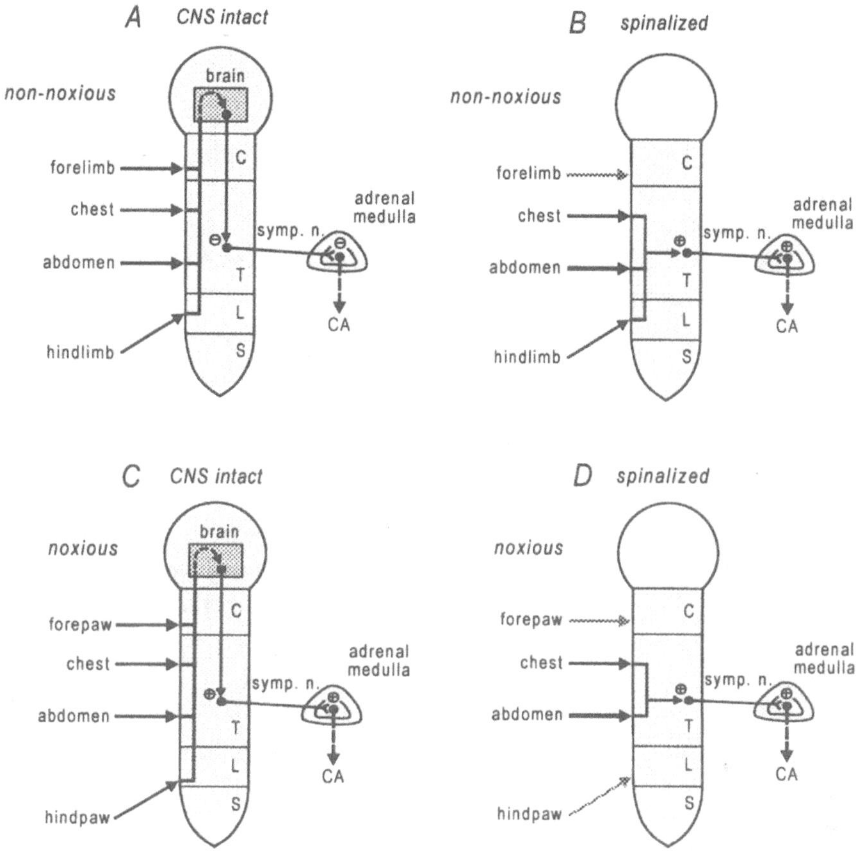

Fig. 87A–D. Reflex pathway for the adrenal medulla with **A,B** innocuous and **C,D** noxious stimulation in **A,C** CNS-intact and **B,D** spinalized animals. **A,C** Innocuous brushing of various segmental skin areas produces an inhibitory effect on the adrenal medullary function through the sympathetic nerves via a supraspinal reflex pathway. On the other hand, noxious pinching of various segmental skin areas produces an excitatory effect on the adrenal medullary function via a supraspinal reflex pathway. **B,D** After spinal transection, both innocuous and noxious somatic stimulation at certain areas can produce the segmental excitatory spinal reflex. Other details are the same as in Fig. 51

B cells, glucagon from the A cells, and somatostatin from D cells. It is well known that there are parasympathetic and sympathetic nerve endings within the islet structures, and that the secretion of pancreatic hormones is in part regulated by the autonomic nervous system.

4.5.2.2.1
Insulin and Somatostatin

The secretion of insulin increases in response to suckling in lactating conscious mammals, e.g., in dogs (Uvnäs-Moberg and Eriksson 1983; Eriksson et al. 1987), humans (Widström et al. 1984), pigs (Uvnäs-Moberg et al. 1984) and rats (Eriksson et al. 1994). A similar increase in insulin was even observed in nursing human infants (Marchini et al. 1987).

Uvnäs-Moberg et al. (1986) found a significant relationship between sciatic nerve stimulation (10 V, 0.2 ms, 3 Hz, for 20 min) and basal levels of insulin in anesthetized cats. Electrical stimulation of sciatic afferent nerve fibers produced two opposite responses in secretion of insulin depending upon the different basal levels of insulin; when the basal level was high, insulin secretion decreased, and when the basal level was low, insulin secretion increased. Vissing et al. (1994) electrically stimulated muscle branches of the femoral nerves in anesthetized cats for 10 min at various intensities. Stimulation at 20 times motor threshold, recruiting group III afferents, elicited slight increases in plasma glucose, but did not change plasma insulin. Stimulation at 140 times motor threshold, exciting group III and IV afferents, elicited increases in plasma glucose and decreases in insulin.

Electrical stimulation of sciatic afferent nerve fibers increased secretion of somatostatin in anesthetized cats (Uvnäs-Moberg et al. 1986), whereas suckling in lactating conscious pigs (Uvnäs-Moberg et al. 1984) and dogs (Linden et al. 1987) had inconsistent effects.

The somatically induced responses of insulin and somatostatin secretion appear to be reflex responses, where the afferent limb is a somatic afferent nerve and the efferent limb is an autonomic efferent nerve. The autonomic efferent limb may act on the pancreatic B and D cells either directly or indirectly, via plasma catecholamines secreted from the adrenal medulla or from other sympathetic nerves not concerned with pancreatic innervation. The contribution of the vagal nerve was recently investigated. The suckling-induced insulin response persisted after vagotomy. In contrast, somatostatin levels that rose significantly in sham-operated rats during suckling rose even more in vagotomized animals. This study shows that vagal nerve activity is of importance for the regulation of somatostatin secretion but not insulin during suckling (Eriksson et al. 1994).

4.5.2.2.2
Glucagon

Suckling of lactating conscious dogs increased plasma glucagon (Eriksson et al. 1987). Electrical stimulation of the sciatic nerve (10 V, 0.2 ms, 3 Hz, for 20 min) in anesthetized cats also increased the plasma glucagon concentration (Uvnäs-Moberg et al. 1986).

In anesthetized rats, noxious and innocuous mechanical stimulation of the skin by pinching and brushing, were delivered to various segmental areas (Kurosawa et al. 1994). Cutaneous pinching for 3.5 min of the face, forepaw, abdomen, or hindpaw increased the plasma level of immunoreactive glucagon (IRG). The increase was larger following pinching of the abdomen or hindpaw than following pinching of the face or forepaw (Fig. 88B). Brushing for 3.5 min of the face, forelimb, abdomen, or hindlimb did not significantly affect the plasma IRG. The increase in plasma IRG following skin pinching was abolished when both parasympathetic vagal and sympathetic nerves to the pancreas were severed bilaterally (Fig. 88A). The increase in plasma IRG was not abolished after bilateral severance of either the parasympathetic vagal or sympathetic nerves alone. These findings indicate that excitation by cutaneous nociceptive afferent information from various spinal segments can regulate glucagon secretion from the pancreas as a reflex response via an efferent limb composed of both sympathetic and parasympathetic nerves. Stimulation of sympathetic or parasympathetic vagal nerves increased the plasma glucagon concentration or glucagon secretion from the isolated perfused pancreas in animals (Bloom et al. 1973, 1974; Bloom and Edwards 1975; Andersson et al. 1982; Bobbioni et al. 1983; Holst et al. 1986; Nishi et al. 1987; Kurose et al. 1990). Thus both pancreatic sympathetic and parasympathetic vagal nerves appear to be activated by cutaneous stimulation.

4.5.3
Hypothalamo-anterior Pituitary Hormone System

In the 1930s, Hans Selye noted that stressful stimuli, including noxious somatic sensory stimulation, produced atrophy of the thymus, hemorrhage in the gastric mucosa and hypertrophy of the adrenal gland (see review by Selye 1956). Since then there have been many reports along this line. It was found that hypertrophy of the adrenal gland was related to hyperfunction of the adrenal cortex, including excessive secretion of adrenal glucocorticoids. Other hypothalamo-anterior pituitary hormones are also

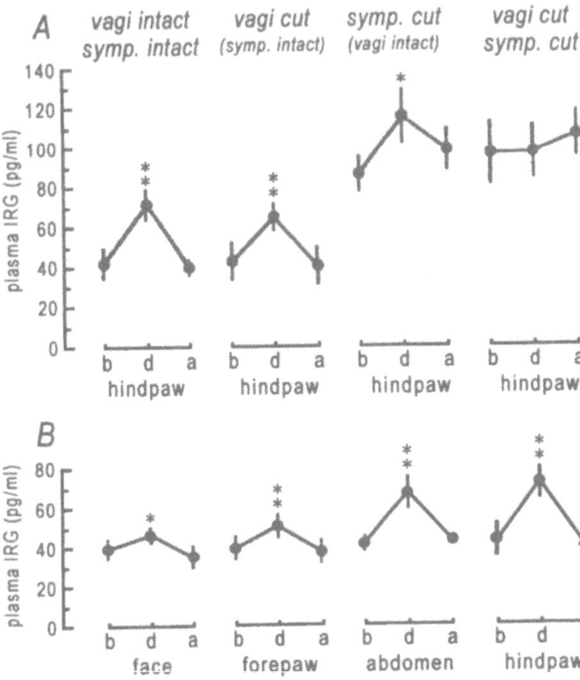

Fig. 88A,B. Effect of cutaneous pinching on plasma immunoreactive glucagon (*IRG*) concentration in chloralose/urethane-anesthetized rats. **A** The effect of bilateral severance of pancreatic autonomic nerves on the increase in the plasma IRG concentration following hindpaw pinching. The increase in the plasma IRG concentration following cutaneous pinching remained after bilateral severance of either sympathetic or vagal parasympathetic nerves innervating the pancreas. However, the response was abolished after bilateral severance of both sympathetic and vagal parasympathetic nerves. **B** Responses of IRG following pinching of various skin regions. Blood samples were collected 30 min before (*b*), during (k), and 20 min after (*a*) cutaneous stimulation for 3.5 min. The increases in plasma IRG concentrations were elicited by stimulation delivered to various segmental skin areas, although abdominal or hindpaw stimulation produced much larger responses than stimulation of the face or forepaw. *P<0.05, **P<0.01; significantly different from prestimulus control values. (Modified from Kurosawa et al. 1994)

known to be influenced by stressful stimuli. The definition of stress, a stressful stimulus or a stressor appears to include many stimuli, and it is difficult to delineate a uniform response, thus appear to be difficult to analyze neural reflex mechanisms of somato-endocrinological reflex responses. To analyze the neural mechanisms of somatosensory modulation of anterior pituitary hormones, it is necessary to use anesthetized animals to eliminate emotional factors. However, there are still few studies of this type.

It was found that the electrical stimulation of somatic afferent nerves increased plasma levels of adrenocorticotropic hormone (ACTH) (Feldman et al. 1981; Hamamura et al. 1986) and corticosterone (Dallman and Jones 1973; Feldman et al. 1975a–c), and that noxious mechanical stimulation of the skin increased plasma corticosterone levels (Tsuchiya et al. 1991b) in anesthetized rats. Furthermore, in anesthetized rats, noxious cutaneous stimulation increased the release of corticotropin-releasing hormone (CRH), which mediates the release of ACTH, from the hypothalamus into hypophysial portal blood (Hotta et al. 1992).

Schanberg et al. (1984) have shown that tactile stimulation is necessary for growth hormone secretion. This type of study using tactile stimulation of the skin can be employed here for discussion because this type of stimulation may not produce stress and will give us a chance to analyze neural reflex mechanism of the response.

4.5.3.1
CRH-ACTH-Cortical Hormones

Secretion of glucocorticoids from the adrenal cortex is regulated by plasma ACTH secreted from the anterior lobe of the hypophysis. ACTH secretion is regulated by CRH secreted from the hypothalamic neurons into the hypothalamo-pituitary portal system, and also by glucocorticoids secreted from the adrenal cortex. The ACTH-glucocorticoid secretion system is known to have a circadian rhythm. The center for this rhythm appears to be in the suprachiasmatic nucleus and to be modulated by various external stimuli, including stimulation of the eyes by light. It has been well documented that this ACTH-glucocorticoid secretion system is influenced by stress, including stimuli such as severe trauma, laparotomy, electroconvulsive treatment, hypoxia or anoxia, pyrogen, acute hypoglycemia, injection of histamine and hemorrhage. Certain stresses are related to somatic sensory stimulation. In addition, plasma corticosterone levels were increased in response to suckling in lactating rats (Voogt et al. 1969) and sows (Okrasa et al. 1989).

Somatosensory regulation of plasma corticosterone has been examined using anesthetized animals (Dallman and Jones 1973; Feldman et al. 1975a,b,c; Gibbs 1969; Greer et al. 1970; Matsuda et al. 1964; Tsuchiya et al. 1991b). Under pentobarbital anesthesia, plasma corticosterone in resting rats was stable and did not show any significant circadian change over a few hours (Tsuchiya et al. 1991b). In anesthetized rats, an increase of plasma corticosterone was observed following electrical stimulation of hindlimb afferents (Dallman and Jones 1973; Feldman et al. 1975a–c, 1988), noxious cutaneous stimulation of the hindpaw (Tsuchiya et al. 1991b), or traumatic fracture of a hindleg bone (Gibbs 1969; Greer et al. 1970), whereas innocuous mechanical stimulation produced no significant change in plasma corticosterone (Tsuchiya et al. 1991b).

It is characteristic that the time course of the corticosterone response to somatosensory stimulation is much longer than the duration of the stimulus. Tsuchiya et al. (1991b) found that bilateral noxious mechanical stimulation of hindpaws by pinching for 10 min significantly increased plasma corticosterone for the following hour (Fig. 89A). The peak of the

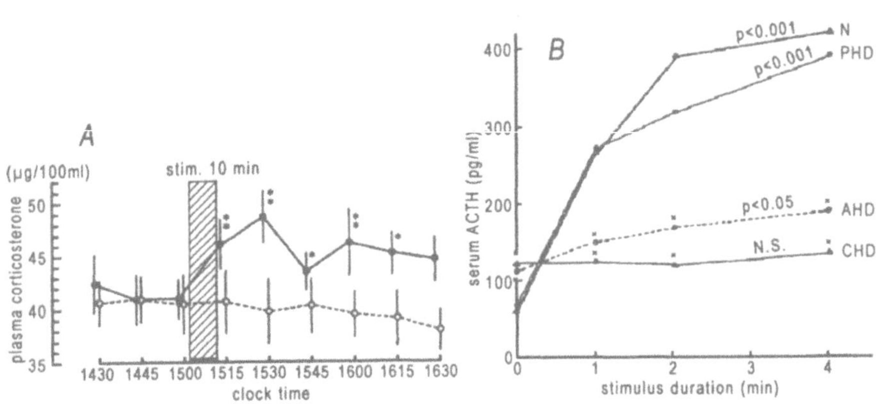

Fig. 89A,B. Effect of somatic stimulation on corticosterone and adrenocorticotropic hormone (*ACTH*) secretion in pentobarbital-anesthetized rats. **A** Effect of bilateral hindpaw pinching (*solid line*) and hindlimb brushing (*broken line*) for 10 min on plasma corticosterone concentration. Noxious pinching stimulation significantly increased plasma corticosterone for the following 1 h, whereas innocuous brushing stimulation produced no significant change in plasma corticosterone. $^{*}P<0.05$, $^{**}P<0.01$; significantly different from prestimulus control values. **B** Effect of sciatic nerve stimulation (0.1 ms, 0.3 mA, 10 Hz, 2 min) on serum ACTH concentration in normal rats (*N*) and rats with anterior (*AHD*), posterior (*PHD*), or complete (*CHD*) hypothalamic deafferentation. Stimulation of the sciatic nerve induced large increases in ACTH concentration in N and PHD. (**A** Modified from Tsuchiya et al. 1991b; **B** From Feldman et al. 1981)

plasma corticosterone response in anesthetized rats occurred 30 min after the onset of noxious mechanical stimulation (Tsuchiya et al. 1991b), 20 min after bone fracture or ischemia by rubberband tourniquet (Greer et al. 1970), and 15 min after electrical stimulation of a thigh for 2 min (Dallman and Jones 1973).

In anesthetized rats and humans, plasma ACTH was shown to increase after surgery (Oyama et al. 1979; Amann and Lembeck 1987) or electrical stimulation of hindlimb afferent nerve fibers (Feldman et al. 1981; Hamamura et al. 1986; Vissing et al. 1994). Electrical stimulation of group III muscle afferents of the femoral nerve increased plasma ACTH, and further stimulation of group IV muscle afferents in addition to group III afferents elicited a even larger increase in ACTH (Vissing et al. 1994). Serum ACTH levels were determined following sciatic nerve stimulation, in intact rats and in rats subjected to anterior, posterior or complete hypothalamic deafferentation. In both intact and posterior hypothalamic deafferentation groups, serum ACTH concentrations were markedly elevated following stimulation (0.3 mA, 0.1 ms, 10 Hz, 1–4 min). In the complete hypothalamic deafferentation group, this response was completely eliminated, and in the anterior hypothalamic deafferentation group, only a very slight and marginally significant increase in serum ACTH occurred upon stimulation (Fig. 89B). These results demonstrate that sciatic nerve stimulation increases ACTH secretion via the hypothalamus, and confirm that this somatosensory stimulation impinges upon the hypothalamus from the rostral direction (Feldman et al. 1981).

CRH in the hypophyseal portal blood is secreted mainly by parvocellular neurosecretory cells in the paraventricular nucleus, which project to the median eminence (Swanson et al. 1983; Merchenthaler et al. 1984; Plotsky and Vale 1984). The activity of the paraventricular nucleus, which was antidromically activated by median eminence stimulation but not by posterior pituitary stimulation, was shown to increase after electrical stimulation of the saphenous nerve (3 mA, 1 ms, 4 Hz, 50–150 s) in anesthetized rats (Hamamura et al. 1986).

In anesthetized rats, Hotta et al. (1992) examined the effects of noxious and innocuous mechanical stimulation of various segmental skin areas on the secretion of immunoreactive CRH (iCRH) from the hypothalamus into hypophysial portal blood. Pinching of the forepaws or hindpaws and brushing of the hindlimbs for 20 min increased hypothalamic iCRH secretion during stimulation (Fig. 90). In contrast, pinching of the face or abdomen and brushing of the face, forelimbs, or abdomen for 20 min did

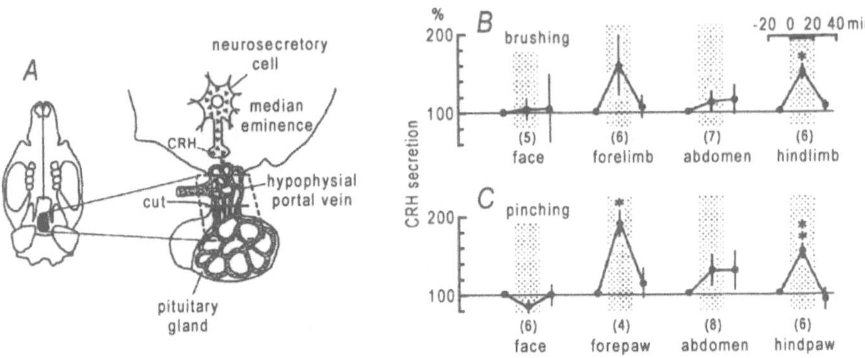

Fig. 90A–C. **A** Hypophysial portal blood sampling by the parapharyngeal approach. After exposing the pituitary gland, the pituitary stalk was transected and the hypophysial portal blood was collected. Responses of hypothalamic corticotropin-releasing hormone (*CRH*) secretion rate to **B** bilateral cutaneous brushing or **C** pinching of various segmental sites of the body in halothane-anesthetized rats for 20 min. Pinching of the forepaws or hindpaws and brushing of the hindlimbs increased hypothalamic immunoreactive CRH secretion. *$P<0.05$, **$P<0.01$; significantly different from prestimulus control values. (Modified from Hotta et al. 1992)

not significantly influence iCRH. These results indicate that cutaneous mechanical sensory stimulation contributes to the reflex regulation of CRH secretion from the hypothalamus into hypophysial portal blood, and also that this effect is highly dependent on the site of stimulation.

4.5.3.2 Gonadotropin-Gonadal Sex Hormones

The hypothalamus releases gonadotropin-releasing hormone (GnRH), which causes the anterior pituitary to release the gonadotropins, luteinizing hormone (LH) and follicle-stimulating hormone (FSH). LH and FSH have stimulative effects on estrogen, progesterone and testosterone secretion.

A number of studies have suggested that, in conscious males of several species, including rat, sensory stimuli associated with mating can elevate LH and testosterone, but not FSH levels (Kamel et al. 1977; Oaknin et al. 1989).

In conscious lactating female animals, the effects of suckling on plasma LH, FSH, estrogen and progesterone have been studied. Plasma LH was increased in ewes (Schirar et al. 1990), decreased in rats (Isherwood and

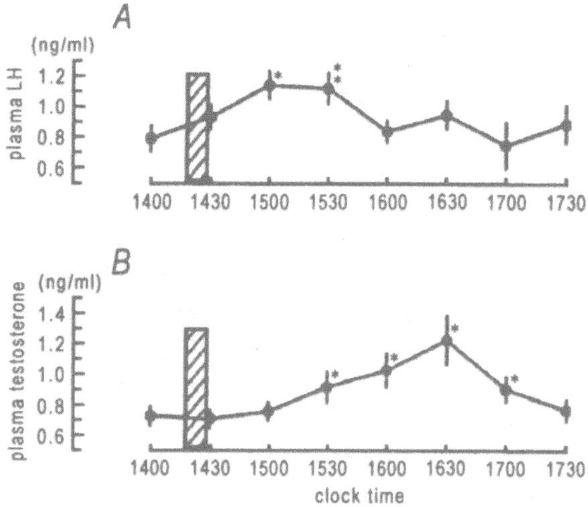

Fig. 91A,B. Effects of cutaneous pinching on **A** luteinizing hormone (*LH*) and **B** testosterone levels in plasma collected from the femoral veins in pentobarbital-anesthetized rats. Bilateral pinching of the hindpaws for 10 min significantly increased plasma LH (30 and 60 min after stimulation) and testosterone (60–150 min after stimulation). *$P<0.05$, **$P<0.01$; significantly different from prestimulus control values. (From Tsuchiya et al. 1992)

Cross 1980) and cows (Chang et al. 1981; Williams et al. 1983), and unchanged in humans (Dawood et al. 1981). Plasma FSH was decreased (Williams et al. 1983), or unchanged (Schiar et al. 1990; Dawood et al. 1981). Plasma progesterone was increased (Bridges and Goldman 1976) or unchanged (Dawood et al. 1981). Plasma estradiol was unchanged (Dawood et al. 1981).

In pentobarbital-anesthetized rats, cutaneous nociceptive information led to increased secretion of LH from the anterior pituitary, resulting in an increase in testosterone secretion from the testes into the plasma (Tsuchiya et al. 1992). Under resting conditions, without somatic sensory stimulation, plasma LH and testosterone measured every 30 min revealed no significant fluctuations over a few hours. Bilateral noxious mechanical stimulation of the hindpaws by pinching for 10 min significantly increased plasma LH and testosterone (Fig. 91). Plasma LH was increased 30 and 60 min after stimulation, while plasma testosterone was increased 60–150 min after stimulation. However, bilateral innocuous mechanical stimulation of the hindlimbs by brushing for 10 min did not significantly change either plasma LH or testosterone. These results indicate that, following pinching stimulation to the skin, there is some latent period, in the order of 30 min, before LH, released from the anterior pituitary gland, begins its action of inducing the secretion of testosterone from the testes.

4.5.3.3 Growth Hormone

The secretion of growth hormone (GH) is regulated by GH-releasing factor and GH-inhibiting factor secreted from hypothalamic neurons into the hypothalamo-pituitary portal system. The secretion of GH is increased during sleep (particularly during slow wave sleep), hypoglycemia, exercise, and stressful stimuli. Stressful stimuli include surgical operations, electrical shock therapy, hemorrhage, ether anesthesia, and emotional stress.

Gentle cutaneous stimulation has been shown to facilitate gain in body weight in weaning rats (Weininger 1954; McClelland 1956; Ruegamer and Silverman 1956) and in human preterm neonates (Field et al. 1986). The preterm neonate who was given tactile/kinesthetic stimulation consisting of body stroking and passive movements of the limbs for three 15-min periods per day for 10 days averaged a 47% greater weight gain per day (Field et al. 1986). Suckling stimulation to the nipple has been shown to increase the concentration of plasma GH in lactating rats (Saunders et al. 1976).

It is known that the concentration of plasma GH decreases when the neonatal rat is separated from the mother, even for as short a time as 2 h (Kuhn et al. 1978; Evoniuk et al. 1979; Schanberg et al. 1984; Pauk et al. 1986; Schanberg and Field 1987; Kacsoh et al. 1989, 1990). It is interesting that vigorous tactile stimulation of the rat pups' skin, i.e., fairly heavy strokes on the back and head areas using a moistened brush (20–30 s stimulation was given at 5 min intervals for 2 h), completely prevented this effect of maternal deprivation, while other forms of tactile stimulation, light stroking with a brush and tail pinching, were ineffective (Evoniuk et al. 1979; Pauk et al. 1986).

Ornithine decarboxylase (ODC) is a sensitive biochemical index of cell growth and differentiation. ODC activity also declined rapidly in newborn rats separated from their mother (Butler and Schanberg 1977; Butler et al. 1978; Kuhn et al. 1978; Pauk et al. 1986), and ODC in the maternally deprived rats recovered if the brushing stimulation mentioned above was delivered to the skin (Pauk et al. 1986). GH is a well-known regulator of ODC activity in tissues. GH was completely ineffective in inducing ODC activity in the livers or brains of maternally deprived rats, although GH caused a significant induction in normal rats. Tactile stimulation of maternally deprived rat pups also restored tissue sensitivity to exogenous GH (Schanberg et al. 1984).

In neonatal rats separated from their mothers, Kacsoh et al. (1990) examined the effects of manipulations of ambient temperature as well as stimulation of the oral or anogenital regions on serum GH. Exposing rat pups to 37°C (nest temperature) during the 6-h separation period prevented the separation-induced decrease in serum GH levels in 2-day-old pups. Stimulation of the oral zone (feeding from a soft cannula) or anogenital zone (inducing urination and defecation with a wet brush) of the pups significantly decreased neonatal serum GH values. The painful stimulus of subcutaneous administration of hyperosmotic saline was without effect on serum GH levels. They concluded that provision of a warm ambient temperature is a critical component for regulation of GH secretion.

All of these results suggest that some cutaneous sensory stimulation can facilitate secretion of GH and tissue sensitivity to GH, and has a functional meaning for growth.

4.5.3.4 Prolactin

The secretion of prolactin (PRL) is regulated by PRL-releasing factor and PRL-inhibiting factor secreted from hypothalamic neurons. PRL initiates lactation in mammals at parturition. In the postpartum lactating animal, stimulation of the nipple by suckling increases the plasma concentration of PRL.In anesthetized lactating rats, plasma PRL was increased by suckling (Burnet and Wakerley 1976), and by repetitive electrical stimulation of the mammary nerve (Mena et al. 1980; de Greef and Visser 1981).

The effects of noxious and innocuous mechanical stimulation of various segmental skin areas (face, forelimb, abdomen, and hindlimb) on the plasma concentrations of PRL were examined in anesthetized female rats. Pinching the hindpaw for 6 min significantly increased plasma PRL concentration during stimulation, to about 280% of the prestimulation value. Pinching the face, the forepaw or the abdomen had no significant effect. Brushing of any skin area for 6 min did not significantly change plasma PRL concentrations (Fig. 92) (Hotta et al. 1993). These results indicate that cutaneous, nociceptive sensory information contributes to the reflex

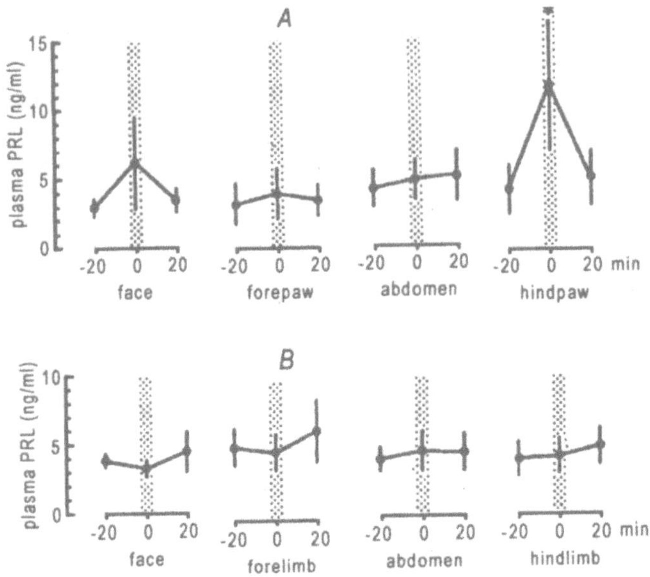

Fig. 92A,B. Plasma prolactin (*PRL*) concentrations before and after **A** pinching and **B** brushing of various skin regions for 6 min (*shaded bar*) in urethane-anesthetized rats. Pinching the hindpaw significantly increased plasma PRL concentration during stimulation, while pinching the face, the forepaw, or the abdomen and brushing of any skin area had no significant effect. (Modified from Hotta et al. 1993)

regulation of PRL secretion from the anterior pituitary, and that this effect is highly dependent on the site of stimulation.

Some of the PRL-releasing factors, such as vasoactive intestinal peptide (VIP), thyrotropin-releasing hormone (TRH), and oxytocin, and some PRL-inhibiting factors such as dopamine (Ben-Jonathan 1990) must be responsible for the somatically induced PRL response. Oxytocin, which stimulates PRL secretion, has been reported to increase with somatic stimulation in anesthetized rats (see Sect. 4.5.4.1; Stock and Uvnäs-Moberg 1988). However, the effects of somatic stimulation on other PRL-regulating factors are still unknown.

4.5.4
Posterior Pituitary Hormone System

It has been established that milk is ejected by contraction of the myoepithelial cells lining the alveoli and small ducts of the mammary gland.

Normally, this milk ejection is dependent upon an increase in oxytocin secretion from the posterior pituitary gland resulting from cutaneous stimulation of the breast (Tindal 1974; Bisset 1974; Cross 1966). Oxytocin is released from the axonal terminals of hypothalamic oxytocin-releasing neurons. Wakerley and Lincoln (1973) first recorded the activity of the hypothalamic oxytocin-releasing neurons. In anesthetized rats, the activity of the hypothalamic oxytocin-releasing neurons was increased by suckling (Wakerley and Lincoln 1973; Lincoln and Wakerley 1974; Freund-Mercier and Richard 1981; Poulin and Wakerley 1982; Moos and Richard 1988). When suckling or a similar kind of innocuous mechanical stimulation is applied to the nipple or breast, the cutaneous sensory information is transmitted through mammary somatic afferent nerves to the spinal cord, then ascends the lateral column of the cord, first to the lateral tegmentum of the contralateral midbrain, and then to the hypothalamic oxytocin-releasing magnocellular neurons in the paraventricular and supraoptic nuclei. This indicates that hypothalamic oxytocin-secretory neuronal activity is modulated by cutaneous afferent stimulation of the breast. In fact, in anesthetized rats, various segmental cutaneous afferent stimuli can change the activity of hypothalamic oxytocin-releasing neurons as a reflex response (Akaishi et al. 1988).

4.5.4.1 Oxytocin

The milk ejection reflex to suckling of the nipples has been demonstrated in conscious and anesthetized lactating animals of various species. Lincoln and Wakerley (1974) succeeded in recording the activity of what they identified as neuroendocrine cells in anesthetized lactating female rat while simulating suckling by rat pups. It was found that a continuous suckling stimulus produced periodic bursts of action potentials in many of the identified neuroendocrine cells. Approximately 13 s after the burst, there was an increase in intra-mammary pressure, indicating the arrival of a pulse of oxytocin to the mammary glands. The effects on hypothalamic-oxytocin releasing neurons of noxious and innocuous mechanical stimulation and electrical stimulation of sciatic nerves were investigated in anesthetized, ovariectomized, estrogen-treated female rats. Extracellular action potentials were recorded from antidromically identified, tonically firing cells in the hypothalamic paraventricular nucleus (PVN). Tonic PVN cells responded to noxious somatosensory stimuli, as well as to electrical stimu-

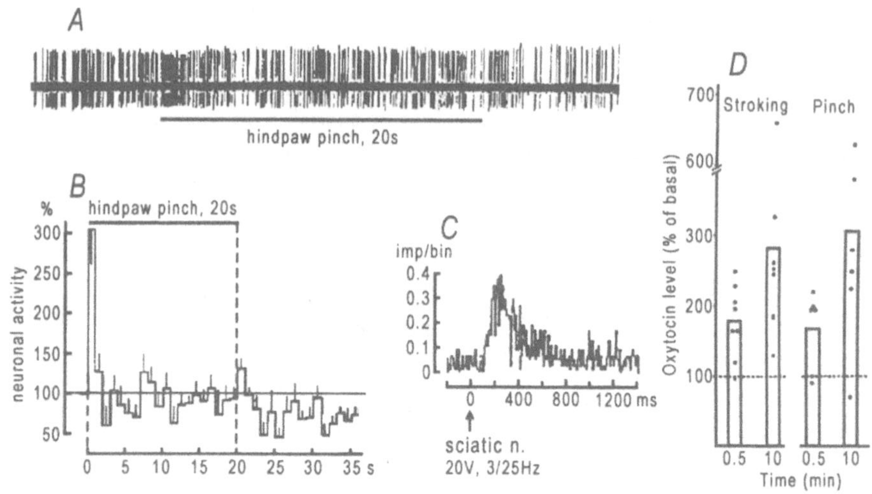

Fig. 93A–D. Effect of somatic stimulation on neuronal activity of the hypothalamic paraventricular nucleus (PVN) and oxytocin secretion in urethane-anesthetized rats. **A,B** Effect of hindpaw pinching on PVN neuronal activity. **A** Representative example of the responses in a tonically firing PVN neuron and **B** histogram of averaged responses for seven neurons. **C** Effect of electrical stimulation of sciatic nerve on PVN neuronal activity. Peristimulus-time histograms constructed for a PVN cell with tonic activity in response to stimulation. Hindpaw pinching or electrical stimulation of the sciatic nerve activated the PVN neurons. **D** Oxytocin levels in male rats increased after stroking the back or pinching one foot for 1 min. The elevation persisted and was even higher 10 min later. (**A–C** Modified from Akaishi et al. 1988; **D** from Stock and Uvnäs-Moberg 1988)

lation of a somatic nerve (Fig. 93A–C) (Akaishi et al. 1988). Pinching of the hindpaw excited 39% of the neurons tested. However, the response was very brief, generally occurring during the initial 1–2 s of the 20-s stimulation period. Pinching other body locations, such as the forepaw or nipple was also effective in a certain number of PVN neurons. In a small percentage of cases, pinching stimuli inhibited the neuronal activity. Innocuous brushing at any site did not produce a response. The sciatic nerve stimulation (20 V, three pulses at 25 Hz) produced excitation in 80% of the cells tested.

In anesthetized male rats, an increase in oxytocin release occurred after stroking stimulation to back skin or pinching stimulation of a hindlimb for 1 min (Fig. 93D), or after afferent electrical stimulation of the sciatic nerve (0.2–2 ms, 5 V, 3–10 Hz, for 1 min) (Stock and Uvnäs-Moberg 1988).

4.5.4.2
Vasopressin

Electrophysiological studies have shown that electrical activities of vasopressin-producing neurons in the supraoptic nucleus (SON) and PVN are related to the amount of vasopressin released from the posterior pituitary gland into the circulation (Bicknell and Leng 1981; Dutton and Dyball 1979).

During muscular exercise, the plasma concentration of vasopressin increases and antidiuresis increases (Kozlowski et al. 1967; Beardwell et al. 1975). In anesthetized cats, electrical stimulation of group I afferent fibers from the gastrocnemius muscle did not change SON neuron discharges, while activation of group III and IV afferent fibers excited them (Yamashita et al. 1984; Kannan et al. 1988). Injection of chemicals (sodium chloride, potassium chloride, bradykinin) into arteries supplying the muscle excited SON neurons. The excitation disappeared after section of the muscle nerves. These results indicate that activation of polymodal fibers of small diameter from muscle excites the SON neurons, leading to an increase in vasopressin secretion during exercise (Yamashita et al. 1984; Kannan et al. 1988). In anesthetized rats, Hamamura et al. (1984) found that the activity of 22 of 91 neurosecretory neurons tested in the SON was increased by tail pinching, and nine of the 22 neurons exhibited a phasic pattern of spontaneous discharge which is known to characterize certain vasopressin-secreting neurons in rats, indicating that these excited neurons are, at least in part, vasopressinergic. In addition, noxious heat stimuli (44°–63°C) applied to the hindlimb produced a transient excitation in 26% of 23 cells tested.

Shibuki and Yagi (1986) also found in anesthetized Wistar rats that electrical stimulation of saphenous afferent nerve fibers with sufficient intensity to produce a flexor reflex of the ipsilateral hindlimb transiently excited 46 of 62 phasically discharging SON neurons (most probably vasopressinergic). The threshold of the saphenous nerve stimulation for exciting the SON neurons was higher than for producing the flexor reflex.

4.5.5
Conclusion

1. Since Cannon's work, there have been many studies in conscious animals on the effects of stressful somatic sensory stimulation on catecho-

lamine secretion from the adrenal medulla via sympathetic efferent nerve activation. Some nonstressful sensory stimuli have been shown to regulate the oxytocin secretion involved in the milk ejection reflex, and to regulate the secretion of growth hormone in conscious animals and humans. In conscious humans, the involvement of emotional factors following somatic sensory stimulation is significant, so that anesthetized animals have been used to examine neural mechanisms of somatically induced hormonal responses.

2. In anesthetized rats, it was found that sympathetic nerve activity and associated catecholamine secretion were reflexly excited or inhibited, depending on the nature of the receptors of the somatic afferent stimulated. Innocuous brushing of the skin inhibited sympatho-adrenal functions, whereas noxious pinching of the skin produced an excitatory reflex response in nerve activity and catecholamine secretion. In CNS-intact rats, a single electrical shock to somatic afferent nerves elicited inhibition of the spontaneous sympathetic discharges before any excitatory discharges appeared. This inhibition, called early depression, is unique in the somato-adrenal sympathetic reflex system, and seems to be responsible for eliciting decreasing sympathetic efferent activity and consequent inhibition in catecholamine secretion during brushing stimulation of the skin. In spinalized preparations, only excitatory responses could be elicited by segmental input. The segmental organization of the excitatory reflexes within the spinal cord is modified into a nonsegmental or generalized inhibitory or excitatory response by supraspinal structures.
3. In anesthetized animals, the secretion of pancreatic hormones has been demonstrated to be reflexly regulated by natural somatic afferent stimulation via autonomic efferent nerves. The secretion of some hypothalamo-pituitary hormones has also been shown to be reflexly influenced by somatic afferent stimulation in anesthetized animals. Further study is necessary to analyze the central neural mechanisms or pathways to the hypothalamic neurons responsible for the hormonal secretions following somatic afferent stimulation. In conscious animal preparations and in humans, well-documented changes in hormonal concentrations in the blood following somatic sensory stimulation seem to involve these somato-hormonal reflex responses. However, it is difficult to distinguish these reflex responses from the responses evoked as a consequence of emotion following somatic sensory stimulation. It is important to clarify the somato-hormonal reflex responses in anes-

thetized animals, because it may be possible to extrapolate the physiological relevance of these reflexes in anesthetized animals to the responses in conscious preparations, even though such responses may include an emotionally elicited component.

4.6 Somatosensory Modulation of the Immune System

4.6.1 Introduction

In the 1930s, Hans Selye found that various harmful or stressful stimuli produced atrophy of the thymus, hemorrhage of the gastric mucosa, and hypertrophy of the adrenal glands in conscious rats. This evidence suggested that stressful stimuli might reduce immune function by inducing atrophy of the thymus. Stress suppresses the immune system, and for many years it was widely believed that this effect was mediated by glucocorticoids. However, with the demonstration that stress-induced immunosuppression occurs not only in normal but in adrenalectomized rats as well, it became clear that other factors must also be involved.

Some immune organs such as the thymus, the lymph nodes and spleen are innervated by autonomic efferent nerves (see Bulloch 1987; Felten et al. 1987). These autonomic nerve fibers have been demonstrated to influence blood flow and immune function itself in these immune-related organs. Besedovsky et al. (1979) found that in rats subjected to bilateral adrenalectomy antibody production increased in response to sheep red blood cells following surgical dissection of splenic sympathetic nerves or chemical denervation of sympathetic nerves using 6-hydroxydopamine. Electrical stimulation of splenic sympathetic efferent fibers decreases cytotoxic activity of the natural killer (NK) cells in the spleen (Katafuchi et al. 1993; Kimura et al. 1994a). Administration of CRH (Irwin 1993) and cytokine (interferon α) (Katafuchi et al. 1993) into the intracerebral ventricles causes decreases in NK cell activity, and this effect is abolished after splenic sympathetic denervation. Electrical stimulation of vagal efferent nerves increases the release of lymphocytes from the thymus, and this release is decreased by severing the vagal nerves (Antonica et al. 1994).

There have been many reports concerning immune function in conscious animals during various stressful stimuli including noxious stimuli to the somatosensory receptors (Fujiwara and Orita 1987; Shavit et al. 1990; Irwin 1993). Acupuncture stimulation enhanced immune responses in conscious mice (Lundeberg et al. 1991) and rats (T. Sato et al. 1996). The contribution of the sympathetic nervous system to these stimulus-induced immune responses in conscious animals has been analyzed by cutting sympathetic nerves to the immune-related organs or by applying blockers of sympathetic neurotransmitters.

Kimura et al. (1994a), using anesthetized animals to eliminate emotional factors, succeeded in proving that somatic afferent stimulation produces a reflex effect on immune function, with autonomic nerves acting as the efferent pathway.

4.6.2
Cutaneous Stimulation

Kimura et al. (1994a) examined the effects of innocuous and noxious mechanical stimulation of the skin on cytotoxic activity of splenic NK cells and splenic blood flow in anesthetized rats. Bilateral brushing of the body surface areas between the lateral chest and hindlimb for 30 min did not significantly influence splenic blood flow, cytotoxic activity of splenic NK cells or splenic sympathetic efferent nerve activity. Bilateral pinching stimulation of the skin of the hindpaws for 30 min reduced cytotoxic activity of splenic NK cells and splenic blood flow to 67% and 82% of the control values, respectively (Fig. 94A). Pinching the hindpaws for 30 min increased splenic sympathetic efferent nerve activity to 143% of control (Fig. 94A). Electrical stimulation (10 V, 10 Hz) of the splanchnic nerve for 20 min produced a decrease in splenic blood flow and cytotoxic activity of splenic NK cells. The hindpaw pinching-induced suppression of cytotoxic activity of NK cells was abolished after surgical transection of the splenic sympathetic nerve or the spinal cord at the cervical level. These results indicate that the suppression of splenic blood flow and cytotoxic activity of NK cells following pinching of the hindpaw skin is a supraspinal reflex response mediated via the splenic sympathetic nerve (Fig. 94B). On the other hand, pinching stimulation of the abdominal skin for 30 min reduced cytotoxic activity of splenic NK cells by increasing splenic sympathetic nerve activity as reflex responses in spinalized rats. These indicate that spinal and supraspinal pathways are involved in these

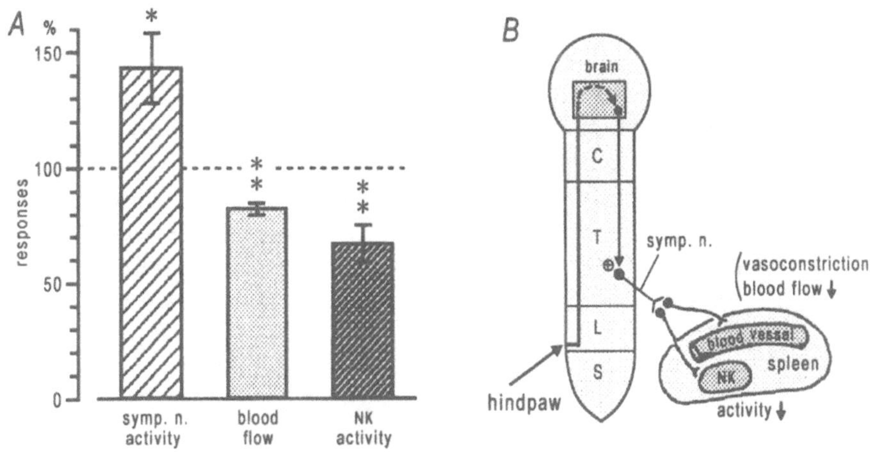

Fig. 94A,B. The effect of hindpaw pinching on splenic function in halothane-anesthetized rats. **A** Summary of the effect of hindpaw pinching on splenic sympathetic nerve activity (*left bar*), splenic blood flow (*middle bar*), and cytotoxic activity of NK cells (*right bar*). Pinching the hindpaws for 30 min increased splenic sympathetic efferent nerve activity and reduced cytotoxic activity of splenic NK cells and splenic blood flow. **B** Reflex pathway for splenic NK cell activity and blood flow responses elicited by hindpaw pinching. *$P<0.05$, **$P<0.01$; significantly different from prestimulus level or nonstimulated group. (Modified from Kimura et al. 1994a)

reflex responses depending on cutaneous segments stimulated, i.e., spinal reflex pathway acts when the abdomen is stimulated, and supraspinal reflex pathway does when hindlimbs are stimulated (Kagitani et al. 1996).

4.6.3 Conclusion

The study of the regulation of immune function under stressful conditions has been developing. However, our knowledge of reflex control of immune functions by somatic afferents is still limited. Using anesthetized animals, cytotoxic activity of the natural killer cell in the spleen was shown to be reflexly influenced by somatic afferent stimulation via sympathetic efferent nerve activity. It seems that there are dominant supraspinal reflex pathways whenever stimulation is delivered to the limb or body, but spinal reflex pathways seem to be used when stimulation is delivered to the lower thoracic segments.

Inhibitory responses of NK activity were demonstrated to be achieved via excitation of sympathetic nerves following noxious stimulation of the skin. It will be necessary to establish which somatic afferents at various segments decrease sympathetic nerve activity and consequently influence the activation of NK cells.

It appears that immune function can be inhibited or facilitated by somatic sensory stimulation in conscious humans. It is expected that the immune functions of the thymus and lymph nodes will be studied in the near future from the point of view of the effects of somatic afferent stimulation.

5 Concluding Remarks

It is readily apparent from the experiences of daily life that stimulation of various types of somatosensory receptors produces not only conscious sensations but also physiological responses. These bodily responses engage the skeletomotor, autonomic, endocrine and immune systems, as well as other organ systems. In traditional medicine, such responses evoked by sensory stimulation have been employed for many hundreds if not thousands of years in the therapy of various kinds of diseases. In modern medicine, too, many forms of physical treatment are used against a variety of diseases without knowing the exact mechanisms by which these therapies bring about their effects. There have been many theories explaining these effects. In humans, somatosensory stimulation provokes conscious sensation and consequent emotional responses that secondarily evoke various bodily responses. As emotional responses vary between individuals, analysis of these responses has been extremely difficult, particularly from the point of view of resolving reflex mechanisms. In fact, many studies have been carried out on humans using modern techniques such as microneurography for recording autonomic efferent activity, or Positron Emission Tomography for measuring regional cerebral blood flow. However, accurate analysis of the reflex mechanisms has not been successful. Sometimes paraplegic patients have been helpful subjects for investigating those spinal reflex pathways that are independent of emotional responses. Total understanding of reflex mechanisms, including spinal and supraspinal reflex pathways, absolutely requires the study of animals which have been anesthetized to eliminate emotional responses.

As discussed in this review, in the last three decades many reflex mechanisms of somatically induced autonomic responses have been clarified using anesthetized animals. In the early stages, most animal experiments were carried out on cats, dogs and rabbits. Studies in rats would have been more convenient from several points of view, however, it was generally accepted that it was impractical to conduct complex studies in anesthetized rats due to the difficulty of maintaining the animals under prolonged anesthesia. Through the early 1970s we developed the appropriate protocols to monitor and sustain physiological parameters, and in 1975 we first published our results from research into somato-autonomic reflexes in anesthetized rats whose respiration, blood pressure and temperature were strictly maintained at normal physiological levels. Since then, an increasingly large amount of research has been conducted using anesthestized rats. Experiments on primates appear to be much less common than those on cats and rats. However, it is expected that in the future much more primate research will be carried out in order to relate the results of animal experiments to humans.

It is now clear that there are a multitude of reflex responses of visceral function following somatic afferent stimulation. The afferent limbs of these reflexes involve group II, III, and IV afferent fibers entering the spinal cord via the spinal nerves, and the brain stem via the trigeminal nerve. Their efferent pathways are autonomic efferent fibers, i.e., sympathetic and parasympathetic efferent fibers. In some cases efferent pathways can be considered to include hormones themselves or hormones regulated by autonomic efferent nerves.

Among the most important findings have been that somato-autonomic or hormone reflexes are characterized by either segmental or nonsegmental organizations, depending upon the visceral organs involved. Furthermore, excitatory and inhibitory reflex responses have been demonstrated depending, again, sometimes on the particular visceral organ's condition and sometimes on the somatic afferent modalities employed. An understanding of all of these properties of the somato-autonomic reflexes appears to be essential for explaining the neural mechanisms by which the majority of physical treatments affect diseases. In this context, we would draw the reader's attention to the fact that the format of this review has been such that investigations pertaining to particular organ systems have been assembled into separate chapters to give the reader a perspective which is biased from the point of view of the effector organs of somato-autonomic reflexes. Secondarily, within the consideration of each target organ system we have discussed the effects of stimulation of different classes of somatic tissues, for example cutaneous or articular, thereby reviewing somato-autonomic reflexes from the point of view of the receptors involved. In synthesizing our concluding remarks we would like to propose a third perspective, which we might term axio-appendicular.

Time and again, in this review of the research on somato-autonomic reflexes, we have been compelled to contrast the supraspinal and spinal components, and to remark upon the segmental organization of the spinal reflex. These are quite exciting findings, since the existence of spinal centers for somato-autonomic reflexes had been denied until quite recently. This is in sharp contrast to, and partly because of, the success of Sherrington's model of the spinal motor reflex. From clinical and even everyday experience, it is known that stimulation of limb afferents, such as those in muscles and tendons, gives rise to motor reflexes. The so-called "patellar tendon reflex" is a familiar example. The model of stimulating somatic limb afferents to elicit somatomotor responses has been so successful that it has necessarily influenced the study of autonomic reflex responses.

Stimulation of somatic limb afferents does indeed elicit broad reflex visceral responses, which are normally abolished in the acute spinal preparation. This led to two widespread generalizations which have impeded the development of studies in this area of physiology. The first was that somatic afferent stimulation produced only a generalized autonomic response, for example of the "fight or flight" variety. The second was the belief that somato-autonomic reflexes were mediated exclusively at the supraspinal level. These conclusions were correct within a limited experimental construct, and in fact were dictated by the bias of experimental design which focused on stimulation of limb afferents. Sensory input from limb afferents enters the spinal cord at the cervical and lumbar enlargements. These regions are richly invested with somatic motor neurons to the virtual exclusion of autonomic motor neurons. Hence, limb afferents readily synapse with somatic motor efferents such that the appropriate stimulation of receptors in the limbs elicits potent somatic motor responses. Necessarily, therefore, any autonomic reflex responses to arise from limb stimulation must be mediated primarily at the supraspinal level and would therefore be abolished in the acute spinal preparation (Fig. 95).

It was not until a sufficient body of evidence was slowly accumulated from the stimulation of spinal segmental (thoracolumbar and sacral) nerves, that the existence of the spinally mediated somato-autonomic reflex could be established. In humans, preganglionic sympathetic neurons form the intermediolateral and intermediomedial columns between, approximately, the first thoracic and the second or third lumbar segments. Preganglionic parasympathetic neurons form the intermediate columns in the second to fourth sacral segments. Somatic afferents entering the

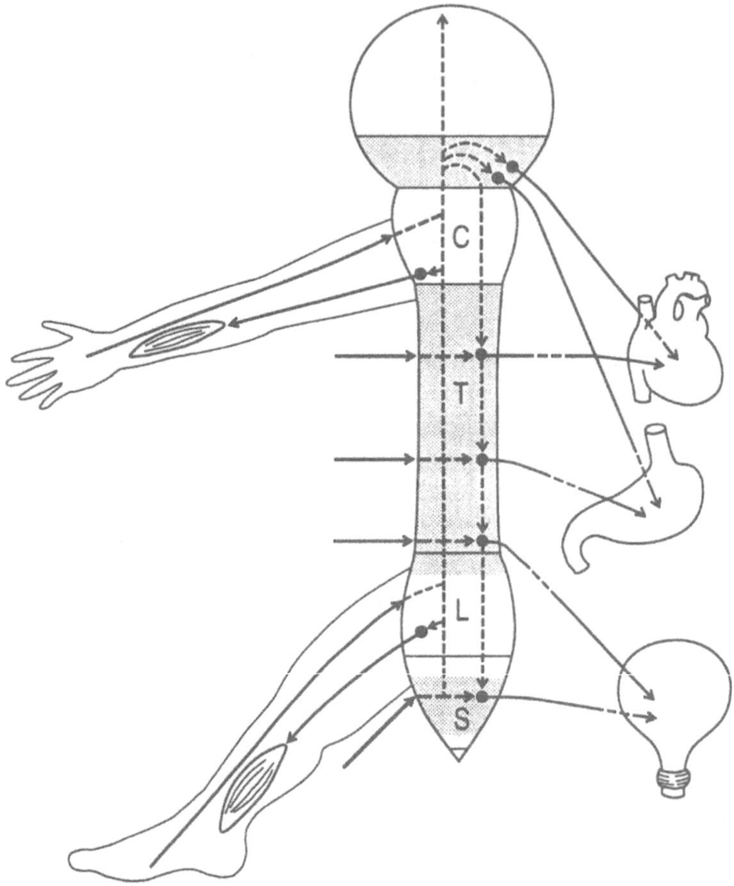

Fig. 95. Reflex pathways for the somato-somatic and somato-autonomic reflexes. Somatic motor neurons innervating limbs are contained in the spinal cord at the cervical and lumbar enlargements. Limb afferents readily synapse with somatic motor efferents to elicit potent spinal reflexes. These spinal segments, however, are essentially devoid of autonomic preganglionic neurons. Thus somato-autonomic reflexes elicited by stimulation of limb afferents appear to be mediated mainly at the supraspinal level. On the other hand, stimulation of segmental afferents may elicit responses from both supraspinal and spinal reflex centers. Spinally mediated somato-autonomic reflexes may show a very strong segmental organization and, under the appropriate conditions, the effects on target organs may be quite specific. Often, however, in the CNS-intact animal, these spinal reflexes are masked by descending influences from the brain. *Broken lines* in this figure indicate synaptic pathways, not direct axonal pathways. *Shadowed areas* of the CNS indicate regions which contain the autonomic preganglionic neurons

spinal cord at these segmental levels have the opportunity, therefore, to synapse with local autonomic motor neurons, as well as to synapse with projections to higher supraspinal centers. Thus stimulation of the appropriate segmental spinal nerves does indeed elicit spinally mediated autonomic responses, and these responses, rather than being broad and generalized, are most likely to be specific for organs served by the local autonomic efferents.

Hence, while in general certain types of stimulation may produce stereotypical responses in particular effector organs or organ systems, the anatomical (axio-appendicular) location of the stimulated receptor organs may have an important and previously under-appreciated influence on the quality or magnitude of the response which is expressed.

Most studies of somato-autonomic reflexes have been carried out using acute experimental procedures. Recent progress in sensory neurophysiological studies of inflammation of the joints appears to be opening a new door to study long-term ("plastic") modifications of somato-autonomic reflexes. In the future more prolonged stimulation, in the order of hours, and repetitive stimulation over a period of days will probably be necessary in order to obtain experimental results which are applicable to the physical treatment of various human diseases.

In this review, diverse autonomic functions were discussed from the point of view of somatic afferent reflex regulation. As an example, neural regulation of regional cerebral blood flow is a completely new finding. This type of neural modulation has never been considered a part of autonomic nervous regulation. However, as mentioned in this review, this type of neural reflex mechanism closely resembles autonomic vascular regulation elsewhere in the body, and there should be no difficulty in considering this type of modulation within a broader definition of autonomic regulation. Within our bodies we may have many other neural reflex mechanisms that have never previously been recognized as autonomic nervous functions, and such new types of nervous regulation can be expected to be studied from the perspective of somato-autonomic reflexes. Along these lines, there is an urgent need to study somato-autonomic immune reflexes.

The main subject of this review has been the somato-autonomic reflexes. However, whenever somatic afferent stimulation is applied clinically for the treatment of visceral disorders, it must be remembered that somatic afferent stimulation may have analgesic effects on various types of pain, for example by releasing endogenous endorphins in the CNS. As this subject has been reviewed repeatedly elsewhere, the present review did not address the important analgesic effects of somatic afferent stimulation. Nonetheless, the combined effects of somatic afferent stimulation on visceral reflexes and analgesia must be kept in mind, in addition to subjective sensory processes, whenever somatic afferents are stimulated, particularly when these afferents are stimulated for medical therapy.

In closing, the authors of this review have enjoyed a long collaboration in which they have endeavored to develop a scientific approach to the study of somato-autonomic reflexes. We believe that some success has been achieved in establishing a neurophysiological rationale for further research. If so, this success has only been possible due to the sincere efforts of many investigators in laboratories around the world. We therefore hope that the present review will provide thoughtful stimulation to the next generation of researchers who will carry forward the study of this important subject.

Acknowledgments. This work was generously supported by a Max Planck Research Award from the Alexander von Humboldt Foundation and the Max Planck Society (1991) to R.F. Schmidt and A. Sato. We are grateful to Dr. Harumi Hotta, Dr. Atsuko Kimura, Dr. Atsuko Suzuki, and Dr. Harue Suzuki (Nakamura) for their devotion in helping to prepare this article and carry it to completion. We would also like to thank Dr. Brian S. Budgell for his assistance in improving the English. Finally, all of the authors would like to express their deep gratitude to Mrs. Takako Matsuzaki (Noguchi) for her excellent secretarial assistance over the past 17 years.

Appendix: A Personal Note

The friendship and collaboration of the authors has spanned the last 30 years. A. Sato worked in R.F. Schmidt's laboratory at Heidelberg for more than 3 years (1966, 1969–1972). Later A. Sato repeatedly visited R.F. Schmidt's laboratories in Kiel and Würzburg to continue our collaborative work. Y. Sato also worked in R.F. Schmidt's laboratories in Kiel and Würzburg for more than 6 months (1982). All three of the current authors have worked together at A. Sato's laboratory in Tokyo for a total of more than 1 year on six occasions since 1977. Our close, long-lasting relationship has made it possible to write this article. One of us (A. Sato) is required by governmental regulations to retire from his present research institute this year (1997) at the age of 63 years, and we assume that there may not be many further opportunities for us to continue our investigative collaboration. We believe that science in this area must be carried forward by many researchers worldwide.

During our collaboration, and in addition to the other authors, A. Sato had the opportunity to host many guest scientists from abroad. These visiting researchers have worked with us in the laboratory for periods of at least 1 month, and often longer. They have made invaluable contributions to many of the studies cited in this review. In alphabetical order and without reference to title, these colleagues are:

Dodo G. Baramidze (Tbilisi, Georgia), Karen J. Berkley (Tallahassee, USA), Dietmar Biesold (Leipzig, Germany), Chandler McC Brooks (New York, USA), Brian S. Budgell (Toronto, Canada), Franco R. Calaresu (London, Canada), Wei-Hua Cao (Hu Nan, China), Samuel H.H. Chan (Taipei, Republic of China), John C. Coote (Birmingham, UK), Bernard T. Engel (Baltimore, USA), Salvatore J. Fidone (Salt Lake City, USA), Giorgio Gabella (London, UK), Zheng-Zhong Gu (Shanghai, China), Åse Hallström (Stockholm, Sweden), Antti Hervonen (Tampere, Finland), Albert Kaufman (New York, USA), Kiyomi Koizumi (New York, USA), Wei-Min Li (Shanghai, China), John F.B. Morrison (Leeds, UK), Dariusz Nowicki (Warsaw, Poland), Ann Robbins (New York, USA), Peter Sandor (Budapest, Hungary), Hans-Georg Schaible (Würzburg, Germany), Albert Simpson (USA), Rand S. Swenson (New Hampshire, USA), Andrzej Trzebski (Warsaw, Poland), Urban Ungerstedt (Stockholm, Sweden), Kerstin Uvnäs-Moberg (Stockholm, Sweden), Ahmad P.M. Yusof (Penang, Malaysia), Wei Zhow (Hu Nan, China).

Furthermore, over the last 23 years A. Sato has had the following Japanese students and coworkers in his laboratory to investigate somato-autonomic reflexes. In alphabetical order, their names are:

Takehiko Adachi, Takao Akaishi, Yasuichiro Fukuda, Toshiro Hamamoto, Harumi Hotta, Osamu Inanami, Tadashi Isa, Kenichi Ito, Fusako Kagitani, Hideki Kametani, Masakazu Kaseda, Jitsu Kato, Pyo Kim, Hitomi Kashiwagi, Atsuko Kimura, Mieko Kurosawa, Konosuke Kumakura, Nobuo Nagai, Osamu Nagata, Takashi Nagayama, Shoichiro Nosaka, Kikuro Ohno, Hideo Ohsawa, Kaoru Okada, Yuka Saeki, Yojiro Sakiyama, Mitsuyoshi Sasaki, Keiichi Shimamura, Atsuko Suzuki, Harue Suzuki (Nakamura), Hisako Sugimoto, Fumiyo Shimada, Yukio Takano, Naoto Terui, Yuko Torigata, Toru Tsuchiya, Sae Uchida, Kikuko Ueki, Yoshiko Yamauchi.

References

Abrahams VC, Hilton SM, Zbrozyna A (1960) Active muscle vasodilatation produced by stimulation of the brain stem: its significance in the defence reaction. J Physiol (Lond) 154:491–513

Abram SE, Kostreva DR, Hopp FA, Kampine JP (1983) Cardiovascular responses to noxious radiant heat in anesthetized cats. Am J Physiol 245:R576–R580

Adachi T, Meguro K, Sato A, Sato Y (1990) Cutaneous stimulation regulates blood flow in cerebral cortex in anesthetized rats. Neuroreport 1:41–44

Adachi T, Chan SHH, Sato A, Yamamoto M (1991) Regulation of norepinephrine release and cerebral blood flow in the parietal cortex by locus coeruleus in the rat. Biogenic Amines 8:19–31

Adachi T, Baramidze DG, Sato A (1992a) Stimulation of the nucleus basalis of Meynert increases cortical cerebral blood flow without influencing diameter of the pial artery in rats. Neurosci Lett 143:173–176

Adachi T, Sato A, Sato Y, Schmidt RF (1992b) Depending on the mode of application morphine enhances or depresses somatocardiac sympathetic A- and C-reflexes in anesthetized rats. Neurosci Res 15:281–288

Adriaensen H, Gybels J, Handwerker HO, van Hees J (1983) Response properties of thin myelinated (A-delta) fibers in human skin nerves. J Neurophysiol 49:111–122

Adrian ED, Bronk DW, Phillips G (1932) Discharges in mammalian sympathetic nerves. J Physiol (Lond) 74:115–133

Akaishi T, Robbins A, Sakuma Y, Sato Y (1988) Neural inputs from the uterus to the paraventricular magnocellular neurons in the rat. Neurosci Lett 84:57–62

Akaishi T, Kimura A, Sato A, Suzuki A (1990) Responses of neurons in the nucleus basalis of Meynert to various afferent stimuli in rats. Neuroreport 1:37–39

Alexander RS (1946) Tonic and reflex functions of medullary sympathetic cardiovascular centers. J Neurophysiol 9:205–217

Allison DJ, Powis DA (1971) Adrenal catecholamine secretion during stimulation of the nasal mucous membrane in the rabbit. J Physiol (Lond) 217:327–339

Amann R, Lembeck F (1987) Stress induced ACTH release in capsaicin treated rats. Br J Pharmacol 90:727–731

Amenta F, Mione MC, Napoleone P (1983) The autonomic innervation of the vasa nervorum. J Neural Transm 58:291–297

Andersson P-O, Holst J, Järhult J (1982) Effects of adrenergic blockade on the release of insulin, glucagon and somatostatin from the pancreas in response to splanchnic nerve stimulation in cats. Acta Physiol Scand 116:403–409

Andres KH, v. Düring M (1973) Morphology of cutaneous receptors. In: Iggo A (ed) Handbook of sensory physiology, vol II. Springer, Berlin Heidelberg New York

Andres KH, v. Düring M, Schmidt RF (1985) Sensory innervation of the achilles tendon by group III fibers. Anat Embryol (Berl) 172:145–156

Andres KH, v. Düring M, Muszinski K, Schmidt RF (1987a) The innervation of the dura mater encephali of the rat. In: Schmidt RF, Schaible H-G, Vahle-Hinz C (eds) Fine afferent nerve fibers and pain. Edition Medizin, Weinheim, pp 15–25

Andres KH, v. Düring M, Muszynski K, Schmidt RF (1987b) Nerve fibres and their terminals of the dura mater encephali of the rat. Anat Embryol (Berl) 175:289–301

Antonica A, Magni F, Mearini L, Paolocci N (1994) Vagal control of lymphocyte release from rat thymus. J Auton Nerv Syst 48:187–197

Aoki M, Yamamura T (1977) Functional properties of peripheral sensory units in hairy skin of a cat's forelimb. Jpn J Physiol 27:279–289

Appenzeller O, Dhital KK, Cowen T, Burnstock G (1984) The nerves to blood vessels supplying blood to nerves: the innervation of vasa nervorum. Brain Res 304:383–386

Araki T, Hamamoto T, Kurosawa M, Sato A (1980) Response of adrenal efferent nerve activity to noxious stimulation of the skin. Neurosci Lett 17:131–135

Araki T, Ito K, Kurosawa M, Sato A (1981) The somato-adrenal medullary reflexes in rats. J Auton Nerv Syst 3:161–170

Araki T, Ito K, Kurosawa M, Sato A (1984) Responses of adrenal sympathetic nerve activity and catecholamine secretion to cutaneous stimulation in anesthetized rats. Neuroscience 12:289–299

Arita H, Kogo N, Ichikawa K (1988) Locations of medullary neurons with non-phasic discharges excited by stimulation of central and/or peripheral chemoreceptors and by activation of nociceptors in cat. Brain Res 442:1–10

Asmussen E (1956) Observations on experimental muscular soreness. Acta Rheumat Scand 2:109–116

Aung-Din R, Mitchell JH, Longhurst JC (1981) Reflex α-adrenergic coronary vasoconstriction during hindlimb static exercise in dogs. Circ Res 48:502–509

Babkin BP, Kite WCJ (1950) Central and reflex regulation of motility of pyloric antrum. J Neurophysiol 13:321–334

Bahr R, Blumberg H, Jänig W (1981) Do dichotomizing afferent fibers exist which supply visceral organs as well as somatic structures? A contribution to the problem of referred pain. Neurosci Lett 24:25–28

Bainbridge FA (1914) On some cardiac reflexes. J Physiol (Lond) 48:332–340

Barker D (1974) The morphology of muscle receptors. In: Hunt CC (ed) Handbook of sensory physiology. Muscles receptors. Springer, Berlin Heidelberg New York, pp 1–90

Barman SM, Wurster RD (1975) Visceromotor organization within descending spinal sympathetic pathways in the dog. Circ Res 37:209–214

Barman SM, Wurster RD (1978) Interaction of descending spinal sympathetic pathways and afferent nerves. Am J Physiol 234:H223–H229
Barman SM, McCaffrey TV, Wurster RD (1976) Cardiovascular and electrophysiological responses to sympathetic pathway stimulation. Am J Physiol 230:1095–1100
Barrington FJF (1914) The nervous mechanism of micturition. Q J Exp Physiol 8:33–71
Barrington FJF (1925) The effect of lesions of the hind- and mid-brain on micturition. Q J Exp Physiol 15:81–102
Barrington FJF (1928) The central nervous control of micturition. Brain 51:209–220
Barron DH, Matthews BHC (1938) Dorsal root reflexes. J Physiol (Lond) 94:26P–27P
Barron W, Coote JH (1973) The contribution of articular receptors to cardiovascular reflexes elicited by passive limb movement. J Physiol (Lond) 235:423–436
Baum T, Becker FT (1983) Suppression of a somatosympathetic reflex by the gamma-aminobutyric acid agonist muscimol and by clonidine. J Cardiovasc Pharmacol 5:121–124
Baum T, Shropshire AT (1979) Interaction of cardiopulmonary chemoreceptor (Bezold-Jarisch) and somatosympathetic reflexes. Arch Int Pharmacodyn 239:99–108
Baumann TK, Simone DA, Shain CN, Lamotte RH (1991) Neurogenic hyperalgesia: the search for the primary cutaneous afferent fibers that contribute to capsaicin-induced pain and hyperalgesia. J Neurophysiol 66:212–227
Beacham WS, Perl ER (1964a) Background and reflex discharge of sympathetic preganglionic neurones in the spinal cat. J Physiol (Lond) 172:400–416
Beacham WS, Perl ER (1964b) Characteristics of a spinal sympathetic reflex. J Physiol (Lond) 173:431–448
Beardwell C, Geelen G, Palmer H, Roberts D, Salamonson L (1975) Radioimmunoassay of plasma vasopressin in physiological and pathological states in man. J Endocrinol 67:189–202
Beaty O III (1985a) Arterial blood pressure control during hindlimb and forelimb contraction in the dog. Am J Physiol 248:H678–H687
Beaty O III (1985b) Carotid sinus and blood pressure control during hindlimb and forelimb contractions. Am J Physiol 248:H688–H694
Bell JA, Sharpe LG, Berry JN (1980) Depressant and excitant effects of intraspinal microinjections of morphine and methionine-enkephalin in the cat. Brain Res 196:455–465
Ben-Jonathan N (1990) Prolactin releasing and inhibiting factors in the posterior pituitary. In: Mÿller EE, MacLeod RM (eds) Neuroendocrine perspectives, vol 8. Springer, Berlin Heidelberg New York, pp 1–38
Berberich P, Hoheisel U, Mense S (1988) Effects of a carrageenan-induced myositis on the discharge properties of group III and IV muscle receptors in the cat. J Neurophysiol 59:1395–1409
Bernthal PJ, Koss MC (1984) Evidence for two distinct sympathoinhibitory bulbo-spinal systems. Neuropharmacology 23:31–36
Besedovsky HO, del Rey A, Sorkin E, da Prada M, Keller HH (1979) Immunoregulation mediated by the sympathetic nervous system. Cell Immunol 48:346–355

Besson JM, Chaouch A (1987) Peripheral and spinal mechanisms of nociception. Physiol Rev 67:67–186
Bessou P, Burgess PR, Perl ER, Taylor CB (1971) Dynamic properties of mechanoreceptors with unmyelinated fibers. J Neurophysiol 34:116–131
Bicknell RJ, Leng G (1981) Relative efficiency of neural firing patterns for vasopressin release in vitro. Neuroendocrinology 33:295–299
Biesold D, Inanami O, Sato A, Sato Y (1989a) Stimulation of the nucleus basalis of Meynert increases cerebral cortical blood flow in rats. Neurosci Lett 98:39–44
Biesold D, Kurosawa M, Sato A, Trzebski A (1989b) Hypoxia and hypercapnia increase the sympathoadrenal medullary functions in anesthetized, artificially ventilated rats. Jpn J Physiol 39:511–522
Bisgard JD, Nye D (1940) The influence of hot and cold application upon gastric and intestinal motor activity. Surg Gynecol Obstet 71:172–180
Bisset GW (1974) Milk ejection. In: Greep RO, Astwood EB, Knobil E, Sawyer WH, Geiger SR (eds) Handbook of physiology. Section 7, endocrinology, vol IV, part 1. American Physiological Society, Washington, pp 493–520
Blair RW (1991) Convergence of sympathetic, vagal, and other sensory inputs onto neurons in feline ventrolateral medulla. Am J Physiol 260:H1918–H1928
Bligh J (1973) Temperature regulation in mammals and other vertebrates. Elsevier, Amsterdam
Bloom S, Edwards A (1975) The release of pancreatic glucagon and inhibition of insulin in response to stimulation of the sympathetic innervation. J Physiol (Lond) 253:157–173
Bloom S, Edwards A, Vaughan N (1973) The role of the sympathetic innervation in the control of plasma glucagon concentration in the calf. J Physiol (Lond) 233:457–466
Bloom S, Edwards A, Vaughan N (1974) The role of the autonomic innervation in the control of glucagon release during hypoglycaemia in the calf. J Physiol (Lond) 236:611–623
Bobbioni E, Marre M, Helman A, Assan R (1983) The nervous control of rat glucagon secretion in vivo. Horm Metab Res 15:133–138
Boyd IA, Davey MR (1968) Composition of peripheral nerves. Livingstone, Edinburgh
Boyd IA, Kalu KU (1979) Scaling factor relating conduction velocity and diameter for myelinated afferent nerve fibres in the cat hind limb. J Physiol (Lond) 289:277–297
Bradley K, Eccles JC (1953) Analysis of the fast afferent impulses from thigh muscles. J Physiol (Lond) 122:462–473
Bradley WE, Teague CT (1968) Spinal cord organization of micturition reflex afferents. Exp Neurol 22:504–516
Bridges RS, Goldman BD (1976) Suckling and LH-induced progesterone secretion in lactating hamsters (Mesocricetus auratus). Neuroendocrinology 21:20–30
Brody MJ (1978) Histaminergic and cholinergic vasodilator systems. In: Vanhoutte PM, Leusen I (eds) Mechanics of vasodilation. Karger, Basel, pp 266–277

Bromm B, Treede RD (1991) Laser-evoked cerebral potentials in the assessment of cutaneous pain sensitivity in normal subjects and patients. Rev Neurol (Paris) 147:625–643

Brooks CMC (1933) Reflex activation of the sympathetic system in the spinal cat. Am J Physiol 106:251–266

Brown AC, Brengelmann GL (1970) The temperature regulation control system. In: Hardy JD, Gagge PA, Stolwijk JAJ (eds) Physiological and behavioral temperature regulation. Charles C. Thomas, Springfield, Illinois, pp 684–702

Brown AG (1977) Cutaneous axons and sensory neurones in the spinal cord. Br Med Bull 33:109–112

Brown AG (1981) The terminations of cutaneous nerve fibres in the spinal cord. TINS March:64–67

Brown AG, Iggo A (1967) A quantitative study of cutaneous receptors and afferent fibers in the cat and rabbit. J Physiol (Lond) 193:707–733

Brown AG, Rose PK, Snow PJ (1977) The morphology of hair follicle afferent fibre collaterals in the spinal cord of the cat. J Physiol (Lond) 272:779–797

Brown AG, Rose PK, Snow PJ (1978) Morphology and organization of axon collaterals from afferent fibres of slowly adapting type I units in cat spinal cord. J Physiol (Lond) 277:15–27

Brown AG, Fyffe REW, Noble R (1980) Projections from Pacinian corpuscles and rapidly adapting mechanoreceptors of glabrous skin to the cat's spinal cord. J Physiol (Lond) 307:385–400

Brück K, Wünnenberg W (1970) "Meshed" control of two effector systems: non-shivering and shivering thermogenesis. In: Hardy JD, Gagge PA, Stolwijk JAJ (eds) Physiological and behavioral temperature regulation. Charles C. Thomas, Springfield, Illinois, pp 562–580

Budgell B, Sato A (1994) Somatoautonomic reflex regulation of sciatic nerve blood flow. J Neuromuscul Syst 2:170–177

Budgell B, Hotta H, Sato A (1995) Spinovisceral reflexes evoked by noxious and innocuous stimulation of the lumbar spine. J Neuromuscul Syst 3:122–131

Bülbring E, Burn JH (1935) The sympathetic dilator fibres in the muscles of the cat and dog. J Physiol (Lond) 83:483–501

Bulloch K (1987) The innervation of immune system tissues and organs. In: Cotman CW, Brinton RE, Galaburda A, McEwen B, Schneider DM (eds) The neuro-immune-endocrine connection. Raven, New York, pp 33–47

Burgess PR, Perl ER (1973) Cutaneous mechanoreceptors and nociceptors. In: Iggo A (ed) Handbook of sensory physiology. Vol II somatosensory system. Springer, Berlin Heidelberg New York, pp 29–78

Burgess PR, Petit D, Warren RM (1968) Receptor types in cat hairy skin supplied by myelinated fibers. J Neurophysiol 31:833–848

Burnet FR, Wakerley JB (1976) Plasma concentrations of prolactin and thyrotrophin during suckling in urethane-anesthetized rats. J Endocrinol 70:429–437

Butler SR, Schanbeg SM (1977) Effect of maternal deprivation on polyamine metabolism in preweanling rat brain and heart. Life Sci 21:877–884

Butler SR, Suskind MR, Schanberg SM (1978) Maternal behavior as a regulator of polyamine biosynthesis in brain and heart of the developing rat pup. Science 199:445–446

Calaresu FR, Pearce JW (1965) Electrical activity of efferent vagal fibres and dorsal nucleus of the vagus during reflex bradycardia in the cat. J Physiol (Lond) 176:228–240

Calaresu FR, Yardley CP (1988) Medullary basal sympathetic tone. Annu Rev Physiol 50:511–524

Calaresu FR, Kim P, Nakamura H, Sato A (1978) Electrophysiological characteristics of renorenal reflexes in the cat. J Physiol (Lond) 283:141–154

Camilleri M, Malagelada JR, Kao PC, Zinsmeister AR (1984) Effect of somatovisceral reflexes and selective dermatomal stimulation on postcibal antral pressure activity. Am J Physiol 247:G703–G708

Campbell JN, Meyer RA, Lamotte RH (1979) Sensitization of myelinated nociceptive afferents that innervate monkey hand. J Neurophysiol 42:1669–1679

Cannon WB (1919) Studies on the conditions of activity in endocrine glands. V. The isolated heart as an indicator of adrenal secretion induced by pain, asphyxia and excitement. Am J Physiol 50:399–432

Cannon WB (1929) Bodily changes in pain, hunger, fear and rage. 2nd edn. Appleton-Century, New York

Cannon WB, Hoskins RG (1912) The effects of asphyxia, hyperpnea, and sensory stimulation on adrenal secretion. Am J Physiol 29:274–279

Cannon WB, Lewis JT, Britton SW (1926) Studies on the conditions of activity in endocrine glands. XVII. A lasting preparation of the denervated heart for detecting internal secretion, with evidence for accessory accelerator fibers from the thoracic sympathetic chain. Am J Physiol 77:326–352

Cao W-H, Inanami O, Sato A, Sato Y (1989) Stimulation of the septal complex increases local cerebral blood flow in the hippocampus in anesthetized rats. Neurosci Lett 107:135–140

Cao W-H, Sato A, Sato Y, Zhou W (1992a) Somatosensory regulation of regional hippocampal blood flow in anesthetized rats. Jpn J Physiol 42:731–740

Cao W-H, Sato A, Yusof APM, Zhou W (1992b) Regulation of regional blood flow in cerebral cortex by serotonergic neurons originating in the dorsal raphe nucleus in the rat. Biogenic Amines 8:351–360

Capra NF (1987) Localization and central projections of primary afferent neurons that innervate the temporomandibular joints in cats. Somatosens Res 4:201–213

Cavanagh PR (1988) On "muscle action" vs "muscle contraction". J Biomech 21:69

Celander O, Folkow B (1953) The nature and the distribution of afferent fibres provided with the axon reflex arrangement. Acta Physiol Scand 29:359–370

Celesia GG, Jasper HH (1966) Acetylcholine released from cerebral cortex in relation to state of activation. Neurology 16:1053–1063

Cervero F (1985) Visceral nociception: peripheral and central aspects of visceral nociceptive systems. In: Iggo A, Iversen LL, Cervero F (eds) Nociception and pain. Royal Society, London, pp 107–119

Cervero F, McRitchie HA (1981) Neonatal capsaicin and thermal nociception: a paradox. Brain Res 215:414–418

Chambers MR, Andres KH, v. Düring M, Iggo A (1972) The structure and function of the slowly adapting type II mechanoreceptor in hairy skin. Q J Exp Physiol 57:417–445

Chang CH, Gimenez T, Henricks DM (1981) Modulation of reproductive hormones by suckling and exogenous gonadal hormones in young beef cows post partum. J Reprod Fertil 63:31–38

Chiu DTJ, Cheng K-K (1974) A study of the mechanism of the hypotensive effect of acupuncture in the rat. Am J Chin Med 2:413–419

Chung JM, Wurster RD (1976) Ascending pressor and depressor pathways in the cat spinal cord. Am J Physiol 231:786–792

Chung JM, Wurster RD (1978) Neurophysiological evidence for spatial summation in the CNS from unmyelinated afferent fibers. Brain Res 153:596–601

Chung JM, Webber CLJ, Wurster RD (1979) Ascending spinal pathways for the somatosympathetic A and C reflexes. Am J Physiol 237:H342–H347

Ciriello J, Calaresu FR (1977) Lateral reticular nucleus: a site of somatic and cardiovascular integration in the cat. Am J Physiol 233:R100–R109

Clark FJ (1972) Central projections of sensory fibers from the cat knee joint. J Neurobiol 3:101–110

Clement DL (1976) Neurogenic influences on blood pressure and vascular tone from peripheral receptors during muscular contraction. Cardiology 61[Suppl 1]:65–68

Clement DL, Shepherd JT (1974) Influence of muscle afferents on cutaneous and muscle vessels in the dog. Circ Res 35:177–183

Clement DL, Pelletier CL, Shepherd JT (1973) Role of muscular contraction in the reflex vascular responses to stimulation of muscle afferents in the dog. Circ Res 33:386–392

Clemente CD (ed) (1985) Gray's anatomy. 30th edn. Lea and Febiger, Philadelphia

Coffman JD (1966) The effect of aspirin on pain and hand blood flow responses to intra-arterial injection of bradykinin in man. Clin Pharmacol Ther 7:26–37

Coote JH (1975) Physiological significance of somatic afferent pathways from skeletal muscle and joints with reflex effects on the heart and circulation. Brain Res 87:139–144

Coote JH, Dodds WN (1976) The baroreceptor reflex and the cardiovascular changes associated with sustained muscular contraction in the cat. Pflugers Arch 363:167–173

Coote JH, Downman CBB (1966) Central pathways of some autonomic reflex discharges. J Physiol (Lond) 183:714–729

Coote JH, Downman CBB (1969) Supraspinal control of reflex activity in renal nerves. J Physiol (Lond) 202:161–170

Coote JH, Macleod VH (1975) The spinal route of sympatho-inhibitory pathways descending from the medulla oblongata. Pflugers Arch 359:335–347
Coote JH, Perez-Gonzalez JF (1970) The response of some sympathetic neurones to volleys in various afferent nerves. J Physiol (Lond) 208:261–278
Coote JH, Sato A (1978) Supraspinal regulation of spinal reflex discharge into cardiac sympathetic nerves. Brain Res 142:425–437
Coote JH, Westbury DR (1979) Intracellular recordings from sympathetic preganglionic neurones. Neurosci Lett 15:171–175
Coote JH, Downman CBB, Weber WV (1969) Reflex discharges into thoracic white rami elicited by somatic and visceral afferent excitation. J Physiol (Lond) 202:147–159
Coote JH, Hilton SM, Perez-Gonzalez JF (1971) The reflex nature of the pressor response to muscular exercise. J Physiol (Lond) 215:789–804
Craig AD (1993) Propriospinal input to thoracolumbar sympathetic nuclei from cervical and lumbar lamina I neurons in the cat and the monkey. J Comp Neurol 331:517–530
Craig AD, Heppelmann B, Schaible H-G (1988) The projection of the medial and posterior articular nerves of the cat's knee to the spinal cord. J Comp Neurol 276:279–288
Crayton SC, Aung-Din R, Fixler DE, Mitchell JH (1979) Distribution of cardiac output during induced isometric exercise in dogs. Am J Physiol 236:H218–H224
Crayton SC, Mitchell JH, Payne FC III (1981) Reflex cardiovascular response during injection of capsaicin into skeletal muscle. Am J Physiol 240:H315–H319
Cross BA (1966) Neural control of oxytocin secretion. In: Martini L, Ganong WF (eds) Neuroendocrinology. Vol 1. Academic, New York, pp 217–259
Czlonkowski A, Stein C, Herz A (1993) Peripheral mechanisms of opioid antinociception in inflammation: involvement of cytokines. Eur J Pharmacol 242:229–235
Dahlöf L-G, Larsson K (1976) Interactional effects of pudendal nerve section and social restriction on male rat sexual behavior. Physiol Behav 16:757–762
Dail WG, Minorsky N, Moll MA, Manzanares K (1986) The hypogastric nerve pathway to penile erectile tissue: histochemical evidence supporting a vasodilator role. J Auton Nerv Syst 15:341–349
Dallman MF, Jones MT (1973) Corticosteroid feedback control of ACTH secretion: effect of stress-induced corticosterone secretion on subsequent stress responses in the rat. Endocrinology 92:1367–1375
Darian-Smith I (1984a) Thermal sensibility. In: Darian-Smith I (ed) Handbook of physiology, section 1: the nervous system, vol 3. Sensory processes, part 2. American Physiological Society, Bethesda, pp 879–913
Darian-Smith I (1984b) The sense of touch: perfomance and peripheral neural processes. In: Darian-Smith I (ed) Handbook of physiology, section 1: the nervous system, vol 3. Sensory processes, part 2. American Physiological Society, Bethesda, pp 739–788
Dawood MY, Khan-Dawood FS, Wahi RS, Fuchs F (1981) Oxytocin release and plasma anterior pituitary and gonadal hormones in women during lactation. J Clin Endocrinol Metab 52:678–683

Dawson NJ, Schmid H, Pierau F-K (1992) Pre-spinal convergence between thoracic and visceral nerves of the rat. Neurosci Lett 138:149–152

De Greef WJ, Visser TJ (1981) Evidence for the involvement of hypothalamic dopamine and thyrotrophin-releasing hormone in suckling-induced release of prolactin. J Endocrinol 91:213–223

De Groat WC, Ryall RW (1969) Reflexes to sacral parasympathetic neurones concerned with micturition in the cat. J Physiol (Lond) 200:87–108

De Groat WC, Douglas JW, Glass J, Simonds W, Weimer B, Werner P (1975) Changes in somato-vesical reflexes during postnatal development in the kitten. Brain Res 94:150–154

De Groat WC, Nadelhaft I, Milne RJ, Booth AM, Morgan C, Thor K (1981) Organization of the sacral parasympathetic reflex pathways to the urinary bladder and large intestine. J Auton Nerv Syst 3:135–160

Dembowsky K, Czachurski J, Amendt K, Seller H (1980) Tonic descending inhibition of the spinal somato-sympathetic reflex from the lower brain stem. J Auton Nerv Syst 2:157–182

Dembowsky K, Lackner K, Czachurski J, Seller H (1981) Tonic catecholaminergic inhibition of the spinal somato-sympathetic reflexes originating in the ventrolateral medulla oblongata. J Auton Nerv Syst 3:277–290

Dembowsky K, Czachurski J, Seller H (1985) An intracellular study of the synaptic input to sympathetic preganglionic neurones of the third thoracic segment of the cat. J Auton Nerv Syst 13:201–244

Denny-Brown D, Robertson EG (1933) The state of the bladder and its sphincters in complete transverse lesions of the spinal cord and cauda equina. Brain 56:397–463

Dhital KK, Appenzeller O (1988) Innervation of vasa nervorum. In: Burnstock G, Griffith SG (eds) Nonadrenergic innervation of blood vessels, vol 2. CRC, Boca Raton, pp 191–211

Diederichs W, Lue T, Tanagho EA (1991) Clitoral responses to central nervous stimulation in dogs. Int J Impot Res 3:7–13

Diehl B, Hoheisel U, Mense S (1988) Histological and neurophysiological changes induced by carrageenan in skeletal muscle of cat and rat. Agents Actions 25:210–213

Dixson AF, Kendrick KM, Blank MA, Bloom SR (1984) Effects of tactile and electrical stimuli upon release of vasoactive intestinal polypeptide in the mammalian penis. J Endocrinol 100:249–252

Donoghue JP, Carroll KL (1987) Cholinergic modulation of sensory responses in rat primary somatic sensory cortex. Brain Res 408:367–371

Dorn T, Schaible H-G, Schmidt RF (1991) Response properties of thick myelinated group II afferents in the medial articular nerve of normal and inflamed knee joints of the cat. Somatosens Mot Res 8:127–136

Dorward PK, Burke SL, Jänig W, Cassell J (1987) Reflex responses to baroreceptor, chemoreceptor and nociceptor inputs in single renal sympathetic neurones in the rabbit and the effects of anaesthesia on them. J Auton Nerv Syst 18:39–54

Downing SE, Siegel JH (1963) Baroreceptor and chemoreceptor influences on sympathetic discharge to the heart. Am J Physiol 204:471–479

Dun NJ, Mo N (1989) Inhibitory postsynaptic potentials in neonatal rat sympathetic preganglionic neurones in vitro. J Physiol (Lond) 410:267–281

Dutar P, Lamour Y, Jobert A (1985) Activation of identified septo-hippocampal neurons by noxious peripheral stimulation. Brain Res 328:15–21

Dutton A, Dyball R (1979) Phasic firing enhances vasopressin release from the rat neurohypophysis. J Physiol (Lond) 290:433–440

Eliasson S, Folkow B, Lindgren P, Uvnäs B (1951) Activation of sympathetic vasodilator nerves to the skeletal muscles in the cat by hypothalamic stimulation. Acta Physiol Scand 23:333–351

Engel BT, Sato A, Sato Y (1992) Responses of sympathetic nerves innervating blood vessels in interscapular, brown adipose tissue and skin during cold stimulation in anesthetized C57BL/6J mice. Jpn J Physiol 42:549–559

Eriksen BC, Mjölneröd OK (1987) Changes in urodynamic measurements after successful anal electrostimulation in female urinary incontinence. Br J Urol 59:45–49

Eriksson M, Lindén A, Uvnäs-Moberg K (1987) Suckling increases insulin and glucagon levels in peripheral venous blood of lactating dogs. Acta Physiol Scand 131:391–396

Eriksson M, Bjorkstrand E, Smedh U, Alster P, Matthiesen AS, Uvnäs-Moberg K (1994) Role of vagal nerve activity during suckling. Effects on plasma levels of oxytocin, prolactin, VIP, somatostatin, insulin, glucagon, glucose and of milk secretion in lactating rats. Acta Physiol Scand 151:453–459

Erlanger J, Gasser HS (1937) Electrical signs of nervous activity. University Pennsylvania Press, Philadelphia

Ermirio R, Ruggeri P, Molinari C, Weaver LC (1993) Somatic and visceral inputs to neurons of the rostral ventrolateral medulla. Am J Physiol 265:R35–R40

von Euler US, Folkow B (1953) Einfluß verschiederner afferenter Nervenreize auf die Zusammensetzung des Nebennierenmarkinkretes bei der Katze. Arch Exp Path Pharm 219:242–247

Evoniuk GE, Kuhn CM, Schanberg SM (1979) The effect of tactile stimulation on serum growth hormone and tissue ornithine decarboxylase activity during maternal deprivation in rat pups. Communications Psychopharm 3:363–370

Fedina L, Katunskii AY, Khayutin VM, Mitsanyi A (1966) Responses of renal sympathetic nerves to stimulation of afferent A and C fibres of tibial and mesenterial nerves. Acta Physiol Acad Sci Hung 29:157–176

Feldberg W, Guertzenstein PG (1972) A vasodepressor effect of pentobarbitone sodium. J Physiol (Lond) 224:83–103

Feldman S, Conforti N, Chowers I (1975a) Complete inhibition of adrenocortical responses following sciatic nerve stimulation in rats with hypothalamic islands. Acta Endocrinol 78:539–544

Feldman S, Conforti N, Chowers I (1975b) Adrenocortical responses following sciatic nerve stimulation in rats with partial hypothalamic deafferentations. Acta Endocrinol 80:625–629

Feldman S, Conforti N, Chowers I (1975c) Subcortical pathways involved in the mediation of adrenocortical responses following sciatic nerve stimulation. Neuroendocrinology 18:359–365

Feldman S, Conforti N, Chowers I, Siegel RA (1981) Effects of sciatic nerve stimulation of ACTH secretion in intact and in variously hypothalamic deafferentated male rats. Exp Brain Res 42:486–488

Feldman S, Conforti N, Melamed E (1988) Involvement of ventral noradrenergic bundle in corticosterone secretion following neural stimuli. Neuropharmacology 27:129–133

Felten DL, Felten SY, Bellinger DL, Carlson SL, Ackerman KD, Madden KS, Olschowki JA, Livnat S (1987) Noradrenergic sympathetic neural interactions with the immune system: structure and function. Immunol Rev 100:225–259

Fernandez de Molina A, Kuno M, Perl ER (1965) Antidromically evoked responses from sympathetic preganglionic neurones. J Physiol (Lond) 180:321–335

Fidone SJ, Sato A (1969) A study of chemoreceptor and baroreceptor A and C-fibres in the cat carotid nerve. J Physiol (Lond) 205:527–548

Field TM, Schanberg SM, Scafidi F, Bauer CR, Vega LN, Garcia R, Nystrom J, Kuhn CM (1986) Tactile-kinesthetic stimulation effects on preterm neonates. Pediatrics 77:654–658

Fields HL, Meyer GA, Partridge LDJ (1970a) Convergence of visceral and somatic input onto spinal neurons. Exp Neurol 26:36–52

Fields HL, Partridge LDJ, Winter DL (1970b) Somatic and visceral receptive field properties of fibers in ventral quadrant white matter of the cat spinal cord. J Neurophysiol 33:827–837

Fisher ML, Nutter DO (1974) Cardiovascular reflex adjustments to static muscular contractions in the canine hindlimb. Am J Physiol 226:648–655

Fock S, Mense S (1976) Excitatory effects of 5-hydroxytryptamine, histamine and potassium ions on muscular group IV afferent units: a comparison with bradykinin. Brain Res 105:459–469

Folkow B, Uvnäs B (1948) The distribution and functional significance of sympathetic vasodilators to the hind limbs of the cat. Acta Physiol Scand 15:389–400

Folkow B, Uvnäs B (1950) Do adrenergic vasodilator nerves exist? Acta Physiol Scand 20:329–337

Folkow B, Ström G, Uvnäs B (1950) Do dorsal root fibres convey centrally induced vasodilator impulses ? Acta Physiol Scand 21:145–158

Folkow B, Heymans C, Neil E (1965) Integrated aspects of cardiovascular regulation. In: Hamilton WF, Dow P (eds) Handbook of physiology, section 2, circulation, vol 3. American Physiological Society, Washington, pp 1787–1823

Foreman RD, Wurster RD (1973) Localization and functional characteristics of descending sympathetic spinal pathways. Am J Physiol 255:212–217

Foreman RD, Wurster RD (1975) Conduction in descending spinal pathways initiated by somatosympathetic reflexes. Am J Physiol 228:905–908

Freeman MAR, Wyke B (1967) The innervation of the knee joint. An anatomical and histological study in the cat. J Anat 101:502–532

Freund-Mercier MJ, Richard P (1981) Excitatory effects of intraventricular injections of oxytocin on the milk ejection reflex in the rat. Neurosci Lett 23:193–198

Friden J, Sfakianos PN, Hargens AR, Akeson WH (1988) Residual muscular swelling after repetitive eccentric contractions. J Orthop Res 6:493–498

Frost F, Hartwig D, Jaeger R, Leffler E, Wu Y (1993) Electrical stimulation of the sacral dermatomes in spinal cord injury: effect on rectal manometry and bowel emptying. Arch Phys Med Rehabil 74:696–701

Fujino M, Kurosawa M, Saito H, Sato A, Swenson RS (1987) Effects of a substance P analogue with antagonist properties ([D-Arg^{1},D-Trp^{7,9},Leu^{11}]substance P) on spontaneous activity of the adrenal sympathetic nerve and its evoked reflex discharges in response to somatic afferent stimulation. Neurosci Lett 80:315–320

Fujiwara R, Orita K (1987) The enhancement of the immune response by pain stimulation in mice. I. The enhancement effect on PFC production via sympathetic nervous system in vivo and in vitro. J Immunol 138:3699–3703

Fukuda Y, Sato A, Suzuki A, Trzebski A (1989) Autonomic nerve and cardiovascular responses to changing blood oxygen and carbon dioxide levels in the rat. J Auton Nerv Syst 28:61–74

Fukunaga AF, Taniguchi Y, Kikuta Y (1990) Cardiovascular responses to noxious stimuli in experimental animals: "pressor or depressor"? Anesthesiology 72:769–771

Fulton JF (1949) Physiology of the nervous system. Oxford University Press, New York

Fyffe REW (1979) The morphology of group II muscle afferent fibre collaterals. J Physiol (Lond) 296:39–40

Fyffe REW (1984) Afferent fibers. In: Davidoff RA (ed) Handbook of the spinal cord. Dekker, New York, pp 79–136

Gardner E (1944) The distribution and termination of nerves in the knee joint of the cat. J Comp Neurol 80:11–32

Gasser HS (1960) Effect of method leading on the recording of the nerve fibre spectrum. J Gen Physiol 43:927–940

Gaumann DM, Yaksh TL, Tyce GM (1990) Effects of intrathecal morphine, clonidine, and midazolam on the somato-sympathoadrenal reflex response in halothane-anesthetized cats. Anesthesiology 73:425–432

Gebber GL, Taylor DG, Weaver LC (1973) Electrophysiological studies on organization of central vasopressor pathways. Am J Physiol 224:470–481

Gelsema AJ, De Groot G, Bouman LN (1983) Instantaneous cardiac acceleration in the cat elicited by peripheral nerve stimulation. J Appl Physiol 55:703–710

Gelsema AJ, Bouman LN, Karemaker JM (1985) Short-latency tachycardia evoked by stimulation of muscle and cutaneous afferents. Am J Physiol 248:R426–R433

Georgopoulos AP (1976) Functional properties of primary afferent units probably related to pain mechanisms in primate glabrous skin. J Neurophysiol 39:71–83

Gernandt B, Liljestrand G, Zotterman Y (1946) Efferent impulses in the splanchnic nerves. Acta Physiol Scand 11:230–247

Gibbs FP (1969) Central nervous system lesions that block release of ACTH caused by traumatic stress. Am J Physiol 217:78–83

Gibbs NM, Larach DR, Skeehan TM, Schuler HG (1989) Halothane induces depressor responses to noxious stimuli in the rat. Anesthesiology 70:503–510

Gildemeister M (1928) Der galvanische Hautreflex. In: Bethe A, Bergmann GV, Embden G, Ellinger A (eds) Handbuch der normalen und pathologischen Physiologie, vol 8, part 2. Springer, Berlin Heidelberg New York, pp 657–1095

Gottschaldt KM, Iggo A, Young DW (1973) Functional characteristics of mechanoreceptors in sinus hair follicles of the cat. J Physiol (Lond) 235:287–315

Gray JAB (1966) The representation of information about rapid changes in a population of receptor units signalling mechanical events. In: Reuck AVSD, Knight J (eds) Touch, heat and pain. Ciba Foundation Symposium. Churchill, London, pp 299–315

Grayson J (1949) Vascular reactions in the human intestine. J Physiol (Lond) 109:439–447

Green JH (1959) Cardiac vagal efferent activity in the cat. J Physiol (Lond) 149:47–49

Greer MA, Allen CF, Gibbs FP, Gullickson C (1970) Pathways at the hypothalamic level through which traumatic stress activates ACTH secretion. Endocrinology 86:1404–1409

Grigg P, Schaible H-G, Schmidt RF (1986) Mechanical sensitivity of group III and IV afferents from posterior articular nerve in normal and inflamed cat knee. J Neurophysiol 55:635–643

Guertzenstein PG, Silver A (1974) Fall in blood pressure produced from discrete regions of the ventral surface of the medulla by glycine and lesions. J Physiol (Lond) 242:489–503

Guilbaud G, Iggo A (1985) The effect of lysine acetylsalicylate on joint capsule mechanoreceptors in rats with polyarthritis. Exp Brain Res 61:164–168

Häbler H-J, Jänig W, Koltzenburg M (1988) A novel type of unmyelinated chemosensitive nociceptor in the acutely inflamed bladder. Agents Actions 25:219–221

Häbler H-J, Jänig W, Koltzenburg M (1990) Activation of unmyelinated afferent fibres by mechanical stimuli and inflammation of the urinary bladder in the cat. J Physiol (Lond) 425:545–562

Häbler H-J, Jänig W, Krummel M, Peters OA (1994) Reflex patterns in postganglionic neurons supplying skin and skeletal muscle of the rat hindlimb. J Neurophysiol 72:2222–2236

Hagbarth H-E, Hallin RG, Hongell A, Torebjörk HE, Wallin BG (1972) General characteristics of sympathetic activity in human skin nerves. Acta Physiol Scand 84:164–176

Hagbarth KE, Vallbo B (1968) Pulse and respiratory grouping of sympathetic impulses in human muscle nerves. Acta Physiol Scand 74:96–108

Hallin RG, Torebjörk HE (1974) Single unit sympathetic activity in human skin nerves during rest and various manoeuvres. Acta Physiol Scand 92:303–317

Hallström Å, Sato A, Sato Y, Ungerstedt U (1990) Effect of stimulation of the nucleus basalis of Meynert on blood flow and extracellular lactate in the cerebral cortex with special reference to the effect of noxious stimulation of skin and hypoxia. Neurosci Lett 116:227–232

Hamamura M, Shibuki K, Yagi K (1984) Noxious inputs to supraoptic neurosecretory cells in the rat. Neurosci Res 2:49–61

Hamamura M, Onaka T, Yagi K (1986) Parvocellular neurosecretory neurons: Converging inputs after saphenous nerve and hypovolemic stimulations in the rat. Jpn J Physiol 36:921–933

Handwerker HO, Kobal G (1993) Psychophysiology of experimentally induced pain. Physiol Rev 73:639–671

Handwerker HO, Kilo S, Reeh PW (1991) Unresponsive afferent nerve fibres in the sural nerve of the rat. J Physiol (Lond) 435:229–242

Hartman FA, Hartman WB (1923) Influence of temperature changes on the secretion of epinephrine. Am J Physiol 65:612–622

He X, Schmidt RF, Schmittner H (1988) Effects of capsaicin on articular afferents of the cat's knee joint. Agents Actions 25:221–224

He X, Schepelmann K, Schaible H-G, Schmidt RF (1990) Capsaicin inhibits responses of fine afferents from the knee joint of the cat to mechanical and chemical stimuli. Brain Res 530:147–150

Hellon R (1983) Thermoreceptors. In: Shepherd JT, Abboud FM, Geiger SR (eds) Handbook of physiology, section 2, the cardiovascular system, vol 3, part 2. American Physiological Society, Bethesda, pp 659–673

Henry JL, Calaresu FR (1974a) Excitatory and inhibitory inputs from medullary nuclei projecting to spinal cardioacceleratory neurons in the cat. Exp Brain Res 20:485–504

Henry JL, Calaresu FR (1974b) Pathways from medullary nuclei to spinal cardioacceleratory neurons in the cat. Exp Brain Res 20:505–514

Henry JL, Calaresu FR (1974c) Origin and course of crossed medullary pathways to spinal sympathetic neurons in the cat. Exp Brain Res 20:515–526

Hensel H (1973) Cutaneous thermoreceptors. In: Iggo A (ed) Handbook of sensory physiology. Springer, Berlin Heidelberg New York, pp 79–110

Hensel H (1974) Thermoreceptors. Annu Rev Physiol 36:233–249

Hensel H (1981) Thermoreception and temperature regulation. Academic, New York

Hensel H, Brück K, Raths P (1973) Homeothermic organisms. In: Precht H, Christophersen J, Hensel H, Larcher W (eds) Temperature and life. Springer, Berlin Heidelberg New York, pp 503–761

Heppelmann B (1990) Morphologische Grundlagen peripherer und spinaler Prozesse beim Gelenkschmerz. Habilitationsschrift, Medizinische Fakultät der Universität Würzburg, Würzburg

Heppelmann B, Schaible H-G, Schmidt RF (1985) Effects of prostaglandins E1 and E2 on the mechanosensitivity of group III afferents from normal and inflamed cat knee joints. In: Fields HL, Dubner R, Cervero F (eds) Advances in pain research and therapy, vol. 9. Raven, New York, pp 91–101

Heppelmann B, Pfeffer A, Schaible H-G, Schmidt RF (1986) Effects of acetylsalicylic acid and indomethacin on single groups III and IV sensory units from acutely inflamed joints. Pain 26:337–351

Heppelmann B, Herbert MK, Schaible H-G (1987) Morphological and physiological characteristics of the innervation of cat's normal and arthritic knee joint. In: Pubols LM , Sessle B (eds) Effects of injury on trigeminal and spinal somatosensory systems. Liss, New York, pp 19–27

Heppelmann B, Heuss C, Schmidt RF (1988) Fiber size distribution of myelinated and unmyelinated axons in the medial and posterior articular nerves of the cat's knee joint. Somatosens Mot Res 5: 273–281

Heppelmann B, Messlinger K, Schmidt RF (1989) Three-dimensional reconstruction of fine afferent nerve fibres in the knee joint capsule. Eur J Cell Biol 25:115–118

Heppelmann B, Messlinger K, Neiss WF, Schmidt RF (1990a) Ultrastructural three-dimensional reconstruction of group III and group IV sensory nerve endings ("free nerve endings") in the knee joint capsule of the cat: evidence for multiple receptive sites. J Comp Neurol 292:103–116

Heppelmann B, Messlinger K, Neiss WF, Schmidt RF (1990b) The sensory terminal tree of "free nerve endings" in the articular capsule of the knee. In: Zenker W, Neuhuber WL (eds) The primary afferent neuron. A survey of recent morpho-functional aspects. Plenum, New York, pp 73–85

Heppelmann B, Messlinger K, Neiss WF, Schmidt RF (1995) Fine sensory innervation of the knee joint capsule by group III and group IV nerve fibers in the cat. J Comp Neurol 351:415–428

Hervonen A, Kimura A, Sato A, Sato Y (1990) Responses of cortical cerebral blood flow produced by stimulation of cervical sympathetic trunks are well maintained in aged rats. Neurosci Lett 120:55–57

Heymans C, Neil E (1958) Reflexogenic areas of the cardiovascular system. Churchill, London

Hodes R (1940) Reciprocal innervation in the small intestine. Am J Physiol 130:642–650

Hoheisel U, Lehmann-Willenbrock E, Mense S (1989) Termination patterns of identified group II and III afferent fibres from deep tissues in the spinal cord of the cat. Neuroscience 28:495–507

Hoheisel U, Mense S, Simons DG, Yu X-M (1993) Appearance of new receptive fields in rat dorsal horn neurons following noxious stimulation of skeletal muscle: a model for referral of muscle pain. Neurosci Lett 153:9–12

Holst J, Schwartz T, Knuhtsen S, Jensen S, Nielsen O (1986) Autonomic nervous control of the endocrine secretion from the isolated, perfused pig pancreas. J Auton Nerv Syst 17:71–84

Holzer P (1988) Local effector functions of capsaicin-sensitive sensory nerve endings: involvement of tachykinins, calcitonin gene-related peptide and other neuropeptides. Neuroscience 24:739–768

Holzer P, Lippe IT, Amann R (1992) Participation of capsaicin-sensitive afferent neurons in gastric motor inhibition caused by laparotomy and intraperitoneal acid. Neuroscience 48:715–722

Horeyseck G, Jänig W (1974a) Reflexes in postganglionic fibres within skin and muscle nerves after mechanical non-noxious stimulation of skin. Exp Brain Res 20:115–123

Horeyseck G, Jänig W (1974b) Reflexes in postganglionic fibres within skin and muscle nerves after noxious stimulation of skin. Exp Brain Res 20:125–134

Horeyseck G, Jänig W, Kirchner F, Thämer V (1976) Activation and inhibition of muscle and cutaneous postganglionic neurons to hindlimb during hypothalamically induced vasoconstriction and atropine-sensitive vasodilation. Pflugers Arch 361:231–240

Hotta H, Nishijo K, Sato A, Sato Y, Tanzawa S (1991) Stimulation of lumbar sympathetic trunk produces vasoconstriction of the vasa nervorum in the sciatic nerve via α-adrenergic receptors in rats. Neurosci Lett 133:249–252

Hotta H, Sato A, Sumitomo T (1992) Hypothalamic corticotropin-releasing hormone (CRH) secretion into hypophysial portal blood is regulated by cutaneous sensory stimulation in anesthetized rat. Jpn J Physiol 42:515–524

Hotta H, Sato A, Sato Y, Uvnäs-Moberg K (1993) Somatic afferent regulation of plasma prolactin in anesthetized rats. Jpn J Physiol 43:501–509

Hotta H, Sato A, Sato Y, Uchida S (1996) Stimulation of saphenous afferent nerve produces vasodilatation of the vasa nervorum via axon reflex-like mechanism in the sciatic nerve of the anesthetized rats. Neurosci Res 24:305–308

Hough T (1902) Ergographic studies in muscular soreness. Am J Physiol 7:76–92

Hunt CC (1954) Relation of function to diameter in afferent fibres of muscle nerves. J Gen Physiol 38:117–131

Hunt CC (ed) (1974) Handbook of sensory physiology, muscle receptors. Springer, Berlin Heidelberg New York

Hunt CC, McIntyre AK (1960) Properties of cutaneous touch receptors in cat. J Physiol (Lond) 153:88–98

Hunt R (1895) The fall of blood-pressure resulting from the stimulation of afferent nerves. J Physiol (Lond) 18:381–410

Hursh JB (1939) Conduction velocity and diameter of nerve fibres. Am J Physiol 127:131–139

Hussain SNA, Chatillon A, Comtois A, Roussos C, Magder S (1991) Chemical activation of thin-fiber phrenic afferents. 2. Cardiovascular responses. J Appl Physiol 70:77–86

Iggo A, Andres KH (1982) Morphology of cutaneous receptors. Annu Rev Neurosci 5:1–31

Iggo A, Muir AR (1969) The structure and function of a slowly adapting touch corpuscle in hairy skin. J Physiol (Lond) 200:763–796

Ikeda H, Sunakawa M, Suda H (1995) Three groups of afferent pulpal feline nerve fibres show different electrophysiological response properties. Arch Oral Biol 40:895–904

Illert M, Gabriel M (1972) Descending pathways in the cervical cord of cats affecting blood pressure and sympathetic activity. Pflugers Arch 335:109–124

Ingvar DH (1976) Functional landscapes of the dominant hemisphere. Brain Res 107:181–197

Inokuchi H, Yoshimura M, Trzebski A, Polosa C, Nishi S (1992) Fast inhibitory postsynaptic potentials and responses to inhibitory amino acids of sympathetic preganglionic neurons in the adult cat. J Auton Nerv Syst 41:53–60

Iriki M, Riedel W, Simon E (1971) Regional differentiation of sympathetic activity during hypothalamic heating and cooling in anesthetized rabbits. Pflugers Arch 328:320–331

Irisawa H, Caldwell WM, Wilson MF (1971) Neural regulation of atrioventricular conduction. Jpn J Physiol 21:15–25

Iriuchijima J, Kumada M (1963) Efferent cardiac vagal discharge of the dog in response to electrical stimulation of sensory nerves. Jpn J Physiol 13:599–605

Iriuchijima J, Kumada M (1964) Activity of single vagal fibers efferent to the heart. Jpn J Physiol 14:479–487

Irwin M (1993) Stress-induced immune suppression. Ann NY Acad Sci 697:203–218

Isa T, Kurosawa M, Sato A, Swenson RS (1985) Reflex responses evoked in the adrenal sympathetic nerve to electrical stimulation of somatic afferent nerves in the rat. Neurosci Res 3:130–144

Isherwood KM, Cross BA (1980) Effect of the suckling stimulus on secretion of prolactin and luteinizing hormone in conscious and anaesthetized rats. J Endocrinol 87:437–444

Ito K, Kaseda M, Sato A, Torigata Y (1978) Excitatory and inhibitory electrodermal reflexes evoked by cutaneous stimulation in acute spinal cats. Jpn J Physiol 28:737–747

Ito K, Kim P, Sato A, Torigata Y (1979) Reflex changes in gastric motility produced by nociceptive stimulation of the skin in anesthetized cats. In: Ito M (ed) Integrative control functions of the brain. Kodansha, Tokyo, pp 255–256

Ito K, Nakamura H, Sato A, Sato Y (1983) Depressive effect of morphine on the sympathetic reflex elicited by stimulation of unmyelinated hindlimb afferent nerve fibers in anesthetized cats. Neurosci Lett 39:169–173

Ito K, Sato A, Shimamura K, Swenson RS (1984) Convergence of noxious and non-noxious cutaneous afferents and baroreceptor afferents onto single adrenal sympathetic neurons in anesthetized rats. Neurosci Res 1:105–116

Iwamoto GA, Botterman BR (1985) Peripheral factors influencing expression of pressor reflex evoked by muscular contraction. J Appl Physiol 58:1676–1682

Iwamoto GA, Kaufman MP, Botterman BR, Mitchell JH (1982) Effects of lateral reticular nucleus lesions on the exercise pressor reflex in cats. Circ Res 51:400–403

Iwamoto GA, Botterman BR, Waldrop TG (1984a) The exercise pressor reflex: evidence for an afferent pressor pathway outside the dorsolateral sulcus region. Brain Res 292:160–164

Iwamoto GA, Parnavelas JG, Kaufman MP, Botterman BR, Mitchell JH (1984b) Activation of caudal brainstem cell groups during the exercise pressor reflex in the cat as elucidated by 2-[^{14}C] deoxyglucose. Brain Res 304:178–182

Iwamoto GA, Waldrop TG, Kaufman MP, Botterman BR, Rybicki KJ, Mitchell JH (1985) Pressor reflex evoked by muscular contraction: contributions by neuraxis levels. J Appl Physiol 59:459–467

Iwamura Y, Uchino Y, Ozawa S, Kudo N (1969) Excitatory and inhibitory components of somato-sympathetic reflex. Brain Res 16:351–358

Iwata K, Tsuboi Y, Toda K, Yagi J, Tsujimoto C, Sumino R (1991) Comparisons of the sensation perceived and intradental nerve activity following temperature changes in human teeth. Exp Brain Res 87:213–217

Izumi H, Karita K (1992) Somatosensory stimulation causes autonomic vasodilatation in cat lip. J Physiol (Lond) 450:191–202

Jänig W (1970) Morphology of rapidly and slowly adapting mechanoreceptors in the hairless skin of the cat's hind foot. Brain Res 28:217–231

Jänig W (1975) Central organization of somatosympathetic reflexes in vasoconstrictor neurones. Brain Res 87:305–312

Jänig W (1985a) Organization of the lumbar sympathetic outflow to skeletal muscle and skin of the cat hindlimb and tail. Rev Physiol Biochem Pharmacol 102:119–213

Jänig W (1985b) Systemic and specific autonomic reactions in pain: efferent, afferent and endocrine components. Eur J Anaesthesiol 2:319–346

Jänig W, Koltzenburg M (1991) Receptive properties of sacral primary afferent neurons supplying the colon. J Neurophysiol 65:1067–1077

Jänig W, Räth B (1977) Electrodermal reflexes in the cat's paws elicited by natural stimulation of skin. Pflugers Arch 369:27–32

Jänig W, Schmidt RF (1970) Single unit responses in the cervical sympathetic trunk upon somatic nerve stimulation. Pflugers Arch 314:199–216

Jänig W, Spilok N (1978) Functional organization of the sympathetic innervation supplying the hairless skin of the hindpaws in chronic spinal cats. Pflugers Arch 377:25–31

Jänig W, Szulczyk P (1980) Functional properties of lumbar preganglionic neurones. Brain Res 186:115–131

Jänig W, Schmidt RF, Zimmermann M (1968) Single unit responses and the total afferent outflow from the cat's food pad upon mechanical stimulation. Exp Brain Res 6:100–115

Jänig W, Sato A, Schmidt RF (1972) Reflexes in postganglionic cutaneous fibres by stimulation of group I to group IV somatic afferents. Pflugers Arch 331:244–256

Jansson G (1969a) Extrinsic nervous control of gastric motility. An experimental study in the cat. Acta Physiol Scand Suppl 326:1–42

Jansson G (1969b) Effect of reflexes of somatic afferents on the adrenergic outflow to the stomach in the cat. Acta Physiol Scand 77:17–22

Jessen C (1985) Thermal afferents in the control of body temperature. Pharmacol Ther 28:107–134

Johansson B (1962) Circulatory responses to stimulation of somatic afferents. Acta Physiol Scand 57[Suppl 198]:1–91

Johansson H, Sjölander P, Sojka P (1991) Receptors in the knee joint ligaments and their role in the biomechanics of the joint. Crit Rev Biomed Eng 18:341–368

Johansson RS, Vallbo AB (1979) Tactile sensibility in the human hand: relative and absolute densities of four types of mechanoreceptive units in glabrous skin. J Physiol (Lond) 286:283–300

Jolesz FA, Cheng-Tao X, Ruenzel PW, Henneman E (1982) Flexor reflex control of the external sphincter of the urethra in paraplegia. Science 216:1243–1245

Jyväsjärvi E, Kniffki KD (1989) Afferent C fibre innervation of cat tooth pulp: confirmation by electrophysiological methods. J Physiol (Lond) 411:663–675

Kacsoh B, Terry LC, Meyers JS, Crowley WR, Grosvenor CE (1989) Maternal modulation of growth hormone secretion in the neonatal rat. I. Involvement of milk factors. Endocrinology 125:1326–1336

Kacsoh B, Meyers JS, Crowley WR, Grosvenor CE (1990) Maternal modulation of growth hormone secretion in the neonatal rat: involvement of mother-offsping interactions. J Endocrinol 124:233–240

Kagitani F, Kimura A, Sato A, Suzuki A (1996) The role of the spinal cord as a reflex center for the somatically induced reflex responses of splenic sympathetic and natural killer cell activity in anesthetized rats. Neurosci Lett 217:109–112

Kamel F, Wright WW, Mock EJ, Frankel A (1977) The influence of mating and related stimuli on plasma levels of luteinizing hormone, follicle stimulating hormone, prolactin, and testosterone in the male rat. Endocrinology 101:421–429

Kametani H, Sato A, Sato Y, Ueki K (1978) Reflex facilitation and inhibition of gastric motility from various skin areas in rats. In: Ito M (ed) Integrative control functions of the brain. Kodansha, Tokyo, pp 285–287

Kametani H, Sato A, Sato Y, Simpson A (1979) Neural mechanisms of reflex facilitation and inhibition of gastric motility to stimulation of various skin areas in rats. J Physiol (Lond) 294:407–418

Kamosinska B, Nowicki D, Szulczyk P (1989) Control of the heart rate by sympathetic nerves in cats. J Auton Nerv Syst 26:241–249

Kannan H, Yamashita H, Koizumi K, Brooks CMC (1988) Neuronal activity of the cat supraoptic nucleus is influenced by muscle small-diameter afferent (groups III and IV) receptors. Proc Natl Acad Sci USA 85:5744–5748

Kanosue K, Matsuo R, Tanaka H, Nakayama T (1986) Effect of body temperature on salivary reflexes in rats. J Auton Nerv Syst 16:233–237

Kao FF (1963) An experimental study of the pathway involved in exercise hyperpnoea employing cross-circulation techniques. In: Cunningham DJC, Lloyd BB (eds) The regulation of human respiration. Blackwell, Oxford, pp 461–502

Karl H, Sato A, Schmidt RF (1975) Electrodermal reflexes induced by activity in somatic afferent fibers. Brain Res 87:145–150

Katafuchi T, Take S, Hori T (1993) Roles of sympathetic nervous system in the suppression of cytotoxicity of splenic natural killer cells in the rat. J Physiol (Lond) 465:343–357

Kato J, Meguro K, Sato A, Sato Y (1992) The effects of morphine administered into the vertebral artery on the somatosympathetic A- and C-reflexes in anesthetized cats. Neurosci Lett 138:207–210

Katunsky AY, Khayutin VM (1968) The reflex latency and the level of mediation of spinal afferent impulses to the cardiovascular sympathetic neurones. Pflugers Arch 298:294–304

Kaufman A, Koizumi K (1971) Spontaneous and reflex activity of single units in lumbar white rami. In: Kao FF, Vasalle M, Koizumi K (eds) Research in physiology, a liber memorials in honor of Prof. Chandler McC Brooks. Aulo Gaggi, Bologna, pp 469–481

Kaufman A, Sato A, Sato Y, Sugimoto H (1977) Reflex changes in heart rate after mechanical and thermal stimulation of the skin at various segmental levels in cats. Neuroscience 2:103–109

Kaufman MP, Iwamoto GA, Longhurst JC, Mitchell JH (1982) Effects of capsaicin and bradykinin on afferent fibers with endings in skleteal muscle. Circ Res 50:133–139

Kaufman MP, Longhurst JC, Rybicki KJ, Wallach JH, Mitchell JH (1983) Effects of static muscular contraction on impulse activity of groups III and IV afferents in cat. J Appl Physiol 55:105–112

Kaufman MP, Rybicki KJ, Waldrop TG, Mitchell JH (1984) Effect on arterial pressure of rhythmically contracting the hindlimb muscles of cats. J Appl Physiol 56:1265–1271

Kaufman MP, Kozlowski GP, Rybicki KJ (1985) Attenuation of the reflex pressor response to muscular contraction by a substance P antagonist. Brain Res 333:182–184

Kaufman MP, Rybicki KJ, Kozlowski GP, Iwamoto GA (1986) Immunoneutralization of substance P attenuates the reflex pressor response to muscular contraction. Brain Res 377:199–203

Kawamura Y, Yamamoto T (1977) Salivary secretion to noxious stimulation of the trigeminal area. In: Anderson DJ, Matthews B (eds) Pain in the trigeminal region. Elsevier, Amsterdam, pp 395–404

Kehl H (1975) Studies of reflex communications between dermatomes and jejunum. J Am Osteopath Assoc 74:667–669

Kerr FWL, Alexander S (1964) Descending autonomic pathways in the spinal cord. Arch Neurol 10:249–261

Khayutin VM, Lukoshkova EV (1970) Spinal mediation of vasomotor reflexes in animals with intact brain studied by electrophysiological methods. Pflugers Arch 321:197–222

Khayutin VM, Lukoshkova EV, Gailans JB (1986) Somatic depressor reflexes: results of specific 'depressor' afferents' excitation or an epiphenomenon of general anesthesia and certain decerebrations? J Auton Nerv Syst 16:35–60

Kimmel DL (1961) Innervation of spinal dura mater and dura mater of the posterior cranial fossa. Neurology 11:800–809

Kimura A, Sato A, Sato Y, Trzebski A (1993) Role of the central and arterial chemoreceptors in the response of gastric tone and motility to hypoxia, hypercapnia and hypocapnia in rats. J Auton Nerv Syst 45:77–85

Kimura A, Nagai N, Sato A (1994a) Somatic afferent regulation of cytotoxic activity of splenic natural killer cells in anesthetized rats. Jpn J Physiol 44:651–664

Kimura A, Okada K, Sato A, Suzuki H (1994b) Regional cerebral blood flow in the frontal, parietal and occipital cortices increases independently of systemic arterial pressure during slow walking in conscious rats. Neurosci Res 20:309–315

Kimura A, Ohsawa H, Sato A, Sato Y (1995) Somatocardiovascular reflexes in anesthetized rats with the central nervous system intact or acutely spinalized at the cervical level. Neurosci Res 22:297–305

Kimura A, Sato A, Sato Y, Suzuki A (1996a) Single electrical shock of a somatic afferent nerve elicits A- and C-reflex discharges in gastric vagal efferent nerves in anesthetized rats. Neurosci Lett 210:53–56

Kimura A, Sato A, Sato Y, Suzuki H (1996b) A- and C-reflexes elicited in cardiac sympathetic nerves by single shock to a somatic afferent nerve include spinal and supraspinal components in anesthetized rats. Neurosci Res 25:91–96

Kirchner F, Sato A, Weidinger H (1971) Bulbar inhibition of spinal and supraspinal sympathetic reflex discharges. Pflugers Arch 326:324–333

Kissin I, Green D (1984) Effect of halothane on cardiac acceleration response to somatic nerve stimulation in dogs. Anesthesiology 61:708–711

Kline RL, Yeung KY, Calaresu FR (1978) Role of somatic nerves in the cardiovascular responses to stimulation of an acupuncture point in anesthetized rabbits. Exp Neurol 61:561–570

Koizumi K, Brooks CMC (1972) The integration of autonomic system reactions: a discussion of autonomic reflexes, their control and their association with somatic reactions. Ergebn Physiol Biol Chem Exp Pharmakol 67:1–68

Koizumi K, Sato A (1972) Reflex activity of single sympathetic fibres to skeletal muscle produced by electrical stimulation of somatic and vago-depressor afferent nerves in the cat. Pflugers Arch 332:283–301

Koizumi K, Sato A, Kaufman A, Brooks CMC (1968) Studies of sympathetic neuron discharges modified by central and peripheral excitation. Brain Res 11:212–224

Koizumi K, Collin R, Kaufman A, Brooks CMC (1970) Contribution of unmyelinated afferent excitation to sympathetic reflexes. Brain Res 20:99–106

Koizumi K, Seller H, Kaufman A, Brooks CMC (1971) Pattern of sympathetic discharges and their relation to baroreceptor and respiratory activities. Brain Res 27:281–294

Koizumi K, Sato A, Terui N (1980) Role of somatic afferents in autonomic system control of the intestinal motility. Brain Res 182:85–97

Koizumi K, Terui N, Kollai M (1983) Neural control of the heart: significance of double innervation re-examined. J Auton Nerv Syst 7:279–294

Kollai M, Koizumi K (1979) Reciprocal and non-reciprocal action of the vagal and sympathetic nerves innervating the heart. J Auton Nerv Syst 1:33–52

Koltzenburg M, Handwerker HO (1994) Differential ability of human cutaneous nociceptors to signal mechanical pain and to produce vasodilatation. J Neurosci 14:1756–1765

Koltzenburg M, Torebjörk H, Wahren LK (1994) Nociceptor modulated central sensitization causes mechanical hyperalgesia in acute chemogenic and chronic neuropathic pain. Brain 117:579–591

Kozelka JW, Wurster RD (1985) Ascending spinal pathways for somatoautonomic reflexes in the anesthetized dog. J Appl Physiol 58:1832–1839

Kozelka JW, Chung JM, Wurster RD (1981) Ascending spinal pathways mediating somato-cardiovascular reflexes. J Auton Nerv Syst 3:171–175

Kozlowski S, Szczepanska E, Zielinski A (1967) The hypothalamo-hypophyseal antidiuretic system in physic exercises. Arch Int Physiol Biochim 75:218–228

Kress M, Koltzenburg M, Reeh PW, Handwerker HO (1992) Responsiveness and functional attributes of electrically localized terminals of cutaneous C-fibers in vivo and in vitro. J Neurophysiol 68:581–595

Kruger L (1987) Morphological correlates of "free" nerve endings – a reappraisal of thin sensory axon classification. In: Schmidt RF, Schaible H-G, Vahle-Hinz C (eds) Fine afferent nerve fibres and pain. VCH Verlagsgesellschaft mbH, Weinheim, pp1–13

Kruse MN, De Groat WC (1990) Micturition reflexes in decerebrate and spinalized neonatal rats. Am J Physiol 258:R1508–R1511

Kruse MN, De Groat WC (1993) Spinal pathways mediate coordinated bladder/urethral sphincter activity during reflex micturition in decerebrate and spinalized neonatal rats. Neurosci Lett 152:141–144

Kuhn CM, Butler SR, Schanberg SM (1978) Selective depression of serum growth hormone during maternal deprivation in rat pups. Science 201:1034–1036

Kumagai Y, Norman J, Whitwam JG (1975) The sympathetic contribution to increase in heart rate evoked by cutaneous nerve stimulation in the dog. J Physiol (Lond) 252:36P–37P

Kumazawa T, Mizumura K (1977) Thin-fiber receptors responding to mechanical, chemical, and thermal stimulation in the skeletal muscle of the dog. J Physiol (Lond) 273:179–194

Kumazawa T, Perl ER (1977) Primate cutaneous sensory units with unmyelinated (C) afferent fibers. J Neurophysiol 40:1325–1338

Kumazawa T, Tadaki E, Kim K (1980) A possible participation of endogenous opiates in respiratory reflexes induced by thin-fiber muscular afferents. Brain Res 199:244–248

Kuno Y (1956) Human perspiration. Thomas, Springfield

Kuntz A (1946) Anatomic and physiologic properties of cutaneo-visceral vasomotor reflex arcs. J Neurophysiol 8:421–429

Kuntz A, Haselwood LA (1940) Circulatory reactions in the gastrointestinal tract elicited by localized cutaneous stimulation. Am Heart J 20:743–749

Kurosawa M, Suzuki A, Utsugi K, Araki T (1982) Response of adrenal efferent nerve activity to non-noxious mechanical stimulation of the skin in rats. Neurosci Lett 34:295–300

Kurosawa M, Saito H, Sato A, Tsuchiya T (1985) Reflex changes in sympatho-adrenal medullary functions in response to various thermal cutaneous stimulations in anesthetized rats. Neurosci Lett 56:149–154

Kurosawa M, Meguro K, Nagayama T, Sato A (1989) Effects of sevoflurane on autonomic nerve activities controlling cardiovascular functions in rats. J Anesthesia 3:109–117

Kurosawa M, Sato A, Sato Y (1992) Cutaneous mechanical sensory stimulation increases extracellular acetylcholine release in cerebral cortex in anesthetized rats. Neurochem Int 21:423–427

Kurosawa M, Okada K, Sato A, Uchida S (1993) Extracellular release of acetylcholine, noradrenaline and serotonin increases in the cerebral cortex during walking in conscious rats. Neurosci Lett 161:73–76

Kurosawa M, Nagai N, Sato A, Uvnäs-Moberg K (1994) Somatic afferent regulation of plasma immunoreactive glucagon in anesthetized rats. Jpn J Physiol 44:221–230

Kurose T, Seino Y, Nishi S, Tsuji K, Taminato T, Tsuda K, Imura H (1990) Mechanism of sympathetic neural regulation of insulin, somatostatin, and glucagon secretion. Am J Physiol 258:E220–E227

Kuru M (1965) Nervous control of micturition. Physiol Rev 45:425–494

Kvetnansky R, Sun CL, Torda T, Kopin IJ (1977) Plasma epinephrine and norepinephrine levels in stressed rats – effect of adrenalectomy. Pharmacologist 19:241–240

Lais LT, Shaffer RA, Brody MJ (1974) Neurogenic and humoral factors controlling vascular resistance in the spontaneously hypertensive rat. Circ Res 35:764–774

Lamotte RH, Lundberg LER, Torebjörk HE (1992) Pain, hyperalgesia and activity in nociceptive C units in humans after intradermal injection of capsaicin. J Physiol (Lond) 448:749–764

Lamour Y, Dutar P, Jobert A (1982) Excitatory effect of acetylcholine on different types of neurons in the first somatosensory neocortex of the rat: laminar distribution and pharmacological characteristics. Neuroscience 7:1483–1494

Langford LA, Schmidt RF (1983) Afferent and efferent axons in the medial and posterior articular nerves of the cat. Anat Rec 206:71–78

Langworthy OR (1965) Innervation of the pelvic organs of the rat. Invest Urol 2:491–511

Langworthy OR, Hesser FH (1937) Reflex vesical contraction in the cat after transection of the spinal cord in the lower lumber region. Bull Johns Hopk Hopkins 60:204–214

Laporte Y, Montastruc P (1957) Rôle des différents types de fibres afférentes dans les réflexes circulatoires généraux d'origine cutanée. J Physiol (Paris) 49:1039–1049

Laporte Y, Bessou P, Bouisset S (1960) Action réflexe des différents types de fibres afférentes d'origine musculaire sur la pression sanguine. Arch Ital Biol 98:206–221

Larsson K, Södersten P (1973) Mating in male rats after section of the dorsal penile nerve. Physiol Behav 10:567–571

Lebedev VP, Krasyukov AV, Nikitin SA (1986) Electrophysiological study of sympathoexcitatory structures of the bulbar ventrolateral surface as related to vasomotor regulation. Neuroscience 17:189–203

Lehmann AV (1913) Studien über reflektorische Darmbewegungen beim Hunde. Pflugers Arch 149:413–433
Levine JD, Clark R, Devor M, Helms C, Moskowitz MA, Basbaum AI (1984) Intraneuronal substance P contributes to the severity of experimental arthritis. Science 226:547–549
Levy MN, Ng ML, Zieske H (1966) Functional distribution of the peripheral cardiac sympathetic pathways. Circ Res 19:650–661
Lewis T, Marvin HM (1927) Observations relating to vasodilatation arising from antidromic impulses, to herpes zoster and trophic effects. Heart 14:27–46
Li P, Sun F-Y, Zhang A-Z (1983) The effect of acupuncture on blood pressure: the interrelation of sympathetic activity and endogenous opioid peptides. Acupunct Electrother Res 8:45–56
Li WM, Sato A, Suzuki A (1995) The inhibitory role of nitric oxide (NO) in the somatocardiac sympathetic C-reflex in anesthetized rats. Neurosci Res 22:375–380
Li WM, Sato A, Sato Y, Schmidt RF (1997) Morphine microinjected into the nucleus tractus solitarius and rostral ventrolateral medullary nucleus enhances somatosympathetic A- and C-reflexes in anesthetized rats. Neurosci Lett (submitted)
Li WM, Sato A, Suzuki A, Trzebski A (1996) Systemic hypoxia facilitates somato-cardiac sympathetic A- and C-reflexes in anesthetized rats. Neurosci Lett 216:175–178
Lincoln DW, Wakerley JB (1974) Electrophysiological evidence for the activation of supraoptic neurones during the release of oxytocin. J Physiol (Lond) 242:533–554
Lindahl O (1961) Experimental skin pain induced by injection of water-soluble substances in humans. Acta Physiol Scand 51[Suppl 179]:1–90
Linden A, Eriksson M, Carlquist M, Uvnäs-Moberg K (1987) Plasma levels of gastrin, somatostatin, and cholecystokinin immunoreactivity during pregnancy and lactation in dogs. Gastroenterology 92:578–584
Lindgren P (1955) The mesencephalon and the vasomotor system. Acta Physiol Scand 35[Suppl 121]:1–189
Lloyd DPC (1943) Neuron patterns controlling transmission of ipsilateral hind limb reflexes in cat. J Neurophysiol 6:293–315
Lodder J, Zeilmaker GH (1976) Effects of pelvic nerve and pudendal nerve transection on mating behavior in the male rat. Physiol Behav 16:745–751
Loewy AD, Spyer KM (1990) Central regulation of autonomic functions. Oxford University Press, New York
Longhurst J, Zelis R (1979) Cardiovascular responses to local hindlimb hypoxemia: relation to the exercise reflex. Am J Physiol 237:H359–H365
Longhurst JC, Mitchell JH, Moore MB (1980) The spinal cord ventral root: an afferent pathway of the hind-limb pressor reflex in cats. J Physiol (Lond) 301:467–476
Longhurst JC, Aung-Din R, Mitchell JH (1981) Static exercise in anesthetized dogs, a cause of reflex alpha-adrenergic coronary vasoconstriction. Basic Res Cardiol 76:530–535

Ludwig C (1847) Beiträge zur Kenntnis des Einflusses der Respirationsbewegungen auf den Blutlauf im Aortensysteme. Joh Müllers Arch Anat Physiol Wiss Med 1847:242

Lundeberg T, Eriksson SV, Theodorsson E (1991) Neuroimmunomodulatory effects of acupuncture in mice. Neurosci Lett 128:161–164

Maggi CA, Santicioli P, Meli A (1986a) Somatovesical and vesicovesical excitatory reflexes in urethane-anesthetized rats. Brain Res 380:83–93

Maggi CA, Santicioli P, Meli A (1986b) Postnatal development of micturition reflex in rats. Am J Physiol 250:R926–R931

Marchini G, Lagercrantz H, Feuerberg Y, Winberg J, Uvnäs-Moberg K (1987) The effect of non-nutritive sucking on plasma insulin, gastrin and somtostatin levels in infants. Acta Paediatr Scand 76:573–578

Marguth H, Raule W, Schaefer H (1951) Aktionsstrsme in zentrifugalen Herznerven. Pflugers Arch 254:224–245

Masuda N, Ootsuka Y, Terui N (1992) Neurons in the caudal ventrolateral medulla mediate the somato-sympathetic inhibitory reflex response via GABA receptors in the rostral ventrolateral medulla. J Auton Nerv Syst 40:91–98

Matsuda K, Duyck C, Kendall JWJ, Greer MA (1964) Pathways by which traumatic stress and ether induce increased ACTH release in the rat. Endocrinology 74:981–985

Matsukawa K, Wall PT, Wilson LB, Mitchell JH (1990) Reflex responses of renal nerve activity during isometric muscle contraction in cats. Am J Physiol 259:H1380–H1388

Matsuo R, Yamamoto T, Yoshitaka K, Morimoto T (1989) Neural substrates for reflex salivation induced by taste, mechanical, and thermal stimulation of the oral region in decerebrate rats. Jpn J Physiol 39:349–357

Matthews PBC (1972) Mammalian muscle receptors and their central actions. Arnold, London

Matthews PBC (1981) Muscle spindles: their messages and their fusimotor supply. In: Brooks VB (ed) Handbook of physiology. Section 1: the nervous system. American Physiological Society, Bethesda, pp 189–228

McAllen RM (1985) Mediation of the fastigial pressor response and a somatosympathetic reflex by ventral medullary neurones in the cat. J Physiol (Lond) 368:423–433

McCall RB, Harris LT (1987) Sympathetic alterations after midline medullary raphe lesions. Am J Physiol 253:R91–R100

McClelland WJ (1956) Differential handling and weight gain in the albino rat. Can J Physiol 10:19–22

McCloskey DI (1978) Kinesthetic sensibility. Physiol Rev 58:763–820

McCloskey DI, Mitchell JH (1972) Reflex cardiovascular and respiratory responses originating in exercising muscle. J Physiol (Lond) 224:173–186

McIntyre AK, Proske U, Tracey DJ (1978) Afferent fibres form muscle receptors in the posterior nerve of the cat's knee joint. Exp Brain Res 33:415–424

McKenna KE, Schramm LP (1983) Sympathetic preganglionic neurons in the isolated spinal cord of the neonatal rat. Brain Res 269:201–210

McLachlan EM, Hirst GDS (1980) Some properties of preganglionic neurons in upper thoracic spinal cord of the cat. J Neurophysiol 43:1251–1265

McMahon S, Koltzenburg M (1990) The changing role of primary afferent neurones in pain. Pain 43:269–272

McMahon SB, Morrison JFB (1982a) Spinal neurones with long projections activated from the abdominal viscera of the cat. J Physiol (Lond) 322:1–20

McMahon SB, Morrison JFB (1982b) Two group of spinal interneurones that respond to stimulation of the abdominal viscera of the cat. J Physiol (Lond) 322:21–34

McMahon SB, Morrison JFB (1982c) Factors that determine the excitability of parasympathetic reflexes to the cat bladder. J Physiol (Lond) 322:35–43

McMahon SB, Morrison JFB, Spillane K (1982) An electrophysiological study of somatic and visceral convergence in the reflex control of the external sphincters. J Physiol (Lond) 328:379–387

McPherson A (1966) The effects of somatic stimuli on the bladder in the cat. J Physiol (Lond) 185:185–196

Mena F, Pacheco P, Grosvenor CE (1980) Effect of electrical stimulation of mammary nerve upon pituitary and plasma prolactin concentrations in anesthetized lactating rats. Endocrinology 106:458–462

Mense S (1977) Nervous outflow from skeletal muscle following chemical noxious stimulation. J Physiol (Lond) 267:75–88

Mense S (1978) Effects of temperature on the discharges of muscle spindles and tendon organs. Pflugers Arch 374:159–166

Mense S (1986) Slowly conducting afferent fibers from deep tissues: Neurobiological properties and central nervous actions. Prog Sens Physiol 6:139–219

Mense S (1991a) Physiology of nociception in muscles. J Manual Med 6:24–33

Mense S (1991b) Considerations concerning the neurobiological basis of muscle pain. Can J Physiol Pharmacol 69:610–616

Mense S (1993) Peripheral mechanisms of muscle nociception and local muscle pain. J Musculoskeletal Pain 1:133–170

Mense S, Meyer H (1985) Different types of slowly conducting afferent units in cat skeletal muscle and tendon. J Physiol (Lond) 363:403–417

Mense S, Meyer H (1988) Bradykinin-induced modulation of the response behaviour of different types of feline group III and IV muscle receptors. J Physiol (Lond) 398:49–63

Mense S, Stahnke M (1983) Responses in muscle afferent fibres of slow conduction velocity to contractions and ischemia in the cat. J Physiol (Lond) 342:383–397

Merchenthaler I, Hynes MA, Vigh S, Schally AV, Petrusz P (1984) Corticotropin releasing factor (CRF): origin and course of afferent pathways to the median eminence (ME) of the rat hypothalamus. Neuroendocrinology 39:296–306

Merzenich MM, Harrington T (1969) The sense of flutter vibration evoked by stimulation on the hairy skin of primates: comparison of human sensory capacity with the responses of mechanoreceptor afferents innervating the hairy skin of monkeys. Exp Brain Res 9:236–269

Messlinger K, Hanesch U, Baumgärtel M, Trost B, Schmidt RF (1993) Innervation of the dura mater encephali of cat and rat: ultrastructure and calcitonin gene-related peptide-like and substance P-like immunoreactivity. Anat Embryol (Berl) 188:219–237

Messlinger K, Hanesch U, Kurosawa M, Pawlak M, Schmidt RF (1995) Calcitonin gene related peptide released from dural nerve fibers mediates increase of meningeal blood flow in the rat. Can J Physiol Pharmacol 73:1020–1024

Meyer RA, Davis KD, Cohen RH, Treede R-D, Campbell JN (1991) Mechanically insensitive afferents (MIAs) in cutaneous nerves of monkey. Brain Res 561:252–261

Milne RJ, Foreman RD, Giesler GJJ, Willis WD (1981) Convergence of cutaneous and pelvic visceral nociceptive inputs onto primate spinothalamic neurons. Pain 11:163–183

Milner P, Appenzeller O, Qualls C, Burnstock G (1992) Differential vulnerability of neuropeptides in nerves of the vasa nervorum to streptozotocin-induced diabetes. Brain Res 574:56–62

Mitchell JF (1963) The spontaneous and evoked release of acetylcholine from the cerebral cortex. J Physiol (Lond) 165:98–116

Mitchell JH (1990) Neural control of the circulation during exercise. Med Sci Sports Exerc 22:141–154

Mitchell JH, Schmidt RF (1983) Cardiovascular reflex control by afferent fibers from skeletal muscle receptors. In: Shepherd JT, Abboud FM, Geiger SR (eds) Handbook of physiology, section 2, the cardiovascular system, vol 3. American Physiological Society, Bethesda, pp 623–658

Mitchell JH, Mierzwiak DS, Wildenthal K, Willis WDJ, Smith AM (1968) Effect on left ventricular performance of stimulation of an afferent nerve from muscle. Circ Res 22:507–516

Mitchell JH, Reardon WC, McCloskey DI (1977) Reflex effects on circulation and respiration from contracting skeletal muscle. Am J Physiol 233:H374–H378

Miyamoto JK (1976) Dorsal root reflex response in sympathetic nerves. Brain Res 111:172–180

Miyamoto J, Alanis J (1970) Reflex sympathetic responses produced by activation of vibrational receptors. Jpn J Physiol 20:725–740

Mizumura K, Kumazawa T (1976) Reflex respiratory response induced by chemical stimulation of muscle afferents. Brain Res 109:402–406

Moos F, Richard P (1988) Characteristics of early- and late-recruited oxytocin bursting cells at the beginning of suckling in rats. J Physiol (Lond) 399:1–12

Morrison JFB, Sato A, Sato Y, Suzuki A (1995a) Long-lasting facilitation and depression of periurethral skeletal muscle following acupuncture-like stimulation in anesthetized rats. Neurosci Res 23:159–169

Morrison JFB, Sato A, Sato Y, Yamanishi T (1995b) The influence of afferent inputs from skin and viscera on the activity of the bladder and the skeletal muscle surrounding the urethra in the rat. Neurosci Res 23:195–205

Morrison JFB, Sato A, Sato Y, Suzuki A (1996a) The nitric oxide synthase inhibitor L-NAME reduces inhibitory components of somato-vesical parasympathetic reflexes in the rat. Neurosci Res 24:195–199

Morrison JFB, Sato A, Sato Y, Suzuki A (1996b) Excitatory and inhibitory A- and C-reflexes in pelvic parasympathetic efferent nerves elicited by single shock to A and C afferent fibers of perineal and limb somatic afferent nerves in anesthetized rats. Neurosci Res Lett 212:25–28

Morrison SF, Reis DJ (1989) Reticulospinal vasomotor neurons in the RVL mediate the somatosympathetic reflex. Am J Physiol 256:R1084–R1097

Murakami T, Ishizuka K, Yoshihara M, Uchiyama M (1983) Reflex responses of single salivatory neurons to stimulation of trigeminal sensory branches in the cat. Brain Res 280:233–237

Murakawa K, Izumi R, Noma K, Tashiro C, Minatogawa T, Amatsu M (1995) Effects of electrical stimulation of cervical sympathetic trunks on microcirculation in the facial nerve. Jpn J Physiol 45:801–809

Nagasaka H, Yaksh TL (1995) Effects of intrathecal μ, δ, and κ agonists on thermally evoked cardiovascular and nociceptive reflexes in halothane-anesthetized rats. Anesth Analg 80:437–443

Nagata O, Li W-M, Sato A (1995) Glutamate *N*-methyl-D-aspartate (NMDA) and non-NMDA receptor antagonists administered into the brain stem depress the renal sympathetic reflex discharges evoked by single shock of somatic afferents in anesthetized rats. Neurosci Lett 201:111–114

Nakamura M, Sakurai T, Tsujimoto Y, Tada Y (1986) Bladder inhibition by electrical stimulation of the perianal skin. Urol Int 41:62–63

Nakazato Y (1968) Reflex potentials of the vagus nerve in various levels evoked by stimulations of some peripheral nerves. J Physiol Soc Jpn 30:172–180 (in Japanese with English abstract)

Neugebauer V, Lücke T, Schaible H-G (1993) N-methyl-D-aspartate (NMDA) and non-NMDA receptor antagonists block the hyperexcitability of dorsal horn neurons during development of acute arthritis in rat's knee joint. J Neurophysiol 70:1–13

Neugebauer V, Lücke T, Grubb B, Schaible H-G (1994a) The involvement of N-methyl-D-aspartate (NMDA) and non-NMDA receptors in the responsiveness of rat spinal neurons with input from the chronically inflamed ankle. Neurosci Lett 170:237–240

Neugebauer V, Lücke T, Schaible H-G (1994b) Requirement of metabotropic glutamate receptors for the generation of inflammation-evoked hyperexcitability in rat spinal cord neurons. Eur J Neurosci 6:1179–1186

Neugebauer V, Rümenapp P, Schaible H-G (1996) Calcitonin gene-related peptide is involved in the spinal processing of mechanosensory input from the rat's knee joint and in the generation and maintenance of hyperexcitability of dorsal horn neurons during development of acute inflammation. Neuroscience 71: 1095–1109

Nicoll RA, Alger BE, Jahr CE (1980) Enkephalin blocks inhibitory pathways in the vertebrate CNS. Nature 287:22–25

Nishi S, Seino Y, Ishida H, Seno M, Taminato T, Sakurai H, Imura H (1987) Vagal regulation of insulin, glucagon, and somatostatin secretion in vitro in the rat. J Clin Invest 79:1191–1196

Nishijo K, Enpin U, Yoshikawa K, Yazawa K, Mori H, Miyamoto T, Katai S (1991) The neural machanism of the response in heart rate induced by acupuncture. In: Yoshikawa M et al (eds) New trends in autonomic nervous system research, Elsevier Science Publishers, Amsterdam, pp 594

Niv D, Whitwam JG (1983) Selective effect of fentanyl on group III and IV somato-sympathetic reflexes. Neuropharmacology 22:703–709

Noguchi E, Hayashi H (1996) Increases in gastric acidity in response to electroacupuncture stimulation of the hindlimb of anesthetized rats. Jpn J Physiol 46:53–58

Norman J, Whitwam JG (1973a) The effect of stimulation of somatic afferent nerves on sympathetic nerve activity, heart rate and blood pressure in dogs. J Physiol (Lond) 231:76P–77P

Norman J, Whitwam JG (1973b) The vagal contribution to changes in heart rate evoked by stimulation of cutaneous nerves in the dog. J Physiol (Lond) 234:89P–90P

Nosaka S, Sato A, Shimada F (1980) Somatosplanchnic reflex discharges in rats. J Auton Nerv Syst 2:95–104

Nunez R, Gross GH, Sachs BD (1986) Origin and central projections of rat dorsal penile nerve: possible direct projection to autonomic and somatic neurons by primary afferents of nonmuscle origin. J Comp Neurol 247:417–429

Nutter DO, Wickliffe CW (1981) Regional vasomotor responses to the somatopressor reflex from muscle. Circ Res 48[Suppl 1]:98–103

Oaknin S, Castillo ARD, Guerra M, Battaner E, Mas M (1989) Changes in forebrain Na, K-ATPase activity and serum hormone levels during sexual behavior in male rats. Physiol Behav 45:407–410

Oberle J, Elam M, Karlsson T, Wallin BG (1988) Temperature-dependent interaction between vasoconstrictor and vasodilator mechanisms in human skin. Acta Physiol Scand 132:459–469

Ogawa S, Saito H, Saeki S, Suzuki H (1994) Reflex sympathetic activities during inhalation of anaesthetics in cats: nitrous oxide. Neurosci Lett 168:16–18

Ohsawa H, Okada K, Nishijo K, Sato Y (1995) Neural mechanism of depressor responses of arterial pressure elicited by acupuncture-like stimulation to a hindlimb in anesthetized rats. J Auton Nerv Syst 51:27–35

Okada H, Nakano O, Nisida I (1960) Effects of sciatic stimulation upon the efferent impulses in the long ciliary nerve of the cat. Jpn J Physiol 10:327–339

Okada H, Okamoto K, Nisida I (1961) The activity of the cardioregulatory and abdominal sympathetic nerves of the cat in the Bainbridge reflex. Jpn J Physiol 11:520–529

Okrasa S, Kotwica G, Ciereszko R, Dusza L, Czarnyszewicz J (1989) Hormonal changes during lactation in sows: influence of spontaneous suckling on prolactin, oxytocin and corticoids concentrations. Exp Clin Endocrinol 93:95–103

Oyama T, Taniguchi K, Ishihara H, Matsuki A, Maeda A, Murakawa T, Kudo T (1979) Effects of enflurane anaesthesia and surgery on endocrine function in man. Br J Anaesth 51:141–148

Patterson TL, Rubright LW (1934) The influence of tonal conditions on the muscular response of the monkey's stomach. Q J Exp Physiol 24:3–21

Pauk J, Kuhn CM, Field TM, Schanberg SM (1986) Positive effects of tactile versus kinesthetic or vestibular stimulation on neuroendocrine and ODC activity in maternally-deprived rat pups. Life Sci 39:2081–2087

Pearce WJ, Scremin OU, Sonnenschein RR, Rubinstein EH (1981) The electroencephalogram, blood flow, and oxygen uptake in rabbit cerebrum. J Cereb Blood Flow Metab 1:419–428

Pedersen HE, Blunck CFJ, Gardner E (1956) The anatomy of lumbosacral posterior rami and meningeal branches of spinal nerves (sinu-vertebral nerves). J Bone Joint Surg 38A:377–391

Perez-Gonzalez JF (1981) Factors determining the blood pressure responses to isometric exercise. Circ Res 48[Suppl 1]:76–86

Perl ER (1984) Pain and nociception. In: Darian-Smith I (ed) Handbook of physiology, section 1, the nervous system, vol 3, part 2. American Physiological Society, Bethesda, pp 915–975

Pescatori ES, Calabro A, Artibani W, Pagano F, Triban C, Italiano G (1993) Electrical stimulation of the dorsal nerve of the penis evokes reflex tonic erections of the penile body and reflex ejaculatory responses in the spinal rat. J Urol 149:627–632

Phillis JW (1968) Acetylcholine release from the cerebral cortex: its role in cortical arousal. Brain Res 7:378–389

Pierau F-K, Fellmer G, Taylor DCM (1984) Somato-visceral convergence in cat dorsal root ganglion neurones demonstrated by double-labelling with fluorescent tracers. Brain Res 321:63–70

Pitetti KH, Iwamoto GA, Mitchell JH, Ordway GA (1989) Stimulating somatic afferent fibers alters coronary arterial resistance. Am J Physiol 256:R1331–R1339

Pitts RF, Larrabee MG, Bronk DW (1941) Hypothalamic cardiovascular control. Am J Physiol 134:359–383

Plotsky PM, Vale W (1984) Hemorrhage-induced secretion of corticotropin-releasing factor-like immunoreactivity into the rat hypophysial portal circulation and its inhibition by glucocorticoids. Endocrinology 114:164–169

Popper CW, Chiueh CC, Kopin IJ (1977) Plasma catecholamine concentrations in unanesthetized rats during sleep, wakefulness, immobilization and after decapitation. J Pharmacol Exp Ther 202:144–148

Poulain DA, Wakerley JB (1982) Electrophysiology of hypothalamic magnocellular neurones secreting oxytocin and vasopressin. Neuroscience 7:773–808

Proske U, Schaible H-G, Schmidt RF (1988) Joint receptors and kinaesthesia. Exp Brain Res 72:219–224

Purinton PT, Fletcher TF, Bradley WE (1976) Innervation of pelvic viscera in the rat. Evoked potentials in nerves to bladder and penis (clitoris). Invest Urol 14:28–32

Rampal G, Mignard P (1975a) Organization of the nervous control of urethral sphincter. A study in the anaesthetized cat with intact central nervous system. Pflugers Arch 353:21–31

Rampal G, Mignard P (1975b) Behaviour of the urethral striated sphincter and of the bladder in the chronic spinal cat. Implications at the central nervous system level. Pflugers Arch 353:33–42

Ranson SW, Billingsley PR (1916) Afferent spinal paths and the vasomotor reflexes. Studies in vasomotor reflex arcs. VI. Am J Physiol 42:16–35

Rausell E, Avendano C (1989) Laminar and segmental termination in the dorsal horn of the spinal cord of an articular nerve of the forepaw in the cat. In: Cervero F, Bennett GJ, Headley PM (eds) Processing of sensory information in the superficial dorsal horn of the spinal cord. Plenum, New York, pp 83–87

Riedel W, Iriki M, Simon E (1972) Regional differentiation of sympathetic activity during peripheral heating and cooling in anesthetized rabbits. Pflugers Arch 332:239–247

Roman C, Gonella J (1981) Extrinsic control of digestive tract motility. In: Johnson LR (ed) Physiology of the gastrointestinal tract, Raven, New York, pp 289–333

Romanes GJ (ed) (1964) Cunningham's textbook of anatomy. 10th edn. Oxford University Press, London

Rosenblueth A, Freeman NE (1931) The reciprocal innervation in reflex changes of heart rate. Am J Physiol 98:430–434

Round JM, Jones DA, Cambridge G (1987) Cellular infiltrates in human skeletal muscle: exercise induced damage as a model for inflammatory disease. J Neurol Sci 82:1–11

Rowell LB (1981) Active neurogenic vasodilation in man. In: Vanhoutte PM , Leusen I (eds) Vasodilatation. Raven, New York, pp 1–17

Ruch TC (1960) Central control of the bladder. In: Field J, Magoun HW, Hall VE (eds) Handbook of physiology, section 1, neurophysiology, vol 2. American Physiological Society, Washington, pp 1207–1223

Ruegamer WR, Silverman FR (1956) Influence of gentling on physiology of the rat. Proc Soc Exp Biol Med 92:170–174

Ruggeri P, Ermirio R, Molinari C, Calaresu FR (1995) Role of ventrolateral medulla in reflex cardiovascular responses to activation of skin and muscle nerves. Am J Physiol 268:R1464–R1471

Ruhmann W (1927) Örtliche Hautreizbehandlung des Magens und ihre physiologischen Grundlagen. Arch Verdauungskr 41:336–350

Russell NJW, Schaible H-G, Schmidt RF (1987) Opiates inhibit the discharges of fine afferent units from inflamed knee joint of the cat. Neurosci Lett 76:107–112

Sachs BD, Garinello LD (1980) Hypothetical spinal pacemaker regulating penile reflexes in rats: evidence from transection of spinal cord and dorsal penile nerves. J Comp Physiol Psychol 94:530–535

Saeki Y, Sato A, Sato Y, Trzebski A (1990) Effects of stimulation of cervical sympathetic trunks with various frequencies on the local cortical cerebral blood flow measured by laser Doppler flowmetry in the rat. Jpn J Physiol 40:15–32

Sakakibara H, Iwase S, Mano T, Watanabe T, Kobayashi F, Furuta M, Kondo T, Miyao M, Yamada S (1990) Skin sympathetic activity in the tibial nerve triggered by vibration applied to the hand. Int Arch Occup Environ Health 62:455–458

Samso E, Farber NE, Kampine JP, Schmeling WT (1994) The effects of halothane on pressor and depressor responses elicited via the somatosympathetic reflex: a potential antinociceptive action. Anesth Analg 79:971–979

Sándor P, Sato A, Sato Y, Swenson RS (1985) The effects of vertebral artery injections of an enkephalin analogue, (D-Met^2, Pro^5)-enkephalinamide, on somatosympathetic reflexes. Neurosci Lett 53:87–93

Saper CB, De Marchena O (1986) Somatosympathetic reflex unilateral sweating and pupillary dilatation in a paraplegic man. Ann Neurol 19:389–390

Sasaki M, Morrison JFB, Sato Y, Sato A (1994) Effect of mechanical stimulation of the skin on the external urethral sphincter muscles in anesthetized cats. Jpn J Physiol 44:575–590

Sato A (1972a) Somato-sympathetic reflex discharges evoked through supramedullary pathways. Pflugers Arch 332:117–126

Sato A (1972b) The relative involvement of different reflex pathways in somato-sympathetic reflexes, analyzed in spontaneously active single preganglionic sympathetic units. Pflugers Arch 333:70–81

Sato A (1972c) Spinal and supraspinal inhibition of somato-sympathetic reflexes by conditioning afferent volleys. Pflugers Arch 336:121–133

Sato A (1973) Spinal and medullary reflex components of the somato-sympathetic reflex discharges evoked by stimulation of the group IV somatic afferents. Brain Res 51:307–318

Sato A (1987) Neural mechanisms of somatic sensory regulation of catecholamine secretion from the adrenal gland. Adv Biophys 23:39–80

Sato A, Sato Y (1992) Regulation of regional cerebral blood flow by cholinergic fibers originating in the basal forebrain. Neurosci Res 14:242–274

Sato A, Schmidt RF (1966) Muscle and cutaneous afferents evoking sympathetic reflexes. Brain Res 2:399–401

Sato A, Schmidt RF (1971) Spinal and supraspinal components of the reflex discharges into lumbar and thoracic white rami. J Physiol (Lond) 212:839–850

Sato A, Schmidt RF (1973) Somatosympathetic reflexes: afferent fibers, central pathways, discharge characteristics. Physiol Rev 53:916–947

Sato A, Schmidt RF (1987) The modulation of visceral functions by somatic afferent activity. Jpn J Physiol 37:1–17

Sato A, Swenson RS (1984) Sympathetic nervous system response to mechanical stress of the spinal column in rats. J Manipulative Physiol Ther 7:141–147

Sato A, Tsushima N, Fujimori B (1965) Reflex potentials of lumbar sympathetic trunk with sciatic nerve stimulation in cats. Jpn J Physiol 15:532–539

Sato A, Sato N, Ozawa T, Fujimori B (1967) Further observations of reflex potentials in the lumbar sympathetic trunk in cats. Jpn J Physiol 17:294–307

Sato A, Kaufman A, Koizumi K, Brooks CMC (1969) Afferent nerve groups and sympathetic reflex pathways. Brain Res 14:575–587

Sato A, Sato Y, Shimada F, Torigata Y (1975a) Changes in gastric motility produced by nociceptive stimulation of the skin in rats. Brain Res 87:151–159

Sato A, Sato Y, Shimada F, Torigata Y (1975b) Changes in vesical function produced by cutaneous stimulation in rats. Brain Res 94:465–474

Sato A, Sato Y, Shimada F, Torigata Y (1976) Varying changes in heart rate produced by nociceptive stimulation of the skin in rats at different temperatures. Brain Res 110:301–311

Sato A, Sato Y, Sugimoto H, Terui N (1977) Reflex changes in the urinary bladder after mechanical and thermal stimulation of the skin at various segmental levels in cats. Neuroscience 2:111–117

Sato A, Sato Y, Schmidt RF (1979a) Effects on reflex bladder activity of chemical stimulation of small diameter afferents from skeletal muscle in the cat. Neurosci Lett 11:13–17

Sato A, Sato Y, Schmidt RF (1979b) Somatic afferents and their effects on bladder function. In: Brooks CMC, Koizumi K, Sato A (eds) Integrative functions of the autonomic nervous system. Universtiy of Tokyo Press, Tokyo, Elsevier, Amsterdam, pp 309–318

Sato A, Sato Y, Schmidt RF (1980) Reflex bladder activity induced by electrical stimulation of hind limb somatic afferents in the cat. J Auton Nerv Syst 1:229–241

Sato A, Sato Y, Schmidt RF (1981) Heart rate changes reflecting modifications of efferent cardiac sympathetic outflow by cutaneous and muscle afferent volleys. J Auton Nerv Syst 4:231–247

Sato A, Sato Y, Schmidt RF (1982) Changes in heart rate and blood pressure upon injection of algesic agents into skeletal muscle. Pflugers Arch 393:31–36

Sato A, Sato Y, Schmidt RF, Torigata Y (1983) Somato-vesical reflexes in chronic spinal cats. J Auton Nerv Syst 7:351–362

Sato A, Sato Y, Schmidt RF (1984) Changes in blood pressure and heart rate induced by movements of normal and inflamed knee joints. Neurosci Lett 52:55–60

Sato A, Sato Y, Swenson RS (1985) Effect of morphine on somatocardiac sympathetic reflexes in spinalized cats. J Auton Nerv Syst 12:175–184

Sato A, Sato Y, Schmidt RF (1986a) Catecholamine secretion and adrenal nerve activity in response to movements of normal and inflamed knee joints in cats. J Physiol (Lond) 375:611–624

Sato A, Sato Y, Suzuki A, Swenson RS (1986b) The effects of morphine administered intrathecally on the somatosympathetic reflex discharges in anesthetized cats. Neurosci Lett 71:345–350

Sato A, Sato Y, Suzuki A (1992) Mechanism of the reflex inhibition of micturition contractions of the urinary bladder elicited by acupuncture-like stimulation in anesthetized rats. Neurosci Res 15:189–198

Sato A, Sato Y, Suzuki A, Uchida S (1993) Neural mechanisms of the reflex inhibition and excitation of gastric motility elicited by acupuncture-like stimulation in anesthetized rats. Neurosci Res 18:53–62

Sato A, Uchida S, Yamauchi Y (1994a) A new method for continuous measurement of regional cerebral blood flow using laser Doppler flowmetry in a conscious rat. Neurosci Lett 175:149–152

Sato A, Sato Y, Uchida S (1994b) Blood flow in the sciatic nerve is regulated by vasoconstrictive and vasodilative nerve fibers originating from the ventral and dorsal roots of the spinal nerves. Neurosci Res 21:125–133

Sato A, Sato Y, Schmidt RF (1995) Modulation of somatocardiac sympathetic reflexes mediated by opioid receptors at the spinal and brainstem level. Exp Brain Res 105:1–6

Sato A, Sato Y, Suzuki A, Uchida S (1996) Reflex modulation of catecholamine secretion and adrenal sympathetic nerve activity by acupuncture-like stimulation in anesthetized rat. Jpn J Physiol 46:411–421

Sato T, Yu Y, Guo SY, Kasahara T, Hisamitsu T (1996) Acupuncture stimulation enhances splenic natural killer cell cytotoxicity in rat. Jpn J Physiol 46:131–136

Sato Y, Terui N (1976) Changes in duodenal motility produced by noxious mechanical stimulation of the skin in rats. Neurosci Lett 2:189–193

Sato Y, Schaible H-G, Schmidt RF (1983) Types of afferents from the knee joint evoking sympathetic reflexes in cat inferior cardiac nerves. Neurosci Lett 39:71–75

Sato Y, Schaible H-G, Schmidt RF (1985) Reactions of cardiac postganglionic sympathetic neurons to movements of normal and inflamed knee joints. J Auton Nerv Syst 12:1–13

Saunders A, Terry LC, Audet J, Brazeau P, Martin JB (1976) Dynamic studies of growth hormone and prolactin secretion in the female rat. Neuroendocrinology 21:193–203

Schaefer H (1960) Central control of cardiac function. Physiol Rev 40[Suppl 4]:213–231

Schaible H-G, Grubb BD (1993) Afferent and spinal mechanisms of joint pain. Pain 55:5–54

Schaible H-G, Schmidt RF (1983a) Activation of groups III and IV sensory units in medial articular nerve by local mechanical stimulation of knee joint. J Neurophysiol 49:35–44

Schaible H-G, Schmidt RF (1983b) Responses of fine medial articular nerve afferents to passive movements of knee joint. J Neurophysiol 49:1118–1126

Schaible H-G, Schmidt RF (1984) Mechanosensibility of joint receptors with fine afferent fibres. In: Creutzfeldt O, Schmidt RF, Willis WD (eds) Sensory-motor integration in the nervous system. Springer, Berlin Heidelberg New York, pp 284–297

Schaible H-G, Schmidt RF (1985) Effects of an experimental arthritis on the sensory properties of fine articular afferent units. J Neurophysiol 54:1109–1122

Schaible H-G, Schmidt RF (1988a) Excitation and sensitization of fine articular afferent units from cat's knee joint by prostaglandin E2 (PG E2). J Physiol (Lond) 403:91–104

Schaible H-G, Schmidt RF (1988b) Direct observation of the sensitization of articular afferents during an experimental arthritis. In: Dubner R, Gebhardt GF, Bond MR

(eds) Pain research and clinical management. Proceedings of the 5th world congress on pain. Elsevier, Amsterdam, pp 44–50
Schaible H-G, Schmidt RF (1988c) Time course of mechanosensitivity changes in articular afferents during a developing experimental arthritis. J Neurophysiol 60:2180–2195
Schaible H-G, Freudenberger U, Neugebauer V, Stiller RU (1994) Intraspinal release of immunoreactive calcitonin gene-related peptide during developement of inflammation in the joint in vivo – a study with antibody microprobes in the cat and rat. Neuroscience 62:1293–1305
Schanberg SM, Field TM (1987) Sensory deprivation stress and supplemental stimulation in the rat pup and preterm human neonate. Child Dev 58:1431–1447
Schanberg SM, Evoniuk G, Kuhn CM (1984) Tactile and nutritional aspects of maternal care: specific regulators of neuroendocrine function and cellular development. Proc Soc Exp Biol Med 175:135–146
Schirar A, Cognié Y, Louault F, Poulin N, Meusnier C, Levasseur MC, Martinet J (1990) Resumption of gonadotrophin release during the post-partum period in sucking and non-sucking ewes. J Reprod Fertil 88:593–604
Schmidt RF (ed) (1995a) Neuro- und Sinnesphysiologie. 2nd edn. Springer, Berlin Heidelberg New York
Schmidt RF, Schönfuss K (1970) An analysis of the reflex activity in the cervical sympathetic trunk induced by myelinated somatic afferents. Pflugers Arch 314:175–198
Schmidt RF, Thews G (eds) (1995b) Physiologie des Menschen. 26th edn. Springer, Berlin Heidelberg New York
Schmidt RF, Weller E (1970) Reflex activity in the cervical and lumbar sympathetic trunk induced by unmyelinated somatic afferents. Brain Res 24:207–218
Schmidt RF, Senges J, Zimmermann M (1967) Excitability measurements at the central terminals of single mechanoreceptor afferents during slow potential changes. Exp Brain Res 3:220–233
Schmidt RF, Schaible H-G, Meblinger K, Heppelmann B, Hanesch U, Pawlak M (1994) Silent and active nociceptors: structure, functions, and clinical implications. In: Anonymous (ed) Proceedings of the 7th World Congress on pain. IASP, Seattle, pp 1–933
Schmidt RF, Schmelz M, Forster C, Ringkamp M, Torebjörk E, Handwerker H (1995) Novel classes of responsive and unresponsive C nociceptors in human skin. J Neurosci 15:333–341
Schramm L, Adair J, Stribling J, Gray L (1975) Preganglionic innervation of the adrenal gland of the rat: a study using horseradish peroxidase. Exp Neurol 49:540–553
Sell R, Erdelyi A, Schaefer H (1958) Untersuchungen über den Einfluß peripherer Nervenreizung auf die sympathische Aktivität. Pflugers Arch 267:566–581
Selye H (1956) The stress of life. McGraw-Hill, New York
Shavit Y, Yirmiya R, Beilin B (1990) Stress neuropeptides, immunity, and neoplasia. In: Freier S (ed) The neuroendocrine-immune network. CRC, Florida, pp 163–175

Shen E, Mo N, Dun NJ (1990) APV-sensitive dorsal root afferent transmission to neonate rat sympathetic preganglionic neurons in vitro. J Neurophysiol 64:991–999
Sherrington CS (1906) The integrative action of the nervous system. Yale University Press, New Haven
Shibuki K, Yagi K (1986) Synergistic activation of rat supraoptic neurosecretory neurons by noxious and hypovolemic stimuli. Exp Brain Res 62:572–578
Sicuteri F (1967) Vasoneuroactive substances and their implication in vascular pain. In: Friedmann AP (ed) Res. clin. stud. headache. Karger, Basel, pp 6–45
Sicuteri F, Franchi G, Fanciullacci M (1964) Bradichinina e dolore da ischemia. Settim Med 52:127–139
Simon E (1974) Temperature regulation: the spinal cord as a site of extrahypothalamic thermoregulatory functions. Rev Physiol Biochem Pharmacol 71:1–76
Simons DG (1987) Myofascial pain syndromes due to trigger points. In: Goodgold J (ed) Rehabilitation medicine. Mosby, St. Louis, pp 686–723
Sjölander P, Johansson H, Sojka P, Rehnholm A (1989) Sensory nerve endings in the cat cruciate ligaments: a morphological investigation. Neurosci Lett 102:33–38
Sjölund B, Terenius L, Eriksson M (1977) Increased cerebrospinal fluid levels of endorphins after electro-acupuncture. Acta Physiol Scand 100:382–384
Skoglund CR (1960) Vasomotor reflexes from muscle. Acta Physiol Scand 50:311–327
Skoglund S (1956) Anatomical and physiological studies of knee joint innervation in the cat. Acta Physiol Scand 36 S124:1–101
Stacey MJ (1969) Free nerve endings in skeletal muscle of the cat. J Anat 105:231–254
Staton WM (1951) New approach to muscle soreness. Athletic J 31:24–61
Stauber WT, Clarkson PM, Fritz VK, Evans WJ (1990) Extracellular matrix disruption and pain after eccentric muscle action. J Appl Physiol 69:868–874
Stebbins CL, Longhurst JC (1985) Bradykinin-induced chemoreflexes from skeletal muscle: implications for the exercise reflex. J Appl Physiol 59:56–63
Stebbins CL, Brown B, Levin D, Longhurst JC (1988) Reflex effect of skeletal muscle mechanoreceptor stimulation on the cardiovascular system. J Appl Physiol 65:1539–1547
Stebbins CL, Ortiz-Acevedo A, Hill JM (1992) Spinal vasopressin modulates the reflex cardiovascular response to static contraction. J Appl Physiol 72:731–738
Steers WD, Mallory B, De Groat WC (1988) Electrophysiological study of neural activity in penile nerve of the rat. Am J Physiol 254:R989–R1000
Stein C (1993) Periphere Opioidrezeptoren und ihre Bedeutung für die postoperative Schmerztherapie. Schmerz 7:4–7
Stein C, Hassan AHS, Przewlocki R, Gramsch C, Peter K, Herz A (1990) Opioids from immunocytes interact with receptors on sensory nerves to inhibit nociception in inflammation. Proc Natl Acad Sci USA 87:5935–5939
Stein C, Comisel K, Haimerl E, Yassouridis A, Lehrberger K, Herz A, Peter K (1991) Analgesic effect of intraarticular morphine after arthroscopic knee surgery. N Engl J Med 325:1123–1126

Stock S, Uvnäs-Moberg K (1988) Increased plasma levels of oxytocin in response to afferent electrical stimulation of the sciatic and vagal nerves and in response to touch and pinch in anaesthetized rats. Acta Physiol Scand 132:29–34

Stornetta RL, Morrison SF, Ruggiero DA, Reis DJ (1989) Neurons of rostral ventrolateral medulla mediate somatic pressor reflex. Am J Physiol 256:R448–R462

Streatfeild KA, Davidson NS, McCloskey DI (1977) Muscular reflex and baroreflex influences on heart rate during isometric contractions. Cardiovasc Res 11:87–93

Sugiura Y, Terui N, Hosoya Y, Kohno K (1989) Distribution of unmyelinated primary afferent fibers in the dorsal horn. In: Cervero F, Bennett GJ, Headley PM (eds) Processing of sensory information in the superficial dorsal horn of the spinal cord. Plenum, New York, pp 15–27

Sugiyama Y, Xue Y-X, Mano T (1995) Transient increase in human muscle sympathetic nerve activity during manual acupuncture. Jpn J Physiol 45:337–345

Sun M-K, Spyer KM (1991) Nociceptive inputs into rostral ventrolateral medulla-spinal vasomotor neurones in rats. J Physiol (Lond) 436:685–700

Sundqvist T, Öberg P, Rapoport SI (1985) Blood flow in rat sciatic nerve during hypotension. Exp Neurol 90:139–148

Swanson LW, Sawchenko PE, Rivier J, Vale WW (1983) Organization of ovine corticotropin-releasing factor immunoreactive cells and fibers in the rat brain: an immunohistochemical study. Neuroendocrinology 36:165–186

Szerb JC (1967) Cortical acetylcholine release and electroencephalographic arousal. J Physiol (Lond) 192:329–343

Szulczyk P (1976) Descending spinal sympathetic pathway utilized by somato-sympathetic reflex and carotid chemoreflex. Brain Res 112:190–193

Takagi K, Sakurai T (1950) A sweat reflex due to pressure on the body surface. Jpn J Physiol 1:22–28

Takahashi H, Izumi H, Karita K (1995) Parasympathetic reflex salivary secretion in the cat parotid gland. Jpn J Physiol 45:475–490

Takahashi Y, Nakajima Y, Sakamoto T, Moriya H, Takahashi K (1993) Capsaicin applied to rat lumbar intervertebral disc causes extravasation in the groin skin: a possible mechanism of referred pain of the intervertebral disc. Neurosci Lett 161:1–3

Tallarida G, Baldoni F, Peruzzi G, Brindisi F, Raimondi G, Sangiorgi M (1979) Cardiovascular and respiratory chemoreflexes from the hindlimb sensory receptors evoked by intra-arterial injection of bradykinin and other chemical agents in the rabbit. J Pharmacol Exp Ther 208:319–329

Tallarida G, Baldoni F, Peruzzi G, Raimondi G, Massaro M, Sangiorgi M (1981) Cardiovascular and respiratory reflexes from muscles during dynamic and static exercise. J Appl Physiol 50:784–791

Tallarida G, Baldoni F, Peruzzi G, Raimondi G, Di Nardo P, Massaro M, Visigalli G, Franconi G, Sangiorgi M (1985) Cardiorespiratory reflexes from muscles during dynamic and static exercise in the dog. J Appl Physiol 58:844–852

Tallarida G, Peruzzi G, Raimondi G (1990) The role of chemosensitive muscle receptors in cardiorespiratory regulation during exercise. J Auton Nerv Syst 30:S155–S162

Tam K-C, Yiu H-H (1975) The effect of acupuncture on essential hypertension. Am J Chin Med 3:369–375

Taylor DG, Brody MJ (1976) Spinal adrenergic mechanisms regulating sympathetic outflow to blood vessels. Circ Res 38[Suppl II]:II10–II20

Terui N, Koizumi K (1984) Responses of cardiac vagus and sympathetic nerves to excitation of somatic and visceral nerves. J Auton Nerv Syst 10:73–91

Terui N, Numao Y, Kumada M, Reis DJ (1981) Identification of the primary afferent fiber group and adequate stimulus initiating the trigeminal depressor response. J Auton Nerv Syst 4:1–16

Terui N, Saeki Y, Kumada M (1987) Confluence of barosensory and nonbarosensory inputs at neurons in the ventrolateral medulla in rabbits. Can J Physiol Pharmacol 65:1584–1590

Thauer R, Simon E (1972) Spinal cord and temperature regulation. In: Itoh S, Ogata K, Yoshimura H (eds) Advances in climatic physiology. Igaku Shoin, Tokyo/Springer, Berlin Heidelberg New York, pp 22–49

Thor KB, Blais DP, De Groat WC (1989) Behavioral analysis of the postnatal development of micturition in kittens. Dev Brain Res 46:137–144

Thorén P, Jones JV (1977) Characteristics of aortic baroreceptor C-fibres in the rabbit. Acta Physiol Scand 99:448–456

Tibes U (1977) Reflex inputs to the cardiovascular and respiratory centers from dynamically working canine muscles: some evidence for involvement of group III or IV nerve fibers. Circ Res 41:332–341

Tindal JS (1974) Stimuli that cause the release of oxytocin. In: Greep RO, Astwood EB, Knobil E, Sawyer WH, Geiger SR (eds) Handbook of physiology. Section 7, endocrinology, vol 4, part 1. American Physiological Society, Washington, pp 257–267

Togashi H, Yoshioka M, Minami M, Shimamura K, Saito H, Kitada C, Fujino. M (1987) Effect of the substance-P antagonist spantide on adrenal sympathetic nerve activity in rats. Jpn J Pharmacol 43:253–261

Torebjörk HE, Hallin RG (1973) Perceptual changes accompanying controlled preferential blocking of A and C fibre responses in intact human skin nerves. Exp Brain Res 16:321–332

Torrens M, Morrison JFB (eds) (1987) The physiology of the lower urinary tract. Springer, Berlin Heidelberg New York

Travell JG, Simons DG (1983) Myofascial pain and dysfunction. Williams and Wilkins, Baltimore

Trzebski A, Kubin L (1981) Is the central inspiratory activity responsible for pCO_2-dependent drive of the sympathetic discharge? J Auton Nerv Syst 3:401–420

Trzebski A, Lipski J, Majcherczyk S, Szulczyk P, Chruscielewski L (1975) Central organization and interaction of the carotid baroreceptor and chemoreceptor sympathetic reflex. Brain Res 87:227–237

Tsuchiya T, Nakayama Y, Ozawa T (1991a) Response of adrenal sympathetic efferent nerve activity to mechanical and thermal stimulations of the facial skin area in anesthetized rats. Neurosci Lett 123:240–243

Tsuchiya T, Nakayama Y, Sato A (1991b) Somatic afferent regulation of plasma corticosterone in anesthetized rats. Jpn J Physiol 41:169–176

Tsuchiya T, Nakayama Y, Sato A (1992) Somatic afferent regulation of plasma luteinizing hormone and testosterone in anesthetized rats. Jpn J Physiol 42:539–547

Tsukayama H, Mori H, Nishijo K (1993) The effects of acupuncture treatment on autonomic function – stimulation to the sympathetic nervous function and paUvnäs B (1960) Central cardiovascular control. In: Field J, Magoun HW, Hall VE (eds) Handbook of physiology, section 1, neurophysiology, vol 2. American Physiological Society, Washington, pp 1131–1162

Uvnäs-Moberg K, Eriksson M (1983) Release of gastrin and insulin in response to suckling in lactating dogs. Acta Physiol Scand 119:181–185

Uvnäs-Moberg K, Eriksson M, Blomquist L-E, Kunavongkrit A, Einarsson S (1984) Influence of suckling and feeding on insulin, gastrin, somatostatin and VIP levels in peripheral venous blood of lactating sows. Acta Physiol Scand 121:31–38

Uvnäs-Moberg K, Posloncec B, Ahlberg L (1986) Influence on plasma levels of somatostatin, gastrin, glucagon, insulin and VIP-like immunoreactivity in peripheral venous blood of anaesthetized cats induced by low intensity afferent stimulation of the sciatic nerve. Acta Physiol Scand 126:225–230

Vallbo ÅB, Hagbarth K-E (1968) Activity from skin mechanoreceptors recorded percutaneously in awake human subjects. Exp Neurol 21:270–289

Vallbo ÅB, Johansson RS (1984) Properties of cutaneous mechanoreceptors in the human hand related to touch sensation. Hum Neurobiol 3:3–14

Vallbo ÅB, Hagbarth K-E, Torebjörk HE, Wallin BG (1979) Somatosensory, proprioceptive, and sympathetic activity in human peripheral nerves. Physiol Rev 59:919–957

Victor RG, Rotto DM, Pryor SL, Kaufman MP (1989) Stimulation of renal sympathetic activity by static contraction: evidence for mechanoreceptor-induced reflexes from skeletal muscle. Circ Res 64:592–599

Vissing J, Wilson LB, Mitchell JH, Victor RG (1991) Static muscle contraction reflexly increases adrenal sympathetic nerve activity in rats. Am J Physiol 261:R1307–R1312

Vissing J, Iwamoto GA, Fuchs IE, Galbo H, Mitchell JH (1994) Reflex control of glucoregulatory exercise responses by group III and IV muscle afferents. Am J Physiol 266:R824–R830

Voogt JL, Sar M, Meites J (1969) Influence of cycling, pregnancy, labor, and suckling on corticosterone-ACTH levels. Am J Physiol 216:655–658

Wada H, Iijima S, Orimo H, Ito H, Sato A (1992) Stimulation of the substantia nigra increases striatal blood flow in rats. Biogenic Amines 9:29–39

Wakerley JB, Lincoln DW (1973) The milk-ejection reflex of the rat: 20- to 40-fold acceleration in the firing of paraventricular neurones during oxytocin release. Endocrinology 57:477–493

Waldrop TG, Mitchell JH (1985) Effects of barodenervation on cardiovascular responses to static muscular contraction. Am J Physiol 249:H710–H714

Walther O-E, Iriki M, Simon E (1970) Antagonistic changes of blood flow and sympathetic activity in different vascular beds following central thermal stimulation. II. Cutaneous visceral sympathetic activity during spinal cord heating and cooling in anesthetized rabbits and cats. Pflugers Arch 319:162–184

Wang C, Chakrabarti MK, Galletly DC, Whitwam JG (1992) Relative effects of intrathecal administration of fentanyl and midazolam on Aδ and C fibre reflexes. Neuropharmacology 31:439–444

Wang GH (1957) The galvanic skin reflex. A review of old and recent works from a physiologic point of view. Part 1. Am J Phys Med 36:295–320

Wang GH (1958) The galvanic skin reflex. A review of old and recent works from a physiologic point of view. Part 2. Am J Phys Med 37:35–57

Wang GH (1964) The neural control of sweating. University of Wisconsin Press, Madison

Weidinger H, Fedina L, Kehrel H, Schaefer H (1961) Über die Lokalisation des "bulbären sympathischen Zentrums" und seine Beeinflussung durch Atmung und Blutdruck. Z Kreisl Forsch 50:229–241

Weininger O (1954) Physiological damage under emotional stress as a function of early experience. Science 119:285–286

Welk E, Fleischer E, Petsche U, Handwerker HO (1984) Afferent C-fibres in rats after neonatal capsaicin treatment. Pflugers Arch 400:66–71

Wiberg M, Widenfalk B (1991) An anatomical study of the origin of sympathetic and sensory innervation of the elbow and knee joint in the monkey. Neurosci Lett 127:185–188

Widenfalk B, Wiberg M (1989) Origin of sympathetic and sensory innervation of the knee joint. A retrograde axonal tracing study in the rat. Anat Embryol (Berl) 180:317–323

Widenfalk B, Wiberg M (1990) Origin of sympathetic and sensory innervation of the temporo-mandibular joint. A retrograde axonal tracing study in the rat. Neurosci Lett 109:30–35

Widenfalk B, Elfvin L-G, Wiberg M (1988) Origin of sympathetic and sensory innervation of the elbow joint in the rat: a retrograde axonal tracing study with wheat germ agglutinin conjugated horseradish peroxidase. J Comp Neurol 271:313–318

Widström AM, Winberg J, Werner S, Hamberger B, Eneroth P, Uvnäs-Moberg K (1984) Suckling in lactating women stimulates the secretion of insulin and prolactin without concomitant effects on gastrin, growth hormone, calcitonin, vasopressin or catecholamines. Early Hum Dev 10:115–122

Williams GL, Talavera F, Petersen BJ, Kirsch JD, Tilton JE (1983) Coincident secretion of follicle-stimulating hormone and luteinizing hormone in early postpartum beef cows: effects of suckling and low-level increases of systemic progesterone. Biol Reprod 29:362–373

Williams T, Mueller K, Cornwall MW (1991) Effect of acupuncture-point stimulation on diastolic blood pressure in hypertensive subjects: a preliminary study. Phys Ther 71:523–529

Willis WD (ed) (1992) Hyperalgesia and allodynia. Raven, New York

Wilson LB, Wall PT, Matsukawa K, Mitchell JH (1992) Effect of spinal microinjections of an antagonist to substance P or somatostatin on the exercise pressor reflex. Circ Res 70:213–222

Wilson LB, Fuchs IE, Matsukawa K, Mitchell JH, Wall PT (1993) Substance P release in the spinal cord during the exercise pressor reflex in anaesthetized cats. J Physiol (Lond) 460:79–90

Winsö O, Biber B, Martner J (1985) Does dopamine suppress stress-induced intestinal and renal vasoconstriction? Acta Anaesthesiol Scand 29:508–514

Wyszogrodski I, Polosa C (1973) The inhibition of sympathetic preganglionic neurons by somatic afferents. Can J Physiol Pharmacol 51:29–38

Yamamoto K, Iwase S, Mano T (1992) Responses of muscle sympathetic nerve activity and cardiac output to the cold pressor test. Jpn J Physiol 42:239–252

Yamashita H, Kannan H, Inenaga K, Koizumi K (1984) The role of cardiovascular and muscle afferent systems in control of body water balance. J Auton Nerv Syst 10:305–316

Yanase K, Meguro K, Sato A, Sato Y (1988) The effect of sevoflurane on somatically induced sympathetic reflexes. J Anesth 2:272–275

Yao T, Andersson S, Thorén P (1982) Long-lasting cardiovascular depression induced by acupuncture-like stimulation of the sciatic nerve in unanaesthetized spontaneously hypertensive rats. Brain Res 240:77–85

Yonei Y, Holzer P, Guth PH (1990) Laparotomy-induced gastric protection against ethanol injury is mediated by capsaicin-sensitive sensory neurons. Gastroenterology 99:3–9

Yoshimura M, Polosa C, Nishi S (1986) Electrophysiological properties of sympathetic preganglionic neurons in the cat spinal cord in vitro. Pflugers Arch 406:91–98

Zanzinger J, Czachurski J, Offner B, Seller H (1994) Somato-sympathetic reflex transmission in the ventrolateral medulla oblongata: spatial organization and receptor types. Brain Res 656:353–358

Zhou L, Chey WY (1984) Electric acupuncture stimulates non-parietal cell secretion of the stomach in dog. Life Sci 34:2233–2238

Zimmermann M (1984) Neurobiological concepts of pain, its assessment and therapy. In: Bromm B (ed) Pain measurement in man. Neurophysiological correlates of pain. Elsevier, Amsterdam, pp 15–35

Zottermann Y (1953) Special senses: thermal receptors. Ann Rev Physiol 15:357–372

Subject Index

Kolibri 360 GmbH
Ferdinand-Jühlke-Straße 7
99095 Erfurt, Deutschland
produktsicherheit@kolibri360.de